Networked Feminisms

Networked Feminisms

Activist Assemblies and Digital Practices

Edited by Shana MacDonald,
Brianna I. Wiens, Michelle MacArthur,
and Milena Radzikowska

LEXINGTON BOOKS
Lanham • Boulder • New York • London

Published by Lexington Books
An imprint of The Rowman & Littlefield Publishing Group, Inc.
4501 Forbes Boulevard, Suite 200, Lanham, Maryland 20706
www.rowman.com

86–90 Paul Street, London EC2A 4NE

British Library Cataloguing in Publication Information Available

Library of Congress Cataloging-in-Publication Data Available

ISBN 978-1-7936-1379-0 (cloth : alk. paper)
ISBN 978-1-7936-1380-6 (electronic)

♾️™ The paper used in this publication meets the minimum requirements of American National Standard for Information Sciences Permanence of Paper for Printed Library Materials, ANSI/NISO Z39.48-1992.

Contents

Acknowledgments

The editorial team would very much like to thank Judith Lakamper at Lexington Books for supporting this collection of essays from day one and for offering so much vital support, patience, and flexibility, especially as we fumbled our way through the COVID-19 pandemic. Every book should be so lucky as to have someone like Judith in their corner. We would also like to offer our gratitude to Sabrina Low and May Allah Nehmat for their editorial assistance in bringing the manuscript to fruition. A special thanks goes to the contributors of this collection for their brilliant feminist work and activism. The world becomes a better place because of the work done by those like the authors in this book. We are eternally grateful to you, and to our future feminist thinkers and activists.

SHANA MACDONALD: I would like to thank the Games Institute at the University of Waterloo for offering the space and institutional support of the qcollaborative (qLab) and the Feminist ThinkTank (FTT) working group. I am also grateful for the support of the Social Sciences and Humanities Research Council of Canada, as well as the University of Waterloo Gender Equity Grant, for helping in the development of the research-creation project Feminists Do Media (Instagram: @aesthetic.resistance), which offered a way to connect with communities of feminist activism online.

BRIANNA WIENS: This editorial team, Shana, Michelle, and Milena, deserves the greatest of thanks. A good deal of labor went into this book and the fiery feminist energy of the team held us together, even amidst a global pandemic. I'm also thankful for the provocative conversations and calls to action from the members of the qLab and Feminist ThinkTank, and for the financial support of the Social Sciences and Humanities Research Council of Canada and the York University Provost Dissertation Scholarship that allowed for the space and time to think more critically and generatively about the relationship between intersectional feminism and digital media and its re-futuring potentials. A special thanks goes, again, to Shana for her academic mentorship, friendship, and magic over the last seven years.

MICHELLE MACARTHUR: I would like to thank the University of Windsor's Humanities Research Group, which supported my work on this project through their 2019–2020 Fellowship and the collegiality and community encouraged by its director, Dr. Kim Nelson. I am also grateful for the support of the Canadian Association of Theatre Research in making space for seminars dedicated to digital feminisms at its 2019 and 2020 conferences. Co-led by myself and Shana, these seminars facilitated networks of feminist scholars working in the discipline of Theatre and Performance Studies and provided an early testing ground for some of the material in this collection. I also extend my gratitude to my co-editors, Shana, Bri, and Milena, for inviting me into their research circle with generosity and warmth.

MILENA RADZIKOWSKA: My biggest thanks to Shana, Bri, and Michelle for including me in this important work, and to the qLab—and Dr. Jennifer Roberts-Smith specifically—for bringing us together. I will be eternally grateful to Dr. Maki Motapanyane for the mentorship and support she's gifted me over the past three years.

Introduction

Feminist Takes on Networking Justice

When this book was coming together in the summer of 2019, we could not have imagined the state of the world that awaited us in 2021 as we finished it or the role that digital media would continue to take within our everyday work and leisure experiences. Hashtag movements like #BlackLivesMatter incited global protests as we mourned the murders of George Floyd and Breonna Taylor among countless others, while the Dakota Access Pipeline was temporarily shut down in July 2020 (and is currently pending an environmental review) after four years of Indigenous and allied resistance, legal action, and digital protest through #NODAPL. The #MeToo movement continued to make waves, initiating its virtual Survivor Healing Series focused on disrupting rape culture and teaching practices for coping with and healing from the trauma of sexual violence. And #StopAAPIHate rallied attention around the rapidly increasing instances of anti-Asian hate crimes and abuse, particularly as a result of racist ideologies surrounding the spread of COVID-19.

Our collection's title, *Networked Feminisms*, took on new meanings during the pandemic, when much activism was relegated online, when digital connections replaced in-person networks, and when social inequities were laid bare. The pandemic exacerbated existing race- and class-based inequities in accessing food, shelter and quarantine areas, stable incomes, and medical care. Moreover, the pandemic made clear the ingrained ableism in our social structures as many institutions quickly and without question transitioned to remote work—an option many disabled and/or chronically ill people have been requesting for a long time. In Canada, where we, the editors of this collection, live and work, a series of constantly shifting stay-at-home orders highlighted the depth of work-based gendered inequities facing those who must balanced a series of full-time jobs: remote working, home schooling for children, care-taking, and home-making. This necessitated the continued recognition that the emotional and physical brunt of pandemic life and care has often come to fall on womxn's shoulders. Evidently, the pandemic underscored an acute need to continue amplifying the decades-long work of racialized, gendered, and otherwise marginalized people, specifically with regard to the foundational work of Black feminists on intersectionality.

To seek greater solidarity and organization against unjust systems of power, like those magnified by the pandemic, as an editorial collective we approach intersectionality, the term famously coined by Kimberlé Crenshaw (1989, 1991), through the words of Patricia Hill Collins (1990, 2015, 2017), who describes it as a broad-based knowledge project and not as the end point of analysis. As a field of study situated within the power relations that it studies, as an analytic strategy that provides new angles of vision on social phenomena, and as a critical praxis that informs social justice projects, intersectionality's "essence lies in its attentiveness to power relations and social inequalities," recognizing how "race, class, gender, sexuality, ethnicity, nation, ability, and age are reciprocally constructing phenomena" (Collins 2015, 3). These mutually constructing categories underlie and shape intersecting system of power that catalyze the "social formations of complex social inequalities that are organized via unequal material realities and distinctive social experiences for people who live within them" (16). bell hooks (2000) reminds us that intersectional feminist work must not forget that a commitment to feminism is a connection to political action, that there is a direct relationship between theory and practice (6). Learning about feminism takes place both inside and outside academic settings, and it is incumbent on scholars to recognize that feminism, "a movement to end sexist oppression," cannot stay in the academy—it should be given back to the communities from which it came in order to renew commitment to political solidarity (6). The now widespread adoption of intersectionality across disciplines, fields, and industries has helped it to flourish in important directions; however with this there is the chance that the grounded, practice based work of intersectionality itself risks being ignored as it circulates with the academy. Inside and outside of the academy, intersectionality has been tokenized by those with the most power. Feminism's calls to action have long been co-opted by white colonial neoliberal forces that benefit from the current unjust status quo. This book resists the siloing of intersectionality within the academic institution that Collins warns against, and instead embraces intersectionality's multifaceted "broad-based knowledge project" (Collins 2015). In different ways, each of our contributors outline intersectional feminist activist practices and theorizations that cut across disciplines, practices, and experiences. It only makes sense, then, to situate ourselves as editors and to pay respects to the land that we work and live on.

Kwe, I am Shana MacDonald, an associate professor and artist-scholar-activist who works and lives on Haldimand Tract, land granted to the Haudenosaunee of the Six Nations, and the territory of the Neutral, and Anishinaabe and Haudenosaunee peoples. I am a straight-passing queer woman and a white, settler of Scottish, French, and Mi'kmaq descent tied the Qalipu First Nation of Western Newfoundland. As a first-generation scholar I have always felt I was a misfit in the elite spaces of academic institutions. This has been a point of shame, but it also defines what matters for me and illuminates where we need to resist, and what kinds of access are needed to ensure community building continues within and beyond the academy despite the constraints placed on those most marginalized by our institutions.

Nĭ hǎo. Guten tag. Olá. I am Brianna Wiens, a postdoctoral fellow at the University of Waterloo, which is situated on the Haldimand Tract, land that was granted to the Haudenosaunee of the Six Nations of the Grand River, and is within the territory of the Neutral, Anishinaabe, and Haudenosaunee peoples. I live and have remotely worked in Stratford, which is the territory of the Anishinaabe, Haudenosaunee, and Ojibway/Chippewa peoples. This territory is covered by the Upper Canada Treaties. I often describe myself as inhabiting the spaces inbetween: mixed-race, the daughter of immigrant-settlers from Malaysia on my mother's side and Brazil on my father's side, bisexual, able-bodied but living with chronic pain from scoliosis and a spinal fusion. Stories, like the ones shared in this collection, have helped to shape and solidify the instability of feeling category-less, to queer the boundaries of the categories themselves, and to articulate the importance of taking up the space of the in-between.

Hello bonjour! I am Michelle MacArthur, a white settler with French and Scottish ancestry. I am assistant professor at the School of Dramatic Art at the University of Windsor. Windsor sits on the traditional territory of the Three Fires Confederacy of First Nations, which includes the Ojibwa, the Odawa, and the Potawatomie. This anthology's focus on the digital realm and our reliance on digital modes of communication as we collaborated on it from across Turtle Island have prompted me to reflect on the ongoing disparities in access to digital technology. I am reflecting on how the digital divide is affecting Indigenous communities during the pandemic, how digital technologies have been used to oppress these communities, and conversely, how they might be used to redress inequities.

Milena Radzikowska is a white settler, refugee Canadian. I work at Mount Royal University, which is situated in an ancient and storied place within the hereditary lands of the Niitsitapi (Blackfoot), Îyârhe Nakoda, Tsuut'ina, and Métis Nations. It is a land steeped in ceremony and history that, until recently, was used and occupied exclusively by peoples indigenous to this place.

As a collective, our own research overlaps at the intersections of media, technology, performance, and design, specifically from an intersectional feminist perspective that is critical of the structures that uphold the academy. This commitment to intersectionality and justice necessarily means a commitment to decentering whiteness and other forms of oppressive operational, institutional, and structural powers, especially within our own editorial team. Within our collective, we take up different academic positions: professor, associate professor, assistant professor, and postdoctoral fellow, each position bringing with it different access points to power, security, and precarity. We're also friends, partners, and, of course, feminists. Some of us are parents, sisters, and daughters. It is because of our individual and collective situatedness and the relationships we have built working together that we have come to think in more complex ways about the concept of networked feminisms and the possibilities that emerge from these digital connections.

Even as we are seeing startling increases in mediated misogyny, racism, and other forms of violence and discrimination online, many of us who are attuned to digital culture are privy to the feminist resurgence that has been building across various spheres of media production. As Sarah J. Jackson, Moya Bailey, and Brooke Foucault Welles (2020) note, "since the 2011 Arab Spring and the upwelling of Occupy movements across the globe, social networks have influenced how both those on the margins and those at the center engage in sociopolitical debate and meaning-making" (xvii). These conversations were made possible through networked activisms, or, as we define it, the various activist forms that take place through online networks and that have material and affective impacts in both mediated and unmediated arenas, from hashtag activism to social media campaigns to hacktivism. Networked feminist activisms have been crucial for inspiring counterpublic formation and maintenance, storytelling, coalition building, and intersectional education and advocacy. Our aim in thinking through networked feminist activisms in this book is to underscore how intersectionality cannot be just another buzzword, but a framework, a practice, and a way of living. As Flavia Dzodan (2011) so eloquently insists, our "feminism must be intersectional or it will be bullshit" (para. 1). To suggest otherwise contradicts the need to fight gendered injustices where they intersect with racism, colonialism, ableism, and queer and transphobias. In embracing an intersectional feminist approach, we must actively rethink feminist research and activism, starting with the canonical texts and histories that are so entrenched in academic and media spaces.

As Tara Conley points out in the opening chapter of this collection, "Black feminists have a long tradition of rejecting white feminism as a liberatory strategy for a select few and for its one-size-fits-all vision for the colonized subject. Black feminists continue this tradition across digital spaces to bring attention to white feminism's ineffectiveness as an organizing strategy as an ethos." Marisa Duarte, a member of the Pasqua Yaqui tribe, outlines in her chapter how Indigenous feminisms' resistance of white colonial practices shares a project with "anti-racist/anti-sexist movements including groups that apply digital tactics toward social change." And we agree—we recognize that the most significant activist movements have come from Black and Indigenous women. Within the last two decades alone, we have seen how the "me too" movement was created by Tarana Burke in 2006 from her organization Just Be Inc. (2003) to help Black women and girls heal from sexual violence before going viral as #MeToo in 2017; Idle No More was created by Jessica Gordon, Sylvia McAdam, Sheelah McLean, and Nina Wilson in 2012 to call attention to Indigenous rights and the protection of the land, air, and water; and Black Lives Matter was created by Alicia Garza, Patrisse Cullors, and Opal Tometi in 2013 in response to the murder of Trayvon Martin and wrongful acquittal of his killer, and has since become a global movement to end white supremacy and seek justice for Black communities. Each of these movements demonstrate the ways in which Black and Indigenous ways of being and knowing are socially and politically transformative, shifting the terms of digital engagement in the process. And these are not the only movements. Recent history has witnessed a range of feminists, including racialized, disabled, queer, and trans feminists, building campaigns, movements, toolkits, and conversations that are setting the agenda for how we work towards greater equities in all aspects of our lives, both online and off.

As editors, we are guided by intersectional and Black feminisms,[1] Native and Indigenous feminisms,[2] as well as crip,[3] queer,[4] and trans[5] feminisms. Grounding ourselves within these feminisms necessitates a commitment to continuously and critically interrogating the insidiousness of white, cis, hetero, ableist neoliberal feminism that seeks to re-center itself time and time again in our academic settings and everyday online digital spaces. Black and Indigenous feminist orientations in particular have served as a blueprint for how we approach scholarship on digital feminist activism and how to intervene into and counter white supremacy. What we outline in this introduction, and what is echoed across the chapters included in this collection, are insights into a series of related activist movements and orientations that are not only about women and gender, but about power more broadly—who holds and wields power and who does not and cannot within the current matrix of domination (D'Ignazio and Klein 2020, drawing on Collins 1990). As such this collection is indebted to the incredible earlier works of the many diverse womxn and gender non-conforming people who have fought, taught, and found joy before us, turning to intersectional feminist scholarly practices of naming and creating historical narratives where none exist in dominant

frames because they were purposefully ignored or erased, as well as supporting practices of imagining better futures, whatever those futures might look
like.

As a collection, the essays gathered here demonstrate the generative and
powerful connections that thrive when we center intersectionality and listen
to disabled experiences, uplift queer and trans voices, and celebrate the work
of Indigenous, Black, and racialized women. Justice is an intersectional endeavour. As editors, we thus bring an overarching commitment to drawing
together a range of experiences and standpoints that reflect the long-standing
notion of "the personal is political" from intersectional perspectives within
the digital moment.

In returning to the cornerstone of "the personal is political" time and time
again, we seek to reaffirm that our everyday lived experiences of constraint
are tied to larger social inequities that are built into the fabric of our institutions and nation-states. Taking our cue from our contributors, as editors we
have oriented this collection and this introduction around the question of
what it means to "take up space" as a networked feminist counterpublic
(Jackson, Bailey, Foucault Welles 2020; Fraser 1990; Warner 2002) within
the white supremacy and misogyny of the digital commons. The chapters in
this book demonstrate a variety of ways that feminists have created support,
solidarity, and community online and highlight the invaluable practices that
have emerged from the nexus of activism, digital media, and technologies. It
is only through these forms of introspection and dialogue that we can begin
to develop scholar-activist spaces of genuine solidarity that are committed to
benefitting, uplifting, and taking up space for and by those most marginalized and targeted by the reach of the imperialist white supremacist capitalist
patriarchy (hooks 2000). And, even as we struggle through these spaces, it is
the joy, hope, imagination, and creativity embodied in the essays in this book
that offer a framework that will change not just the next generation of feminists' futures, but their presents.

FEMINIST NETWORKS, COMMUNITIES, AND CHALLENGES IN THE DIGITAL ERA

As digital media gained popularity in the early 2000s, feminists took notice.
Taking advantage of emerging platforms, feminist activists connected and
amplified their messages through social media sites like YouTube (2005),
Facebook (2006), Twitter (2006), Tumblr (2007), Instagram (2010), and,
more recently, TikTok (2016). The turn to digital media from analog forms
of protest media like zines, film, and guerilla art makes a good deal of sense
for our contemporary moment. The rapid circulation of social media content
has ensured that feminist activist discourses have entered mainstream con-

versations and have seen active debate in public spaces where they were not previously as admissible, in some cases going so far as to enable their virality in hashtag, meme, or video form. This kind of exposure to ideas, practices, and languages of social justice has created important spaces of discussion and political engagement in both mediated and unmediated spheres among the general public and policy makers, and has encouraged greater participation in activist spaces (Anderson, Toor, Rainie, and Smith 2018).

And yet, the uptake in social media use for advocacy and activism has not been without its failures and tensions, from the violent misogyny, white supremacy, colonialism, ableism, queerphobia, and transphobia that run rampant online, to the continual centering of neoliberal white postfeminist brands as the face of #feminism. In the face of these realities, this edited collection and its contributors work to amplify the labor of digital feminists while also holding the worst parts of the internet to account. Three key, interrelated areas inform the conversations this book offers: critiques of postfeminism, the ongoing threat of mediated digital misogyny, and the promises and constraints of networked feminist activism(s) as a field of research and a site of practice. Here, we offer a brief explanation of these areas and their overlapping interests.

Post-Feminism

Feminist media scholarship across the last three decades has grappled with the erasures caused by white feminism, and its most strident iterations: postfeminism and celebrity feminism. As Rosalind Gill (2007; 2017) and Angela McRobbie (2004) have aptly stated, postfeminism is an ideology that displaces the work of feminism as a radical and collective political movement focused on equity and structural change by emphasizing instead neoliberal ideals of individualism, choice, and (self) empowerment. Postfeminism circulates as particular tropes or "sensibilities" (Gill 2007) in popular culture, taking feminism into account only to repudiate it—for example, in the framing of consumerist habits like buying clothing and make-up as driven by individual choice, or in the wellness industry's rebranding of self-surveillance and self-discipline as empowering and fulfilling for women. Particularly pernicious is postfeminism's focus on individualism as it actively resists an intersectional understanding of the experiences of those who have been marginalized, holding instead white, Western, heterosexual, cis-gendered women as its ideal subjects. In a neoliberal climate such as ours, where individuals are required to be increasingly self-reliant and self-governing in the face of privatization and deregulation, where productivity and a "boss babe" ethos invite cheers and pats on the back, postfeminism thrives.

Of concern are the ways longer-standing postfeminist sensibilities proliferate online, particularly on social media platforms where women participate directly in their circulation through their own (often competitive) postfeminist performances. This can be seen anywhere from workout and thigh gap selfies to self-care strategies that include restrictive diet detoxes, to the use of #WifeMaterial when documenting a domestic task like cooking or decorating while spurning other forms and ways of partnership. None of these advance the structural inequalities experienced by women, and all confirm a white-centered set of beauty and feminine ideals. In their own ways, each of the essays in this collection reject the whiteness and elitism of post-feminism through re-centering the racialized and gendered digital labor that diverse networked feminist counterpublics engage in for their own liberatory means.

Mediated Misogyny

These postfeminist discourses have been paralleled by an equally pervasive rise in mediated misogyny and racism (Banet-Weiser 2018; Berridge and Portwood-Stacer 2015; Kendall 2020; Portwood-Stacer and Berridge 2014). Speaking directly to the onslaught of gendered violence online, Sarah Banet-Weiser and Kate Miltner (2016) make clear that "[w]e are in a new era of the gender wars . . ." that are "marked by alarming amounts of vitriol and violence directed at women in online spaces" (171). This "networked misogyny" responds to a perceived threat that feminists are encroaching on men's "rightful place in the social hierarchy" and more specifically "the incursion of women and people of color into what were previously almost exclusively white, male spaces" (Banet-Weiser and Miltner 2016, 172). For instance, groups of 4chan participants, defining themselves as the disenfranchised victims of feminism, employed the platform to "organize a campaign of revenge against women, 'social justice warriors' and the 'alpha males' who had deprived them of sexual success" (Ging 2017, 3). These current forms of popular misogyny continue to ensure that "rape culture is normative, violent threats against women are validated, and rights of the body for women are either under threat or being formally retracted" (Banet-Weiser and Miltner 2016, 172). The reach of misogyny's gendered violence online and the retraction of bodily rights extends to non-binary and trans people, made abundantly clear in the recent rollbacks of transgender health care protections in the United States and related attacks on trans people online.

As just one now well-known example, Gamergate exploded in the summer of 2014 as coordinated public harassments of prominent feminist critics of sexism in gaming culture. Emerging as a backlash to perceived bias within video game journalism, Gamergate quickly became synonymous with a violent form of trolling against women who are vocal about abuse, feminist critics of the games industry, and their allies and supporters. Doxing, rape threats, and death threats used by self-identified members of the gaming community to explicitly silence feminist critiques signal that this outpouring of mediated misogyny has no fear of ramifications, precisely because of the pervasive patriarchal culture we live in and the anonymity offered toabusers online. Every recent feminist happening (e.g., #MeToo, memes and hashtags countering #NotAllMen, #TimesUp, various iterations of the Women's March, #BlackLivesMatter protests, #IdleNoMore, and #MMIWG and the red dress campaign for Missing and Murdered Indigenous women and girls, etc.) has garnered a misogynistic and racist response as a violent reaction to women speaking in public. Despite actions ranging from policy and law to social movements to simple utterances of "no," understanding violence against women, gender non-conforming, racialized, disabled, and otherwise marginalized people as non-normative has been overshadowed by a deep complicity with structures of white supremacy, ableism, necrocapitalism, and misogyny that are so clearly supported within digital culture.

And yet, a growing number of feminist voices online have offered powerful responses to mediated and unmediated harassment by various antifeminist and white supremacist groups. Despite the emotional and physical labor of responding to antifeminist hate online, a hatred that has long existed on the internet, the emergence of networked feminist activisms within the last decade is new and, frankly, exciting. The nature of the conversations unfolding across (un)mediated spheres places greater emphasis on how harassment and violence are manifest in both online and everyday offline spaces in equal measure. Media studies scholars have argued for the last decade that there is no "real" separation between on/offline spheres (Hine 2000). And, as harassment against marginalized groups intensifies in coordinated ways, we cannot dispute this. As such, new forms of critical language are needed to equip ourselves with intersectional feminist tactics in order to face everyday misogyny and white supremacy in a moment when online and offline are inexorably intertwined.

Networked Feminist Activisms

As Rosemary Clark (2016) argues, the shift to the digital includes both a discursive and increasingly inclusive focus within contemporary feminist social movements and activism. For Clark (2016), importantly it is "a hashtag's narrative logic—its ability to produce and connect individual stories" that "fuels its political growth" (2). Drawing on Stacey Young (1987), Frances Shaw (2012) demonstrates how such "discursive activism" understands discourse as always already political, thus enabling new ways of speaking and new social responses and paradigms to emerge. From within the sexist and racist digital spaces outlined above, networked feminist activisms have emerged as movements which are actively "capable of triggering sociopolitical change . . ." (Shaw 2012 quoted in Clark 2016, 792). Shaw thus suggests that the emergence of paradigms and new modes of speech can lead to collective action and movement through the ways in which discursive activism highlights how the sentiment itself is political by virtue of being uttered.

Important research on digital responses to gendered and racialized violence has highlighted these kinds of complexities in the relationship between hashtags and feminist activisms (Berridge and Portwood-Stacer 2015). For example, social media hashtags have become effective ways of talking about Black women's issues when mainstream media outlets will not. In her 2017 essay on processes of decoding as becoming in Black Feminist hashtags, Tara L. Conley describes how hashtags like #WhyIStayed, #YouOkSis, #SolidarityIsForWhiteWomen, and #BlackPowerIsForBlackMen reimagine the human, thinking through how Black women's "encounters, desires, articulations, and bodies" are "entangled among sociopolitical processes of domination and authority," both online and offline, through their "renewal and strategy, mediation and embodiment, and as sites of struggle over representation, as becoming" (23). The work of decoding, Conley argues, contributes to intervening into assemblages of dominant ideologies that uphold "white privilege, racial paternalism, misogyny, and sexual and gender violence" (24). Hashtags as sites of becoming thus offer the potential to address and redress forms of racial, gendered, and sexual violence.

Significantly, Sherri Williams (2015) writes, "Black feminists' use of hashtag activism is a unique fusion of social justice, technology, and citizen journalism. . . . Twitter is often a site of resistance where black feminists challenge violence committed against women of color and they leverage the power of Black Twitter to bring attention and justice to women who rarely receive either" (343). Hashtags in this way enable affective and technological solidarity to express a range of reactions to rape culture and occlusions from mainstream media, including feminist rage, irony, and humor. Feminists are thus able not only to expose rape culture and systems of oppression but share their own experiences with an invitation for response (Rentschler 2014). Moreover, scholarship has recognized emerging opportunities for social justice, commenting on the transnational reach of feminist hashtagging for women's rights activism. "For those who have access to them," observes Eleanor Tiplady Higgs (2015), "social and digital media provide unparalleled opportunities for crossing borders of all kinds, allowing advocates for women's rights to organise around, through, and despite national and cultural divides" (344).

Nevertheless, it is these very acts of "border crossing," both physical and digital, that continue to reveal the kinds of power and privilege inherent to online spaces, even when fighting from intersectional perspectives for justice against sexual violence (Baer 2016; Higgs 2015). Campaigns like #JusticeforLiz and #BringBackOurGirls, though important for bringing awareness and possible change, underscore the prevalence of a white savior complex that plays into dominant stereotypes of Black and Brown bodies perpetuated by white feminist, neoliberal, and colonial frameworks and threatens possibilities for genuine solidarity (Higgs 2015; Khoja-Moolji 2015). Within this context, Williams (2015) underscores how "[w]hen white feminists miss opportunities to stand with their black sisters and mainstream media overlooks the plight of nonwhite women," as has been the case with dominant coverage of the #MeToo campaign, "women of color use social media as a tool to unite and inform" (342). Black feminists on Twitter re-centered the conversation about #MeToo to founder Tarana Burke's work and intentions, countering the ways in which #MeToo became tied to celebrity feminist visibility. As social media and digital technologies enable the circulation of activist work and community-building potentials, they simultaneously constrain such possibilities through perpetuating misogyny, white supremacy, and classism in updated, digital form (Benjamin 2019; Noble 2018; Wiens and MacDonald 2020) and place the burden on those who are most marginalized to provide the necessary correctives.

While the "algorithmic oppression" (Noble 2018) and "new Jim Code" (Benjamin 2018) embedded in search engines and social media platforms must be addressed, there still lies potential for world-building. Despite their shortcomings, including the ways they can distort or misinterpret complex issues, feminist hashtags have been incredibly effective at providing "easy-to-digest shorthands" (Bailey, Jackson, and Welles 2019, para. 37) that speak back to dominant ideologies surrounding violence against women, queer, and trans people and narratives of victimhood. Here, hashtags encourage a specific naming of the challenges that marginalized groups are up against, bringing into broader circulation issues that have long been silenced. Hashtags collectively "name what hurts" (hooks 2012), articulating aloud the harm done in order to draw attention to the matter. What should be underscored here is the significant and necessary role that previous hashtag movements and their creators have played in carving out digital space and sociopolitical awareness and presence for the future of current movements and where they can go, and what impact they can have. As Jackson, Bailey, and Welles (2019, 2020) point out, the important reckoning force of #MeToo could not have been made possible without the work done by its predecessors like #YesAllWomen, #SurvivorPrivilege, #WhyIStayed, and #TheEmptyChair, since each of these hashtag networks was already publicizing the interpersonal and structural violence experienced by women. The compounding efforts of what Deen Freelon, Meredith D. Clark, Sarah J. Jackson, and Lori Lopez (2018) call "Feminist Twitter," a community influenced by Black women, Indigenous women, and women of color, has brought to the forefront subaltern communities and conversations that were originally silenced.

SITUATING *NETWORKED FEMINISMS:*
ACTIVIST ASSEMBLIES AND DIGITAL PRACTICES

Networked Feminisms: Activist Assemblies and Digital Practices illuminates
the myriad interventions feminist assemblies are making online through a
necessarily interdisciplinary approach. This book gathers provocations, analyses, creative explorations, and cases studies of digital feminist practices
from a wide range of disciplinary perspectives including, but not limited to,
media studies, communication studies, critical and cultural studies, gender
and sexuality studies, performance studies, digital humanities, feminist Human-Computer Interaction (HCI), and feminist Science and Technology
Studies (STS). Given our aims, in pulling the book together we actively
sought new ways to articulate scholarly structures in the hopes that, through
the iterative processes of challenging disciplinary boundaries, we find new
modes of meaning-making and more capacious, accessible knowledges. The
result is a series of essays that consider how digital feminist activisms use
conventions of networked assemblies to counter the individualizing forces of
postfeminism, neoliberalism, misogyny, racism, ableism, and colonialism
while foregrounding the types of systemic change so greatly needed, but
often overlooked, in this climate.

Tara L. Conley makes this clear in the opening chapter of this book when
she situates hashtag feminism, a sign of the times, as a conceptual framework: a tool for organizing ideas and stories, collecting information, and
documenting and organizing movements, especially for Black feminists.
Bringing readers through a body of digital feminist research; her website,
Hashtag Feminism, founded in 2013 and its blog posts and infographics; the
feminist storytelling and organizing of #MeToo; and the tensions of competing beliefs and values as seen through Micki Kendall's #SolidarityIsFor-
WhiteWomen, Conley carefully artciulates the relationship between feminist
praxis and proceses used for engaging such praxis, both technical and material. In these ways, hashtag feminism is more than a mere tool; hashtag feminism sparks movement in the world as discourse, as embodied practice, and
as ideology, particularly in the context of documenting historical shifts, and
is thus not neutral.

Similarly, Melissa Brown argues for the important role that Black women
and LGBTQ+ people embody as "virtual sojourners"—digital technology
users who use the same tools that dominant groups may use but for different
means: for self-determination, for self-authorship, and for self-definition,
where possibilities for both visibility and erasure take place. In employing
these tools, Black and queer virtual sojourners establish important social
relationships and discuss shared experiencing, while challenging heternorma-
tive narratives of gender and sexuality and creating their own digital counterpublics and enclaves for Black and queer culture and identity.

Offering a playful cyborgian exploration of what they term "keywords for crip feminists" (keywords: chronic, fem(me)bot, doll, ghost, space, fail, sexy/funny), Adan Jerreat-Poole uses their essay to also carve out overlapping counterpublic spaces, specifically speaking to the ways that keyword searches construct users' relationships to digital information and embodiment. Through grappling with such keywords, Jerreat-Poole intervenes into the algorithmic echo chambers of ableist digital platforms, centering disability and its (in)visibilities, complexities, and affects within feminist activism: "Turning the meaning of these words as we shift registers and positionalities is like turning a door handle," they write, "opening to a room of our (collective, rather than singular) own."

And, indeed, it is in turning words, ideas, and spaces over and in (re)considering the logics that we have been asked to grow accustomed to that Brianna I. Wiens argues we cultivate resistance. In offering a conceptual framework for orienting to hashtag movements, Wiens argues that "virtual dwelling" as a way of thinking and being in digital space encourages feminists to become more attuned to the practices, tools, and communities within reach. Wiens brings the reader through her own journey of virtually dwelling with the accusation that the #MeToo movement was a witch hunt, demonstrating how, in slowing down (a fight in and of itself against the neoliberal capitalism of white supremacy and patriarchy) and lingering with these moments of intrigue and outrage, feminists learn, organize, respond, and (re)purpose openings for digital activism and resistance.

Calling on such resistance and persistence in the face of anti-Indigenous racism and sexism, Marisa Elena Duarte details the digital tactics of cyberdefense that Native and Indigenous women engage with to protect their peoples, waters, and lands. Bringing together stories, histories, and scholarship around Indigenous women's networked practices, Duarte articulates the necessity of framing Indigenous Internet connection, access, and use as issues of human rights and self-determination, rather than merely issues of digital literacy, ISP construction, or ICT use. Drawing on the 2016 #NODAPL movement, network sovereignty, and data-driven tactics for justice, Duarte makes clear that "Indigenous means of production and information distribution are key to Indigenous solidarity."

Controlling the production and circulation of narratives also comes to bear in Ace J. Eckstein's essay on the queer worldmaking potentials of trans intelligibility on YouTube, once again making clear the importance of Brown's virtual sojourner in the quest for self-determination, authorship, and definition. Eckstein closely analyzes the transition channels of five trans men, demonstrating the significance of transmasculine counterpublics on YouTube and how, through these counterpublics, queer worldmaking comes to be. This, he suggests, brings "into focus questions of norms, normativity, and standards of queerness." Ultimately, Eckstein argues that it is precisely the transness of transgender men's self-published stories of manhood that marks them as intelligible and worthy of taking up space and place in their own right—not for their proximity to cismasculinity.

Radhika Gajjala, Sarah Ford, Vijeta Kumar, and Sujatha Subramanian's essay also thinks through Internet use as a practice of place-making and counterpublic formation, specifically in the context of Indian digital diaspora and international outreach. Gajjala, Ford, Kumar, and Subramanian highlight how this is particularly important as white feminism often tokenizes upper-caste brown feminist work while claiming intersectionality, especially within academic settings. As such, when examining activist encounters that affect Indian women and women of Indian descent across digital South Asian diasporas, it is crucial to map the relationalities and tensions of race, caste, gender, sexuality, and geography within hashtag publics. In doing so, the authors ask the important question of how and when the relationship between "BIPOC" and "anticaste" might make political sense, specifically in the context of intersectionality and the material and sociohistorical situatedness of hashtag activism.

In addition to attending to the material and sociohistorical dynamics of hashtag activism, Helena Suárez Val argues for the inclusion of emotional and affective practices in recording and mapping networked feminisms. Suárez Val offers digital cartographies of feminicide as "feminist affect amplifiers" to demonstrate how feminist activists in Ecuador, Mexico, Spain, and Uruguay use their own knowledges and emotions to change mainstream affects of apathy around feminicide, the gender-based murders of women and girls. Because, as she so eloquently argues, feminist activists understand the political strength of publicly displaying emotion, these cartographies illustrate the power of feminist activism in translating grief and rage into public action through collecting, analyzing, processing, and distributing data about site-specific femincides, publicizing the systematic prevalence of these tragedies and modulating the "affective atmosphere" of gender-based violence.

Suárez Val's essay reminds us of the many intersections that are necessary for interrogating gendered violence, and Leandra H. Hernández and Sarah De Los Santos Upton's chapter too argues for the need to approach violence against women from an intersectional perspective, specifically within reproductive contexts. Focusing on what they term "reproductive feminicide," the authors explore social media and digital social justice as tools for combating reproductive and gender-based injustice through analyzing specific hashtags like #ProChoice, #ReproductiveRights, #ReproJustice, and #ReproductiveJustice. By focusing on reproductive justice, the chapter investigates the relationships between reproductive relationships and practices and policies that shape reproductive experiences. The authors highlight and explain intersections of race, class, and gender to transcend the pro-life/pro-choice binary that often prioritizes white women's experiences.

Hernández and De Los Santos Upton's essay emphasizes the need to further engage with discussions of reproductive justice that center matters of intersectionality, inclusivity, and coalition building, themes that are also key to Angela Smith, Ihudiya Finda Williams, and Alexandra To's chapter.

Reflecting on their 2020 paper, "Critical Race Theory for HCI," Smith, Williams, and To discuss the writing process and the reception of the paper, offering personal stories to help form the wider context of the relationship between Critical Race Theory (CRT) and Human-Computer Interaction (HCI), and to shed light on the need for more justice-oriented work that encourages scholars to bring their "full selves to our academic community through a labor of scholarly activism for racial justice." In bringing together their stories and justifications for writing, the authors effectively provide a set of tools for future scholar-activists, contextualizing what scholar-activism can look like and offering calls to action and recommendations for sustainably engaging with CRT and racial justice in technology research and practice.

In the closing chapter, Elizabeth Nathanson asks readers to embrace the complexity of networked meaning-making within digital culture, focusing on the iconic Ruth Bader Ginsburg. Instead of accentuating historical accuracy, Nathanson argues that we might reflect on Ginsburg's popular representations within digital culture as a sign of changing times when it comes to feminine representation, where older women are no longer simply viewed as "symbolic of stasis or obsolescence," but are "agents of change and liberal revolution." At the same time, Nathanson suggests that we might give popular representations the credit they're due, considering the ways that the production and circulation of popular digital artifacts, like memes, of Ginsburg offers "unpredictable and distinctly energetic" interactions with political events. In analyzing RBG's circulation within and across networked publics, Nathanson demonstrates how the Supreme Court justice's popular representations signify and perform important intergenerational feminist politics.

At the beginning of this introduction, we reflected on how much has changed since we started work on this collection in 2019. The digital landscape at the center of this book is, of course, ever evolving, posing challenges for a collection like this one to keep up and ensure its material remains relevant by the time it makes it to print. But if the pandemic and the issues and questions of inequities it magnified have taught us anything, it is that the work of social justice is not over. The central concerns of the networked feminist activisms outlined here extend from the analog past to the digital present, from offline to online spaces, from the pages of this book to the conversations we hope they initiate.

NOTES

1. See, for example, Collins 1990, 2015, 2017; Combahee River Collective 1981; Crenshaw 1989, 1991; Davis 1981, 1990; 2015; hooks 1981, 1984; 1990, 2003; 2012; Lorde 1984, 1997, 2017.

2. See, for example, Anderson 2010, 2011, 2016; Cook-Lynn 1997; Green 2017; Suzack, Huhndorf, Perreault, and Barman 2011; Tallbear 2014, 2015, 2017.

3. See, for example, Bailey and Mobley 2018; Fritsch 2019ab; Garland-Thomson 1997, 2002, 2011, 2016; Hamraie and Fristch 2019; Kafer 2013; Nishida 2016, 2018; Wendell 1989, 1996.

4. See, for example, Ahmed 2006, 2017, 2019; Butler 1990, 1993, 2004; Halberstam 2005, 2011, 2012; Sedgwick 1990, 2003.

5. See, for example, Bettcher and Garry 2009; Bettcher 2006, 2012, 2013; Bornstein 1994; Feinberg 1992, 1996, 1998; Namaste 2000; Stryker 1994, 2004, 2008.

REFERENCES

Ahmed, Sara. 2006. *Queer Phenomenology: Orientations, Objects, and Others*. Duke University Press.

———. 2017. *Living a Feminist Life*. Duke University Press.

———. 2019. *What's the Use? On the Uses of Use*. Duke University Press.

Anderson, Kim. 2010. "Affirmations of an Indigenous Feminist." In *Indigenous Women and Feminism: Politics, Activism, Culture,* edited by Cheryl Suzack, Shari Huhndorf, Jeane Perreault, and Jean Barman. Vancouver: UBC Press.

———. 2011. *Life Stages and Native Women: Memory, Teachings, and Story Medicine*. Winnipeg: University of Manitoba Press.

———. 2016. *A Recognition of Being: Reconstructing Native Womanhood (Second Edition)*. Toronto: Canadian Scholars' Press.

Anderson, Monica, Skye Toor, Lee Rainie, and Aaron Smith. 2018. "Public Attitudes Toward Political Engagement on Social Media." *Pew Research Center: Internet & Technology*. July 11, 2018.

Baer, Hester. 2016. "Redoing Feminism: Digital Activism, Body Politics, and Neoliberalism." *Feminist Media Studies* 16(1): 17–34.

Bailey, Moya and Izetta Autumn Mobley. 2018. "Work in the Intersections: A Black Feminist Disability Framework." *Gender and Society* 33(1): 19–40.

Bailey, Moya, Sarah Jackson, and Brooke Foucault Welles. 2019. "Women Tweet on Violence: From #YesAllWomen to #MeToo." *Ada: A Journal of Gender, New Media, and Technology 15*. Doi: 10.5399/uo/ada.2019.15.6.

Banet-Weiser, Sarah. 2018. *Empowered: Popular Feminism and Popular Misogyny*. Duke University Press.

Banet-Weiser, Sarah and Kate M. Miltner. 2016. "MasculinitySoFragile: Culture, Structure, and Networked Misogyny." *Feminist Media Studies* 16(1): 171–74.

Banet-Weiser, Sarah and Laura Portwood-Stacer. 2017. "The Traffic in Feminism: An Introduction to the Commentary and Criticism on Popular Feminism." *Feminist Media Studies* 17(5): 884–906.

Benjamin, Ruha. 2019. *Race After Technology: Abolitionist Tools for the New Jim Code*. Polity Press.

Berridge, Susan and Laura Portwood-Stacer. 2015. "Introduction: Feminism, Hashtags and Violence Against Women and Girls." *Feminist Media Studies* 15(2): 341.

Bettcher, Talia Mae. 2006. "Understanding Transphobia: Authenticity and Sexual Abuse." In *Trans/Forming Feminisms: Transfeminist Voices Speak Out*, edited by Krista Scott-Dixon. Toronto: Sumach Press, 203–10.

———. 2012. "Trans Women and the Meaning of 'Woman.'" In *Philosophy of Sex: Contemporary Readings (sixth edition)*, edited by Nicholas Power, Raja Halwani, and Alan Soble. New York: Rowan & Littlefield, 233–50.

———. 2013. "Trans Women and 'Interpretive Intimacy': Some Initial Reflections." in *The Essential Handbook of Women's Sexuality (volume two)*, edited by Donna Marie Castañeda. Santa Barbara: Praeger, 51–68.

Bettcher, Talia Mae and Ann Garry. 2009. "Transgender Studies and Feminism: Theory, Politics, and Gender Realities (special issue)." *Hypatia: A Journal of Feminist Philosophy* 24(3).

Bornstein, Kate. 1994. *Gender Outlaw: On Men, Women, and the Rest of Us*. New York: Routledge.

Butler, Judith. 1990. *Gender Trouble: Feminism and the Subversion of Identity*. New York: Routledge.

———. 1993. *Bodies That Matter: On the Discursive Limits of Sex*. New York: Routledge.

———. 2004. *Undoing Gender*. New York: Routledge.

———. 2015. *Notes Toward a Performative Theory of Assembly*. Cambridge MA: Harvard University Press.

Clark, Rosemary. 2016. "'Hope in a Hashtag': The Discursive Activism of #WhyIStayed." *Feminist Media Studies*. *16*(5): 788–804.

Collins, Patricia Hill. 1990. *Black Feminist Thought: Knowledge, Consciousness, and the Politics of Empowerment*. Hyman.

———. 2015. "Intersectionality's Definitional Dilemmas." *Annual Review of Sociology* 41(1): 1–20.

———. 2017. "The Difference that Power Makes: Intersectionality and Participatory Democracy." *Investigating Feminism/Investigaciones Feministas* 8(1): 19–39.

Combahee River Collective. 1981. "A Black Feminist Statement." In *This Bridge Called My Back: Writing by Radical Women of Color*, edited by Cherríe Moraga and Gloria Anzaldúa. New York: Kitchen Table, 210–18.

Conley, Tara. 2017. "Decoding Black Feminist Hashtags as Becoming." *The Black Scholar* 43(3): 22–32.

Cook-Lynn, Elizabeth. 1997. "Who Stole Native American Studies?" *Wicazo Sa Review* 12(1): 9–28.

Crenshaw, Kimberlé Williams. 1989. "Demarginalizing the Intersection of Race and Sex: A Black Feminist Critique of Antidiscrimination Doctrine, Feminist Theory and Antiracist Politics." *University of Chicago Legal Forum*, 139–67.

———. 1991. "Mapping the Margins: Intersectionality, Identity Politics, and Violence Against Women of Color." *Stanford Law Review* 43(6): 1241–300.

D'Ignazio, Catherine and Lauren F. Klein. 2020. *Data Feminism*. MIT Press.

Davis, Angela Y. 1981. *Women, Race and Class*. Penguin Random House.

———. 1990. *Women, Culture and Politics*. Penguin Random House.

———. 2015. *Freedom Is a Constant Struggle: Ferguson, Palestine, and the Foundations of a Movement*. Haymarket Books.

Dzodan, Flavia. 2011. "My Feminism Will Be Intersectional or It Will Be Bullshit." *Tiger Beatdown*. October 10, 2011. http://tigerbeatdown.com/2011/10/10/my-feminism-will-be-intersectional-or-it-will-be-bullshit/.

Feinberg, Leslie. 1992. *Transgender Liberation: A Movement Whose Time Has Come*. New York: World View Forum.

———. 1996. *Transgender Warriors: Making History from Joan of Arc to Dennis Rodham*. Boston: Beacon Press.

———. 1998. *Trans liberation: Beyond Pink or Blue*. Boston: Beacon Press.

Fraser, Nancy. 1990. "Rethinking the Public Sphere: A Contribution to the Critique of Actually Existing Democracy." *Social Text* (25/26): 56–80.

Freelon, Deen, Meredith D. Clark, Sarah J. Jackson, and Lori Lopez. 2018. "How Black Twitter and Other Social Media Communities Interact with Mainstream News." *Knight Foundation*. February 27, 2018. https://knightfoundation.org/reports/how-black-twitter-and-other-social-media-communities-interact-with-mainstream-news.

Fritsch, Kelly. 2019a. "Ramping Up Canadian Disability Culture." In *The Spaces and Places of Canadian Popular Culture*, edited by Victoria Kannen and Neil Shyminsky. Toronto: Canadian Scholars Press, 265–72.

———. 2019b. "Governing Lives Worth Living: The Neoliberal Biopolitics of Disability." In *Governing the Social in Neoliberal Times*, edited by Deborah Brock. Vancouver: UBC Press, 38–62.

Garland-Thomson, Rosemarie. 1997. *Extraordinary Bodies: Figuring Physical Disability in American Culture and Literature*. Columbia University Press.

———. 2002. "Integrating Disability, Transforming Feminist Theory." *National Women's Studies Association Journal* 14(2): 1–32.

———. 2011. "Misfits: A Feminist Materialist Disability Concept." *Hypatia: A Journal of Feminist Philosophy* 26(3): 591–609.

———. 2016. "Becoming Disabled." *The New York Times*. August 19, 2016.

Gill, Rosalind. 2007. "Postfeminist Media Culture: Elements of a Sensibility." *European Journal of Cultural Studies* 10(2): 147–66.

———. 2017. "The Affective, Cultural, and Psychic Life of Postfeminism: A Postfeminist Sensibility 10 Years On." *European Journal of Cultural Studies* 20(6): 606–26.

Greene, Joyce (ed.). 2017. *Making Space for Indigenous Feminism (Second Edition)*. Fernwood Publishing.

Halberstam, J. 2005. *In a Queer Time and Place: Transgender Bodies, Subcultural Lives*. New York University Press.

———. 2011. *The Queer Art of Failure*. Durham: Duke University Press.

———. 2012. *Gaga Feminism*. Boston: Beacon Press.

Hamraie, Aimi and Kelly Fritsch. 2019. "Crip Technoscience Manifesto." *Catalyst: Feminism, Theory, Technoscience* 5(1): 1–33.

Higgs, Eleanor Tiplady. 2015. "#JusticeforLiz: Power and Privilege in Digital Transnational Women's Rights Activism." *Feminist Media Studies* 15(2): 344–47.

Hine, Christine. 2000. *Virtual Ethnography*. Thousand Oaks, California: Sage Publications.

hooks, bell. 1981. *Ain't I a Woman? Black Women and Feminism*. Routledge.

———. 1984. *Feminist Theory: From Margin to Center*. Cambridge, MA: South End Press.

———. 1990. *Yearning: Race, Gender, and Cultural Politics*. Cambridge, MA: South End Press.

———. 2000. *Feminism is for Everybody: Passionate Politics*. Cambridge, MA: South End Press.

———. 2003. *Teaching Community: A Pedagogy of Hope*. New York: Routledge.

———. 2012. *Writing Beyond Race: Living Theory and Practice*. Routledge

Jackson, Sarah J., Moya Bailey, Brooke Foucault Welles. 2020. *#HashtagActivism: Networks of Race and Gender Justice*. MIT Press.

Jane, Emma Alice. 2014. "'Back to the kitchen, cunt': Speaking the Unspeakable About Online Misogyny." *Continuum: Journal of Media and Cultural Studies* 28(4): 558–70.

Kendall, Mikki. 2020. *Hood Feminism: Notes From the Woman That a Movement Forgot*. Bloomsbury.

Khoja-Moolji, Shenila. 2015. "Becoming an 'Intimate Publics': Exploring the Affective Intensities of Hashtag Feminism." *Feminist Media Studies* 15(2): 347–50.

Lorde, Audre. 1984. *Sister Outsider: Essays and Speeches*. Berkeley, CA: Crossing Press.

———. 1997. "Age, Race, Class, and Sex: Women's Redefining Difference." In *Dangerous Liaisons: Gender, Nation, and Postcolonial Perspectives*, edited by Anne McClintock, Aamir Mufti, and Ella Shohat. Minnesota, Minneapolis: University of Minnesota Press, 374–80.

———. 2017. *Your Silence Will Not Protect You: Essays and Poems*. Silver Press.

McRobbie, Angela. 2004. "Post-feminism and Popular Culture." *Feminist Media Studies* 4(3): 255–64.

Namaste, Viviane. 2000. *Invisible Lives: The Erasure of Transsexual and Transgendered People*. University of Chicago Press.

Nishida, Akemi. 2016. "Understanding Politicization Through an Intersectionality Framework: Life Story Narratives of Disability Rights and Justice Activists." *Disability Studies Quarterly* 36(2).

———. 2018. "Critical Disability Praxis." In *Manifestos for the Future of Critical Disability Studies*, edited by Rosemarie Garland-Thomson, Katie Ellis, Mike Kent, and Rachel Robertson. Farnham, UK: Ashgate Publishing.

Noble, Safiya Umoja. 2018. *Algorithms of Oppression: How Search Engines Reinforce Racism*. New York University Press.

Rentschler, Carrie. A. 2014. "Rape Culture and the Feminist Politics of Social Media." *Girlhood Studies* 7(1): 65–82.

Sedgwick, Eve Kosofsky. 1990. *Epistemology of the Closet*. University of California Press.

———. 2003. *Touching Feeling: Affect, Pedagogy, Performativity.* Durham: Duke University Press.

Shaw, Francis. 2012. "'Hottest 100 Women': Cross-Platform Discursive Activism in Feminist Blogging Networks." *Australian Feminist Studies* 27(74): 373–87.

Sills, Sophie, Chelsea Pickens, Karishma Beach, Lloyd Jones, Octavia Calder-Dawe, Paulette Benton-Greig, and Nicola Gavey. 2016. "Rape Culture and Social Media: Young Critics and a Feminist Counterpublic." *Feminist Media Studies* 16(6): 935–51.

Singh, Rianka and Sarah Sharma. 2019. "Platform Uncommons." *Feminist Media Studies* 19(2): 303–2.

Stryker, Susan. 1994. "My Words to Victor Frankenstein Above the Village of Chamounix: Performing Transgender Rage." *GLQ: A Journal of Gay and Lesbian Studies* 1(3): 237–54.

———. 2004. "Transgender Studies: Queer Theory's Evil Twin." *GLQ: A Journal of Lesbian and Gay Studies* 10(2): 212–15.

———. 2008. *Transgender History*. Berkeley: Seal Press.

Suzack, Cheryl, Shari Huhndorf, Jeane Perreault, and Jean Barman (eds.). 2010. *Indigenous Women and Feminism: Politics, Activism, Culture.* Vancouver: UBC Press.

TallBear, Kim. 2014. "Standing With and Speaking as Faith: A Feminist-Indigenous Approach to Inquiry." *Journal of Research Practice* 10(2), Article N17.

———. 2015. "Dossier: Theorizing Queer Inhumanisms: An Indigenous Reflection on Working Beyond the Human/Not Human." *GLQ: A Journal of Lesbian and Gay Studies* 21(2–3): 230–35.

———. 2017. "Beyond the Life/Not Life Binary: A Feminist-Indigenous Reading of Cyropreservation, Interspecies Thinking, and the New Materialisms." In *Cryopolitics*, edited by Joanna Radin and Emma Kowal. MIT Press.

Warner, Michael. 2002. "Publics and Counterpublics." *Quarterly Journal of Speech* 88: 413–25.

Wendell, Susan. 1989. "Toward a Feminist Theory of Disability." *Hypatia: A Journal of Feminist Philosophy* 4(2): 104–24.

———. 1996. *The Rejected Body: Feminist Philosophical Reflections on Disability*. Routledge.

Wiens, Brianna I. and Shana MacDonald. 2020. "Feminist Futures: #MeToo's Possibilities as Poiesis, Techné, and Pharmakon." *Feminist Media Studies*. doi: 10.1080/14680777.2020.1770312.

Williams, Sherri. 2015. "Digital Defense: Black Feminists Resist Violence with Hashtag Activism." *Feminist Media Studies* 15(2): 341–44.

Zarzycka, Marta and Domitilla Olivieri. 2010. "Affective Encounters: Tools of Interruption for Activist Media Practices." *Feminist Media Studies* 17(4): 527–34.

Chapter One

A Sign of the Times

Hashtag Feminism as a Conceptual Framework

Tara L. Conley

INTRODUCTION

Story of the Question

Over the years I have often been asked if I believe online activism is "real." To this day, I find that question jarring because it suggests that organizing work like building coalitions, issue amplification, and storytelling, which provide context for social and political movements, lack merit simply because these practices are digitally mediated. Even before writing about #RenishaMcBride, I always believed that activist and feminist practices can and should take many forms (2013a; 2014a). Though addressing this question is exhausting, it has led me to examine how discourse emerges and what it communicates about the social, political, and cultural contexts in which we live. As a result of looking closely at online organizing and storytelling over the years, I have come to an understanding about the relationship between discourse and practice, and the technologies that carry them across time and space. This chapter reflects years of mulling over this relationship and presents a culmination of my work that describes hashtag feminism as a conceptual framework.

Hashtag feminism describes another enunciation for how to frame feminist theory and practice beginning in the aughts. Hashtags point to things. They locate. They organize information and people. They document. They reflect ever changing language and discourse. The hashtag marks ideas, practices, and ideologies across time and space where one can also locate the politics of representation, the ideological underpinnings of technological infrastructures, and our response to these aspects of social life in the public sphere. Hashtags are sites of becoming (Conley 2017). They are shorthand, sometimes to a fault. #MeToo, for example, is so embedded in popular discourse that its origin story is often lost. The story of #MeToo, as told through white liberal media, begins with Hollywood producer Harvey Weinstein, actress Alyssa Milano, and journalist Ronan Farrow, not with community organizer Tarana Burke, who many years before exposés and tweets sat down with a small group of Black girls and listened to their stories about sexual abuse (Burke n.d.; Me Too Movement 2018; Conley 2018). The more I analyze these moments, the more I see how origin stories around activism and organizing, or "moments of authoring" attached to prominent hashtags get lost overtime (Gray 2013, 108). And yet, this is precisely how language plays out. Language travels and morphs. Language takes work. So too does indexing language reflected in hashtag discourse. I believe the significance of hashtag feminism lies in its utility: While it arguably stands for a type of feminist moniker for the Internet age, more importantly, hashtag feminism points to a corpus of feminist discourse, practices, and ethos that reflect people's everyday lives as well as the tensions that exist around ideas about "how communities ought to be" (Ott 2018).

This chapter examines hashtag feminism according to three constructs: as discourse, as embodied practice, and as ideology. First, in fleshing out an understanding of hashtag feminism-as-discourse (HFAD), I analyze a corpus of digital feminist scholarship along with the website I founded in 2013, Hashtag Feminism, including its blog posts and infographics from 2014–2017. Second, I analyze hashtag feminism-as-embodied-practice (HFAEP) by examining the feminist storytelling and organizing project of #MeToo. I examine these practices through the concept of *homo narrans*, or the storytelling species. This idea comes from Jamaican writer and cultural theorist Sylvia Wynter, who acknowledges that "the human is not only a languaging being but also a storytelling species, a *bios/mythoi*" (McKittrick 2015, 25). Lastly, I conceptualize hashtag feminism-as-ideology (HFAI) as a set of beliefs and values among competing solidarities across contested feminist landscapes.

To carry out this analysis, I begin with a brief literature review on hashtag studies and then move to a cultural history of hashtag feminism, which informs my understanding of hashtag feminism as an intellectual project. I then examine hashtag feminism according to the three themes outlined above. André Brock's critical technocultural discourse analysis (CTDA) influences my approach to explicating hashtag feminism as an "assemblage of artifacts, practices, and cultural beliefs" (2018, 1014). This triadic formulation helps carry out a hermeneutic analysis where culture is positioned as a technological artifact, a view I accept. (Brock 2018, 1016). To conclude, I offer a vision of hashtag feminism as an intellectual project and praxis of our time, one that espouses a Black feminist ethos for justice and liberation.

Hashtag Studies

Research on social tagging and hashtags as sites of analysis to examine culture and society has been a busy space for over a decade. This is expected, since the more hashtags (metadata) are used to locate cultural practices and language (data) across the Internet, the more questions arise for scholars about how to capture and analyze this information as part of our collective memory. In particular, *hashtag activism*, *hashtag feminism*, and *digital feminism* are perhaps the most recognized terms to come out of this scholarship. These terms have been formalized through peer-reviewed research to describe the role hashtags play in documenting social and political life, and in theorizing about power. This body of literature examines modes of activism and feminist practices, namely how Twitter hashtags locate activist campaigns, organize online communities, and amplify social movements.

A more established approach to the study of hashtags comes out of sociolinguistics and corpus studies. This body of literature has been at the forefront of examining social tagging and semiotic mechanisms of hashtags for some time. These studies provide some of the earliest and most exhaustive explanations of social tagging on Twitter, primarily highlighting linguistic metafunctions that point to conversational or "searchable talk," spoken linguistics, and affiliation patterns (Bruns and Stieglitz 2012; Zappavigna 2012, 2015; Zhu 2016; Scott 2018). Research coming out of this area also provides historical frameworks for examining social tagging on Twitter and other social media platforms that predate Twitter (Ames and Naaman 2007; Keho and Gee 2011). Related studies in this field consider hashtags as sites to examine diffusion and emergence in online communities, looking at predictive aspects of hashtags, and the messages they carry overtime (Bastos, Raimundo, and Travitzki 2013; Fox, Cruz, and Lee 2015).

The turn to formalize research about the role hashtags play in movements for gender and racial justice crystalized in 2014 when the academic journal *Feminist Media Studies* published its first special issue on feminist hashtags and Twitter activism, of which my work is part. During the summer of 2014, editors of the *Feminist Media Studies* special issue sent out a call for short essays that reflected on some of the most visible feminist hashtags of the year. By the time these essays were published, stories about hashtag activism had been well covered across the mainstream press and written about in academic press. However, this issue was the first of its kind entirely devoted to reflections on the use of feminist hashtags to "respon[d] to contemporary events and discussions" (Portwood-Stacer and Berridge 2014, 1090). Included in the issue are 14 short pieces that reflect on trending hashtags framed in a transnational context across the United States, Turkey, United Kingdom, and Nigeria. The commentary structure was meant to be short, pithy, and accessible to mirror discourse on Twitter. Though not all trending feminist hashtags were captured in this issues, among those that were included are (Portwood-Stacer and Berridge, 2014b): #AskThicke; #BringBackOurGirls; #RenishaMcBride; #NotBuyingIt; #YesAllWomen; #Direnkahkaha; #OPRollRedRoll and #OccupySteubenville; #EndFathersDay; #YourSlipIsShowing, #MyStealthyFreedom; #KadınKatliamıVar and #TransCinayetleriPolitiktir; #TheVagenda.

Also during this year, Kitsy Dixon published "Feminist online identity: Analyzing the presence of hashtag feminism," in the *Journal of Arts and Humanities* (2014). Alongside this article and the *Feminist Media Studies* special issue, *Ada: A Journal of Gender New Media & Technology* published its fifth issue on Queer Feminist Media Praxis that included an article by Susana Loza (2014) on "Hashtag feminism, #SolidarityIsForWhiteWomen, and the other #Femfuture." These articles and special issues mark some of the first scholarly nods to hashtag feminism and hashtag activism as emerging areas of scholarly inquiry. For instance, Dixon and Loza's articles cite early case studies (most notably #SolidarityIsForWhiteWomen) and the platforms used to engage in feminist praxis across social media, like Tumblr and LiveJournal. Dixon's article, in particular, explores definitions and theories used to explain feminist praxis online. Loza's article takes a more cultural historical approach to explore early feminist praxis, while highlighting the role feminists of color played in "(re)forming mainstream American feminism."

One of the earliest and most widely cited scholarly articles on feminist hashtag activism comes from Sherri Williams' "Digital defense: Black feminists resist violence with hashtag activism," which is part of the 2015 *Feminist Media Studies* second issue. Williams examines Black women's use of Twitter hashtags to advocate for survivors of sexual assault and reframe messages about Black women as victims of sexual assault. Williams considers the use of Twitter hashtags as a media savvy approach to activism. Williams writes: "Black feminists' use of hashtag activism is a unique fusion of social justice, technology, and citizen journalism" (343). Williams considers Black feminism as a strategic approach to spark social action online and offline through digitally born content.

A relatively nascent body of literature considers the technical utility of hashtags and the digital geographic landscapes where hashtags emerge. This literature sets out to examine technology as ideology. These perspectives, typically coming out of code studies, STS (science and technology studies), critical race studies, and digital geography studies, interrogate how beliefs and ideologies are reinscribed through digital artifacts and platforms (Sharma 2013; Brock 2012, 2018). They frame non-visual dimensions of code, in this case U+0023 (the Unicode character for the hashtag), as actors in reconstituting social difference and undermining systems of power (Conley 2017; Elwood and Leszczynski 2018). These perspectives argue that technology platforms and artifacts like hashtags *do* things, they are not neutral.

Perhaps the most burgeoning body of work on hashtags examines publics and counterpublics, and the use of content analysis and social network analysis as primary methodological techniques. Some of this work includes examining, for instance: the socio-political practice of hashtag(ging); the role of hashtags in building and amplifying feminist and anti-racist movements; hashtags as sites to locate affinity spaces; and hashtags as modes of storytelling (Khoja-Moolji 2015; Bonilla and Rosa 2015; Jackson and Foucault Welles 2016; Yang 2016; Walton and Oyewuwo-Gassikia 2017; Kuo 2018; Myles 2018; Conley 2019; Jackson, Foucault Welles, and Bailey 2020). This work spans various disciplines and industries.

Overall, this body of literature offers valuable insights for scholars concerned with the future direction of hashtag research. As more scholarship on hashtags is produced, more questions about methods for capturing data and metadata across platforms will undoubtedly emerge.

A CULTURAL HISTORY OF HASHTAG FEMINISM

Hashtag feminism, as a concept and practice does not exist independent of other discursive forms relative to digital culture in the aughts. Before hashtag feminism and Twitter, Black feminist bloggers were theorizing, organizing, and telling stories across digital spaces like Culture Kitchen, Crunk Feminist Collective, Racialicious, African Diaspora, Ph.D., Diaspora Hypertext, to name a few (Johnson 2013; Bailey and Gumbs 2010). Hashtag feminism also shares a unique history with another like-term, hashtag activism. In 2011, the term "hashtag activism" was coined by journalist Eric Augenbraun from the *Guardian* to describe the #OccupyWallStreet movement. Around this time, people had already been using hashtag vernacular across public discourse. In 2012, *Ms. Magazine* published an article, "Future of Feminism: The Hashtag is Mightier Than the Sword," that signaled a shift towards a new(er) lexicon for feminist activism online and by way of social media (Scott). Other like terms such as Twitter Feminism, an idea amplified by Canadian feminist, and polarizing figure for her transphobic views, Megan Murphy as a critique of online feminism, also signaled toward this new(er) turn. But even before users began to associate online feminist activism with Twitter, hashtags like #fem2 and #femfuture appeared on the platform in 2008 and 2009, respectively, as a way to organize and locate feminist issues and conversations. Early feminist hashtag vernacular finds its roots in feminist blogging and online advocacy work that called out misogyny, misogynoir, and rape culture purported across mainstream media.

Around the same time when hashtag activism and hashtag feminism were becoming recognized, albeit contested forms of activism and feminist practice, data analysis and mining companies saw the potential in profiting from hashtag data. In 2013, Topsy, a free analytics search platform, was available to everyday users to search hashtag data. At the time, it was relatively feasible to locate the origin stories of social movement discourse, political protests, and advocacy campaigns, as well as to map diffusion and trajectories of hashtags like #BlackLivesMatter and #SolidarityIsForWhiteWomen across Twitter. After Apple acquired Topsy in 2015, the platform disappeared, and so did the ability for everyday users to freely explore metadata on Twitter. Topsy was only gone briefly, however. Once analytics companies recognized the profitability of mining and analyzing hashtag data, services like Topsy reappeared, but at-cost.

Fee-for-service hashtag analytics platforms posed challenges for researchers conducting mixed method analyses to examine critical social issues. If researchers did not manually collect tweets in Excel sheets or if they lacked the technical skills to mine hashtag data on their own using Twitter's API[1] then paying for hashtag data analytics was, and continues to be, a more feasible option. The black box conundrum is when data analytics services use their own algorithms and methods to mine data that customers cannot see (Littman 2017). This presents problems for Internet researchers concerned about ethical data collection, the accuracy of mined data, access to historical metadata, and terms of service. When researchers are unable to collect, assess, study, and create repositories for hashtag data, the loss of discurvise histories become apparent. Indeed, the political economy of mining hashtag data is an under-examined area of research that deserves more attention. That said, however, even with the help of platforms, services, and collaborators, questions about the origins of hashtags, attribution for hashtags, and about how hashtags have been used across time and space remain contested. For this reason, as a feminist researcher and mediamaker, I believed it necessary to build a web platform where I could archive digital feminist culture on my own terms.

In mid-December of 2013, roughly a week before Beyoncé released her surprise digital album that told her story about embracing a feminist identity, I was designing a WordPress website called Hashtag Feminism. I wanted the website to house social commentary about digital activism and popular culture led by Black women and women of color on Twitter, and to collect data that would document moments of authoring around a new(er) enunciation of feminism in the Twitter age. Several months prior, hashtags like #SolidarityIsForWhiteWomen (created by Mikki Kendall), #SolidarityIsForBlackMen (created by Jamilah Lemieux), and #NotYourAsianSidekick (created by

Suey Park) went viral and were making national and international headlines. More and more Black women and women of color were publicly calling out the failures of white feminism in the digital age. They were sharing stories about racialized gender violence, and speaking out against erasure in social movements that proclaimed racial justice and gender equality. Also at this time, I was working on a case study about the impact of online organizing led by Black women in response to the murder of 19-year-old Renisha McBride in Dearborn Heights, Michigan (2013a). Looking back, I consider 2013 the year Black women and women of color mediamakers, storytellers, activists, organizers trended. It was a watershed moment for feminist discourse and practice; an origin story that was, for better or worse, fervently documented by Black women and women of color using hashtags.

At present, there appears to be a rhetorical shift to move beyond hashtags in effort to re-center people and community practices in the study of the digital (Florini 2019; Jordan-Zachery and Harris 2019). To be clear, in offering a conceptual framework for hashtag feminism, I do so in order to center people, namely Black women and girls, in an analysis of digital culture while emphasizing that the technological processes for archiving digital and networked cultures are not neutral. The shift to move beyond the rhetoric of hashtags is not so much a gesture towards a *post-hashtag* era in the sense of anti-hashtag discourse, but rather, it appears to be a move to pause, reflect, and further explore hashtags as part of a broader network of communication technologies and sociality. Scholars continue to be critical of neoliberal and techno-utopian discourses that position technologies and *the digital* as neutral and egalitarian. The notion of "beyond the hashtag" also signals a shift away from an exclusive focus on the discursive function of hashtag (and creators) to consider the history of technologies and cybercultures, and social practices. This era of scholarship looks back at watershed moments like 2013, for example, in efforts to formulate newer understandings about theory, praxis, and language.

In the following sections, I analyze hashtag feminism as a conceptual framework through an in depth examination of discursive forms, practices, and ideological underpinnings. I take cues from Brock's seminal work on Black Twitter and his method, Critical Technocultural Discourse Analysis (2012; 2018; 2020)—which enhances decoding-as-method (Conley 2017)—in efforts to unearth a complex assemblage for articulating hashtag feminism.

HASHTAG FEMINISM-AS-DISCOURSE (HFAD)

www.hashtagfeminism.com

In her chapter, "Beyond the Categories of the Master Conception: The Counterdoctrine of the Jamesian Poiesis" Jamaican writer and cultural theorist, Sylvia Wynter writes,

> [t]o be effective, systems of power must be discursively legitimated. This is not to say that power is originally a set of institutional structures that are subsequently legitimated. On the contrary, it is to suggest the equiprimordiality of structure and cultural conceptions in the genesis of power. (65)

Here, Wynter provides insight into the conceptual utility of discourse and its relationship to institutions. Namely, discourse functions as a concept to describe how modes of language are tangled up in relations of power and representation. Wynter's assertion that systems of power legitimize discourse, and discourse legitimizes systems of power through structural and cultural conceptions provides a basis for this section that maps the website Hashtag Feminism (www.hashtagfeminism.com). For this section, I approach the concept of discourse in a way that "focus[es] on representation within and of technology," since technology is a political process for reproducing and re-presenting social hierarchies (Brock 2018, 1016).

Hashtag feminism-as-discourse (HFAD) describes the makeup of discursive forms that characterize a feminist ethos beginning in the aughts. HFAD considers how people engage in ideas of feminism online and considers the role web platforms and indexing processes play in reproducing feminist discourse. Myles (2018) also considers hashtag feminism from a discursive standpoint: "[A]pprehending hashtag feminism from a discursive standpoint makes sense, as it gives new opportunities to understand how citizens engage 'in acts of political creativity, negotiation, dialogue, and productive disagreement' (Shaw, 2016: 3) by enacting both the physical and digital affordances of a myriad of technologies, objects and spaces" (511). In other words, considering hashtag feminism from a discurvise standpoint provides a way to examine the relationship between popular cultural discourse and the technologies that carry discourse across time and space. I have been acutely aware of this relationship as a feminist mediamaker, insomuch as it influenced my approach to build www.hashtagfeminism.org.

The website Hashtag Feminism marks an origin story that represents how communities on Twitter took up feminist issues across national and transnational contexts. The technological aspects of Hashtag Feminism, such as the Wordpress development platform, the HTML codes used to design the platform, and the use of tagging to code content throughout the platform reproduced modes of sociality that first appeared on Twitter. From 2013–2015,

Hashtag Feminism functioned as a corpus of metadata and data that housed metrics and content, including hashtags that located feminist conversations, organizing, and storytelling online. It was also a space for feminist writers to provide commentary about hashtags that garnered national and international attention, such as for example, #SolidarityIsForWhiteWomen and #NotYourAsianSidekick (Conley 2013c).

In 2013 and 2014, I produced two informal analyses on the top feminist hashtags of the year. In 2013, I noted that for each hashtag highlighted, they represented their "own unique entry into conversations dealing with race, gender, sexuality, economic justice, global citizenship, and Ms. Yoncé Carter" (Conley 2013c). At the time, I monitored the top hashtags as determined by Topsy and Keyhole analytics platforms, and also looked at top trends on Twitter and news coverage during a 30-day period. The top hashtags of 2013 were not without social context, as indicated by my informal reference to Beyoncé Knowles Carter, who released her self-titled album that included a sample from Nigerian feminist writer Chimamanda Ngozi Adichie declaring a feminist identity. The design aesthetics of Beyoncé's album, with its black background and bright pink knockout font, influenced the design aesthetics of Hashtag Feminism's website (see figure 1.1 and figure 1.2).

On the surface, it would appear as if feminism was rebranded in 2013. However, I consider 2013 the year when feminist discourse emerged as a strategy to shape popular cultural discourse whilst struggles for representation and visibility were increasingly scrutinized. This is not to suggest that feminist discourse has not been scrutinized in the public sphere before. It is to say, however, that 2013 marks an origin story when racialized feminist discourse went viral through digitization and strategies of diffusion. For example, users employed an indexing process (the hashtag) to archive a message (e.g., white feminism's constant betrayal towards Black women), the framing of that message (e.g. #SolidarityIsForWhiteWomen), and the spread of the message across contexts and platforms. Further, I write (2013d), "[t]hat #NotYourAsianSidekick appeared in *Time*, Al Jazeera America, ABC, and BBC, and that #solidarityisforwhitewomen and #solidarityisforblackmen appeared on NPR, among other outlets, might indicate that broader audiences are paying more attention to our stories than ever before." Portwood-Stacer and Berridge (2014a) further articulate this phenomenon:

> Hashtags specifically related to feminist causes, like #YesAllWomen, #BringBackOurGirls, and #Direnkahkaha, are invoked by social media users worldwide in response to contemporary events and discussions. At their most visible, these terms and their spread are taken up by newspapers, television, and other media outlets as stories of collective public opinion and, sometimes, further action. (1090)

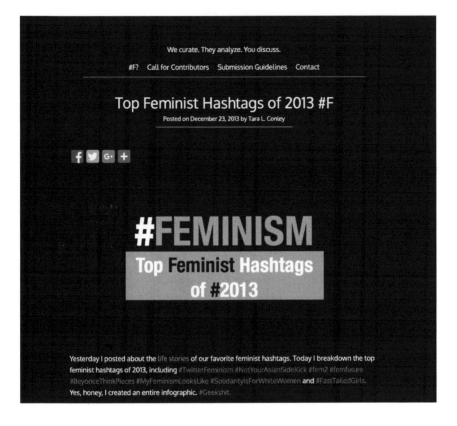

Figure 1.1.   Original Hashtag Feminism webpage from December 23, 2013. Screenshot from The Internet Archive's Wayback Machine, https:// www.archive.org/web/20140902183109/http://www.hashtagfeminism.com:80/ top-feminist-hashtags-0f-2013-f/.

The socio-political and cultural context of 2013 is important to expand on here because it places feminist issues in the context of contemporary discussions and events happening at the time. Tensions among feminists were reaching a tipping point as Black women and other feminists of color continued to call out white feminists' erasure of their histories and online advocacy work (myself included, see Conley 2013f). Mikki Kendall (2013) details aspects of this tipping point in the *Guardian* explaining the origins of #SolidarityIsForWhiteWomen. Jessica Marie Johnson (2013) also presents a detailed cultural history of the #FemFuture report and convening in 2013, funded in part by Barnard College, Columbia University. These moments carried out across Twitter and other online platforms arguably characterize a reckoning for U.S. feminism in the early twenty-first century. Also at this

Figure 1.2.   Beyoncé album cover. By Columbia Records, Public Domain.

time, President Barack Obama was entering his second term, and the political landscape in the U.S. remained divisive, especially online. Twitter also became a publicly traded company in 2013, impacting the platform's operations and capital growth, including the launch of Vine, a short-form video hosting service. The year ended with the surprise release of *Beyoncé*, a non-marketed digital album that generated historic buzz, including 1.2 million tweets in twelve hours on the day of its release. The cultural significance of Beyoncé's digital album coupled with intracommunity tensions, political climate, and the expansion of technology infrastructures shaped how feminist discourse and practices played out across online spaces. To be sure, 2013 was a year digital media landscapes changed and newer modes of sociality emerged with Black women and women of color shaping the frontier.

The year 2014 mirrored 2013 in the way users continued to employ a strategy of rapid diffusion of feminist and anti-racist discourse using hashtags (Conley 2014b). Hashtags like #YouOkSis, #YesAllWomen, #BringBackOurGirls represented transnational feminist discourse that re-centered women and girls, primarily Black women and girls in the face of misogyny and misogynoir (Bailey 2014; Bailey and Trudy 2018). Hashtags archived these moments in time, subsequently opening up an entire area of study that would be written about and researched for years to come.

Roughly 104 hashtags were indexed and 54 articles were published on Hashtag Feminism, which Kelly Ehrenreich and I co-edited from 2013–2015. These articles, some of which are detailed in table 1.1, represent the most visible discourse on Twitter that highlighted racialized gender violence and the failure of liberal white feminism. In explaining the social and political context of hashtag activism, Ehrenreich (2014) outlines four tenets of hashtag feminism to consider based on her analysis of the discourse captured on Hashtag Feminism between 2013 and 2015. She writes:

1. Tweeting about something does not necessarily bring about political change.
2. Hashtag feminism may preach to the choir, but we still have to take our choirs offline and out into the community.
3. Representation in the media is key and hashtags, in many ways, disrupt mainstream media narratives about marginalized communities and unjust legislation.
4. Online communication by way of hashtags can help birth a new generation of understanding, empathy, and acceptance.

The events surrounding Renisha McBride's untimely death marked an origin story of the platform, Hashtag Feminism. My blog post "#RenishaMcBride and online activism #femfuture #fem2 #Twitterfeminism #f" (Conley 2013b) was the second post to appear on Hashtag Feminism the day of its launch on December 20, 2013. This post, along with the online case study I published one month prior (2013a), informed the mission of Hashtag Feminism as an intersectional digital space that reproduced discourse about racialized misogyny and gender violence in the United States. The story of Renisha McBride is also one about how advocates and activists used Twitter and hashtags to organize local protests, push news outlets to cover her death, and pressure local officials to take legal action against her killer. In keeping with Ehrenreich's tenets from above, it was not enough for activists to use #RenishaMcBride and #RememberRenisha to bring about awareness, they also had to take their advocacy offline. They leveraged online communication technologies like hashtags, YouTube videos, and digital news stories to cement Renisha McBride's legacy in our collective memory. What was once an under-

Table 1.1.

Hashtag	Year	Description
#RenishaMcBride	2013	Locates the story about Renisha McBride, who was shot and killed in Dearborn Michigan on November 2, 2013. The hashtag and its adjacent hashtag #RememberRenisha points to grassroots organizing led by activists and mediamakers in protest of her death.
#NotYourAsianSidekick	2013	Created by Suey Park on December 14, 2013 to "engage a critical conversation about Asian American feminism" (Conley 2013e). Park later faced public backlash that scholar Yasmin Nair (2016) documents and critically analyzes.
#TwitterFeminism	2013	Locates contentious debates among online feminist communities about the utility of Twitter and its users to engage in feminist causes.
#MarissaAlexander	2013	Locates the story about Marissa Alexander, a domestic abuse survivor who faced sixty years in prison for defending herself against her husband. The adjacent hashtag #FreeMarissa points to grassroots organizing led by activists and mediamakers in protest of her incarceration and domestic violence laws.
#BlackFemMusic	2013	Created by the PBS program *News and Then* as part of a Twitter chat and inspired by Beyoncé Knowles' album release. The discussion covered issues relating to Black women in music and the role of Black feminism in popular music.
#PGPDVice	2014	Locates public backlash to the Prince George's County Police Department in Maryland when announcing they would live-tweet a prostitution sting.
#YesAllWomen	2014	Created on May 24, 2014 by a user who wished to remain anonymous in response to the Isla Vista killings and in response to the hashtag #NotAllMen. The hashtag locates stories about violence against women, noting the relationship between misogyny and gun violence in the United States. #AllMenCan became an adjacent hashtag deriving from #YesAllWomen.
#HobbyLobby	2014	Locates public backlash to the Supreme Court's decision in *Burwell v. Hobby Lobby* that would allow companies to withhold contraception coverage from their employees on the grounds of religious belief. Adjacent hashtags include: #DrHobbyLobby and #NotyMyBossBusiness.

Hashtag	Year	Description
#YouOkSis	2014	Locates Black women and women of color's stories about experiencing street harassment and bystander intervention, as well as conversations in response to #ThatsWhatHeSaid and #StopStreetHarassment. Feminista Jones started the conversation on Twitter about Black women's encounters of street harassment, in particular, asking "You Ok, sis?" after which Black Girl Danger turned Jones' question into a hashtag (Conley 2017).
#WhyIStayed	2014	Created by Beverly Gooden on September 8, 2014 who started a conversation by posting a series of tweets about domestic abuse after video footage was released online showing NFL Player Ray Rice assaulting his fiancée Janay Palter in an elevator.
#BlackGirlsMatter	2015	Created in response to a report published by Kimberlé Crenshaw and the African American Policy Forum about Black girls and the school-to-prison pipeline. Adjacent hashtags include #WhyWeCantWait, and later #SayHerName.
#BaltimoreUprising	2015	Locates protest stories following Freddy Gray's death in Baltimore, Maryland. Social media users made #BaltimoreUprising to counter the narrative about community protests characterized in #BaltimoreRiots. Adjacent hashtags include: #FreddyGrey, #BreakTheCurfew, #BaltimoreLunch, #BaltimoreCurfew #JusiceforFreddieGray, and #Baltimore.

reported story about a young Black woman who was killed in Dearborn, Michigan became an early marker in hashtag feminism history made possible by Black women's organizing online and offline. Hashtag Feminism also celebrated the organizing work and storytelling surrounding anti-racist issues. #BaltimoreUprising (Springer 2015), for example, was the final hashtag analyzed on Hashtag Feminism. This hashtag locates Ferguson-era racial politics led by young Black activists who were using hashtags as a counter narrative strategy for protesting police violence against Black Americans.

In 2016, Hashtag Feminism was hacked by unknown sources and could not be accessed (Conley 2016). I was able to restore most of the content through a variety of methods, including using the Internet Archive's Wayback Machine and parsing through old emails, Google Documents, and other saved documents that stored older drafts of essays written by former contributors. By 2018, I relaunched Hashtag Feminism primarily as an archive of restored content. The threat of losing discursive content on the web should be an important consideration for anyone documenting and archiving digital cultures. So while I consider Hashtag Feminism a platform "born out of love for Internet culture and passion for social justice" (Conley 2019), I recognize too that preserving feminist discourse on the web is laborious and platforms are vulnerable because technologies remain unstable.

Feminist Digital Media Scholarship

A review of key terms and titles across digital feminist scholarship reveals that the term *hashtag feminism* has been used less often since 2013 than the term *digital feminism* (see table 1.2).[2] Though this review does not encapsulate all research published on feminist hashtags (for instance, there are articles written that cite feminist hashtags without terms being used as keywords or in the title), it does suggest that as far discourse goes, *digital feminist* and a close variation of the term have been the preferred scholarly descriptors. I believe one reason for this is because the term *hashtag feminism*, like *hashtag activism*, has been analyzed in mainstream press according to toxic online discourse and based on a false dichotomy of "real activism versus slacktivism" (Risam 2015; Casey 2016). That said, however, feminist bloggers were among the earliest critics to debunk this false equivalence (Bailey and Gumbs 2010; Conley 2012a; Conley 2012b).

That said, as it relates to digital feminist scholarship, it appears *hashtag feminism* is not considered an all-encompassing term whereas *digital feminism* seems broad enough to encapsulate a study of feminist practices online. While the terms *digital feminist* and *digital feminism* reign supreme in areas of critical studies, the term *hashtag feminism* may have lost discursive relevance as scholars continue to formulate new understandings about feminist digital cultures. A long-term consequence of moving away from "hashtag feminism" as a key term in research could perhaps result in discursive delegitimization across critical studies research—again, as noted by Wynter's words at the beginning of this section (1992), "to be effective, systems of power must be discursively legitimated" (65). Though Wynter was speaking in the context of the European bourgeoisie, her insight can also be applied to how knowledge get constructed and valorized across critical studies scholarship. As both a technical and cultural artifact, the hashtag is a unique indicator for documenting and recording histories of social movements and counternarratives. It also seems necessary to recognize not only the hashtag's discursive and technical function in feminist research, but also its historical context as a culture defining communication technology. As such, I believe hashtag feminism, as an area of inquiry and key term, should continue to be scrutinized and indexed across academic research.

A review of the Hashtag Feminism website and feminist digital media studies database suggests that HFAD characterizes a discursive strategy for 1) locating shared feminist referents across transnational contexts and 2)

Table 1.2.

Terms	Instances of Terms as Keyword	Instances of Terms in Title
Digital feminist; or digital and feminist	5,244	11
Digital feminism; or digital and feminism	2,930	7
Feminist hashtags; or feminist and hashtags	272	5
Hashtag feminism; or hashtag and feminism	176	2

archiving digital feminist cultures. Though hashtag feminism is arguably considered a waning descriptor to examine feminist practices online, as evident in the hashtags analyzed on Hashtag Feminism starting in 2013, hashtag feminism has done well to call attention to worldwide feminist issues and stories, and the social contexts and histories surrounding them. In the case of #RenishaMcBride, for example, Black women and women of color leveraged the hashtag as an indexing process to shape public discourse while calling attention to racialized gender violence. In the case of #SolidarityIsForWhite-Women, among other similar hashtags for example, the hashtag also locates righteous anger about the erasure of Black women and women of color across movements for gender justice.

HASHTAG FEMINISM-AS-EMBODIED-PRACTICE (HFAEP)

By now, the reader might have gathered that I am preoccupied with origin stories. My proclivity to the study of hashtags is likely a result of this preoccupation and why I often recite in my scholarship that hashtags locate stories. In the case of hashtag feminism, I consider origin stories events to critically examine digital cultures that also preserve stories from Black and non-white queer women, femme, and trans people online (Conley 2020). My view of origin stories across digital culture is influenced by Sylvia Wynter's notion that stories reflect human scripts, ones that tell us how we are and what we are across "shared virtual worlds" (Ambroise 2018, 848). Wynter's writes, "[o]ur 'stories' are as much a part of what makes us human . . . as are our bipedalism and the use of our hands" (Wynter, 2015, 217). Stories help organize people lives by narrating what *being* human means. To further elaborate on Wynter's view of the storytelling human, Ambroise writes:

> Wynter proposes that we humans know, feel, and experience ourselves not solely in biocentric terms as a purely biological being but also in pseudospeciating terms as a specific "type," "kind," or "sort" of being human—what she terms a "genre of being hybridly human"—that is derived from each origin

story's/myth's answer to the questions "what is humankind?," "who am I?," "what am I?" (850)

In other words, and this is where it gets murky, the stories we tell ourselves reflect the genre-specific modalities (e.g., race and gender) around which we organize ourselves—and sometimes the stories we tell ourselves can betray us.

Origin stories in the eurocentric context, for example, separate humanity on the basis of genre-specific modalities and through a biocentric knowledge system informed by Darwinian thought. Wynter's vision gives humanness a future beyond biocentric modes of being.[3] Such a vision requires relational and interdisciplinary thinking to understanding "the rhythmic interplay between nature and narrative (McKittrick, O'Shaughnessy, and Witaszek 2018, 867). Given that the eurocentric script of humanness has overwhelmingly privileged the hegemonic man (or normalized genre of the human), I have in my research sought to re-center the humanness of the non-hegemonic man.

Wynter introduces the term *homo narrans* to describe the hybridity of the human as both bios and myths—as a "storytelling species" (McKittrick 2018, 25). Wynter's "anti-colonial project" (Trimble 2017, 277) not only challenges purely biocentric views of humanness, but it also calls into question a reliance on origin stories as events to narrate what it means to be human. For this reason, I often revisit my own conceptualization of origin story to think beyond linearity and hierarchy while maintaining an advocacy for authorial attribution as it concerns the work of Black women and women of color online (Conley 2020). I also consider how multiple origin stories can emerge across time and space, and the implications therein for documenting digital culture, namely feminist storytelling and organizing practices.

Informed by Wynter's concept of *homo narrans*, hashtag feminism-as-embodied-practice (HFAEP) describes a relationship between ontology and technology, between being and technical processes reflected in stories of lived experiences. Rentschler and Thrift consider this relationship through the idea of techne, or "ways of doing" feminism (2015, 242):

> What people actually do emerges from their embodied, situated and practical knowledge, their "ways of doing things" according to anthropological definitions of culture. Techne is an embodied relationship to technology, a learned and socially habituated way of doing things with machines, tools, interfaces, instruments and media. Techne signals more than technical skill; it constitutes embodied habits for acting and doing.

HFAEP suggests that feminist praxis cannot be separated from such things as, for example, memory, corporeal movement, socialization, diaspora, nature, and affective experiences. It also suggests feminist storytelling and organizing that confront sexism, gender violence, and racism, for example, are inextricably tied to ways of knowing, feeling, and being. Cherríe Moraga and Gloria E. Anzaldúa call this ontological entanglement *theory of the flesh* (2002, 21). Patricia Hill-Collins describes it through standpoint theory in Black feminist thought (2000, 270). Wynter's storytelling species questions what it means to *be* among systems of oppression, turning the concept of ontology on its head. Stories, then, "become" encoded modes of ourselves as living beings constantly struggling to disentangle from contained human scripts (Conley 2017). Hashtags, as technical processes, locate this entanglement, providing temporal and spatial ways of seeing the *"symbolic species that we are"* (Ambroise 2018, 848).

#MeToo

In 2017, roughly one month after #MeToo went viral on Twitter, Tarana Burke wrote an op-ed for *The Washington Post* pithily entitled "#MeToo Was Started for Black and Brown Women and Girls. They're Still Being Ignored." The article was written in response to mainstream attention directed towards elite white women after journalist Ronan Farrow broke the story for *The New Yorker* one month prior about Hollywood producer Harvey Weinstein's sexual misconduct. Most interestingly, Burke, who founded the Me Too movement in 2006, begins the article with a story about witnessing a Black waitress sexually harassed by the main cook at a local diner in Montgomery, Alabama. As has always been the case with Burke's advocacy across media, she leveraged her access to mainstream press in efforts to re-center Black women in the movement to fight against sexual assault and harassment. Burke writes:

> From the start of #MeToo going viral and the recognition of my years of work preceding it, I have been happily allowing this wave of attention to shine a much-needed light on the fight to end gender-based violence. I founded the "me too" movement in 2006 because I wanted to find a way to connect with the black and brown girls in the program I ran. But if I am being honest with myself, and you, I often wonder if that sister in the diner has even heard of #MeToo, and if she has, does she know it's for #UsToo?

It is no secret that Black women and girls in the era of Me Too have overwhelmingly been ignored by mainstream press. When Black women and girls are being written about it is usually because other Black women, like Burke are writing about these cases (see also *Rejoinder*'s 2019 special issue). Black women and girls are not the only ones largely excluded; other communities include sex workers (Cooney 2018) and trans people (Talusan 2018), and oftentimes these communities overlap. That said, relative to the entire body of scholarly research on the Me Too movement, examining the lived experiences of Black women and girls continues to be an under researched area, which appears to mirror public discourse.

Burke's framing of #MeToo through a story about witnessing the sexual harassment of Black waitresses in Alabama is as much a mobilizing strategy as it is a discursive one. Oftentimes, when Black women advocates, scholars, and public figures write about, discuss, and analyze #MeToo, we are intent on re-centering Black women and girls in the movement. Doing so creates a narrative strategy that simultaneously disrupts normative discourse surrounding gender-based violence and informs the public, including the academy, about gender-based violence unique to Black women and girls. As an embodied practice, advocates, scholars, and public figures enact Ehrenreich's (2014) third tenet of hashtag feminism, namely to "disrupt mainstream media narratives about marginalized communities and unjust legislation."

Analyzing #MeToo through the construct of HFAEP demonstrates how Black women have organized across online spaces based on shared lived experiences and *"shared resistance"* to erasure in the movement (Sharma 2018, 179). Much like other hashtags that point to Black women and girls' stories of gender-based violence like #WhyIStayed and #YouOkSis, #MeToo locates Wynter's *homo narrans* as well as the mobilizing work involved in re-centering the humanness of Black women and girls. At the same time, #MeToo also locates an origin story across mainstream media, which upholds the idea that sexual harassment and assault are the burden of white upper-class cis women. Further, stories of Me Too and the hashtag that locates them capture a longstanding history of competing solidarities in the fight for gender and racial justice.

HASHTAG FEMINISM-AS-IDEOLOGY (HFAI)

Wynter's once said in an interview, "[g]ood heavens, just as we had to fight Marxism, we're going to have to fight feminism" (Thomas, 2006). In the spirit of Wynter's candid discontentment with ideological markings of feminism, I begin this section citing examples of conflict and confrontation across time and space to illustrate moments when feminist ideologies emerge and collide. In 2013, I wrote an open letter to feminist writer Amanda Marcotte that, in part, asked: "Amanda, I wonder: What is your understanding of online feminism?" (Conley 2013f). The letter was published online for *The Feminist Wire* in the spirit of Audre Lorde's letter to Mary Daly, who in 1979 took Daly to task for her erasure of Black women's experiences and ideas in the book *Gyn/Ecology*. Lorde acknowledged Daly's work for its generative and provocative insights on "the true nature of old female power" (Moraga and Anzaldúa 2002, 101). At the same time, however, Lorde invited Daly to consider how her work promoted an ahistorical view on the ecology of women, which espoused a belief that all women suffer the same oppressions. Lorde called on Daly to explain her misuse of Black women's words to support claims in the book. Lorde writes:

> Mary, do you ever really read the work of black women? Did you ever read my words, or did you merely finger through them for quotations which you thought might valuably support an already-conceived idea concerning some old and distorted connection between us? This is not a rhetorical question. To me this feels like another instance of the knowledge, crone-logy and work of women of color being ghettoized by a white woman dealing only out of a patriarchal western-european frame of reference. (103)

Lorde's letter came during a moment in feminist history when feminists of color were openly challenging second wave feminism in print, on television and radio, in academia, and in the streets. This largely white middle class women's movement did little to speak the issues of working class Black women, Indigenous women, and other women of color.

At the time I published the open letter to Marcotte, I was also referring to white feminists' erasure of Black and Brown women's ideas and grassroots work documented online, namely Marcotte's re-appropriation of feminist blogger Brownfemipower's work on gender-based violence in immigrant communities. A few months later, Mikki Kendall created the hashtag #SolidarityIsForWhiteWomen in response to white feminists' dismissal of women of color who at the time were critical of Hugo Schwyzer, a self-identified feminist who engaged in sexual misconduct with female students and abusive behaviors towards women of color online. Kendall (2013) writes: "Digital feminists" like Jill Filipovic (also a *Guardian* columnist), Jessica Coen, Jessica Valenti and Amanda Marcotte were, in our view, complicit in allowing Hugo Schwyzer to build a platform—which, as he has now confessed, was based partly on putting down women of color and defending white feminism." Kendall argued that white feminists' "brand of solidarity centers on the safety and comfort of white women" (Kendall 2013). This message resonated as #SolidarityIsForWhiteWomen immediately went viral. Kendall's hashtag located contested solidarities in feminism, both generational and ideological across time. Black women's fight for representation and acknowledgment in social movements, in theory, in the home, and in our communities is nothing new. What Kendall and I identified in 2013 was something for which Lorde and Wynter had already been advocating; that is, for white feminists to come to terms with generations of betrayal of Black and Brown women in public and in private spaces.

Implicit in calling attention to the waywardness of white feminism is the struggle to undo scripts of humanness that uphold colonialism, capitalism, and whiteness as ideology. Black feminists have a long tradition of rejecting white feminism as a liberatory strategy for a select few and for its one-size-fits-all vision for the colonized subject. Black feminists continue this tradition across digital spaces to bring attention to white feminism's ineffectiveness as an organizing strategy and as an ethos. Similar to how social media platforms like Twitter and code like U+0022 represent a logic of standardization (Conley 2017), white feminism too is an ideology of standardization. It represents a faction of feminism at odds with a Black feminist ethos for justice and liberation, which seeks to undo the mess colonialism made.

Factions in feminism are perhaps most apparent in the constant struggle for Black and Indigenous women (cis, trans, and queer) to escape otherness in theory and across social movements where race and nativeness have been relegated to footnotes and otherwise ignored. But representation is not enough. A Wynterian view understands representation as "the central mechanism at work" in the process of colonization (Sharma 2015, 169). Black and Indigenous women are trapped within the contours of symbolic representation that have been "designed to institutionalize the new racist orders implemented by different colonial empires" (Sharma 174).

To articulate an escape of *being* colonized should, in part, be a central tenet to feminism, regardless of ideological underpinnings. This requires a reorientation to history and to language. The failure to acknowledge this tenet might also speak to where solidarities lie, whether in the realm of knowledge or practice, solidarities rest in the interpretation (hermeneutic) of what justice and liberation look like. Hashtags point to these interpretations and help us see factions at work. More importantly, however, hashtags locate thresholds where reorienting away from colonized scripts of humanness can take shape.

Hashtag Feminism-as-Ideology relies on technical artifacts and contemporary vernacular to articulate and interpret feminist thought and practice. It is a belief that technical processes reflect shared referents, human scripts, and virtual worlds, and as such, feminist solidarities and resistance must both be scrutinized. HFAI pushes feminist theory to account for documenting feminist ways of knowing, doing, and being online while historicizing the technical processes used to index thought and practice across time and space. HFAI espouses a belief that the practices of indexing and archiving digitally born content can and should respond to ideological standards of feminism that uphold colonialism, capitalism, and whiteness as tenets. As an ideological perspective of feminism at the intersection of technology and culture, hashtag feminism reimagines solidarity, and values the indexing process for locating sites of becoming across feminist ways of knowing, doing and making, and being.

LOCATING BECOMING ACROSS THE EPISTEME, TECHNE, AND ONTOLOGICAL: A VISION FOR FEMINISM

Hashtag feminism as a conceptual framework captures technical processes that locate discursive forms of feminism, feminist practices (like storytelling and organizing), and ideological underpinnings of feminism, as well as locates technical infrastructures that align or malign with a feminist ethos for justice and liberation. The constructs analyzed here manifest in various ways and are informed by ontological renderings of humanness. Hashtags locate these manifestations, namely the tension that exists across feminist ways of knowing (episteme), feminist doing and making (techne), and ways of being (ontological). These constructs, however, ought not be understood in isolation but rather considered for their mutually constitutive function for locating sites of becoming.

When I wrote "Decoding Black feminist hashtags as becoming" (2017), I was demonstrating decoding-as-stance as a critical method for mapping entanglement among feminist discourse, practice, and ideologies of technology. I was also offering this method as a vision for how hashtags locate where and how shared anger can contribute to renewed solidarities. Becoming is understood as a threshold of potentiality where freedom from forms of oppression can take shape. Expanding on that vision here, I consider that in order to locate sites of becoming across ways of knowing and ways of doing, we must reorient and respacialize ourselves to see where "ontological unity" can exist (Sharma 2015, 180). Technical process, while imperfect, can help reorient ourselves to feminist histories across time and space, namely when and where feminists got it right and when and where we got it wrong. For instance, #SolidarityIsForWhiteWomen points not only to the ongoing tensions associated with white feminism in the Twitter age but the hashtag also harkens back to past generations when Black feminists confronted white feminists for their shortsightedness in advocating for gender equality and representation. This hashtag and many others show how ideologies, informed by colonized ways of knowing, move. They also archive and track the ways feminist discourse, storytelling, and ideologies morph overtime.

If a vision of feminism is one of freedom from all forms of oppression, then it must also consider a future that does not rely solely on contained scripts of humanness for justice and liberation. It is not enough to advocate for representation, as evident in some of the hashtags outlined here like #MeToo. A feminist vision of freedom must also consider how ways of knowing and doing can upend systems of oppression that seek to contain and reorganize people as a result of colonization. Locating contested solidarities can be an entry point for transformation. For instance, #YouOkSis was a necessary response to #StopStreetHarassment as #SolidarityIsForWhite-Women was a necessary response to #FemFuture. These hashtags, among others created by Black women, locate both righteous anger and love for communities. They find stories of shared resistance and solidarity. They also provide an opportunity to re-envision freedom as a shared sense of advocacy—one that, in practice, manifests as disrupting fixed systems of thought and process, and disrupting one-size-fits-all ways of *doing* feminism.

CONCLUSION

If it seems trite to theorize feminism on the basis of the hashtag symbol, consider other ways systems of thought have been conceptualized through cultural artifacts. For example, Ford Motor Company's assembly line signifies a structuralist way of thinking about mass production, otherwise understood as Fordism. Similarly, complex technical infrastructures like the Internet might signify a reductionist way of thinking about technologies as at once invisible and agentive, otherwise understood as techno-determinism. Wynter expands on the value of theorizing through cultural conceptions. Wynter (1992) writes:

> Cultural conceptions, encoded in language and other signifying systems, shape the development of political structures and are also shaped by them. The cultural aspects of power are as original as the structural aspects; each serves as a code for the other's development. It is from these elementary cultural conceptions that complex legitimating discourses are constructed. (65)

In other words, there is meaning in the artifacts of our daily lives, and from those artifacts (and what I have attempted to do here) we can produce knowledge and reshape theoretical discourse.

In offering this conceptualization of hashtag feminism, it is my hope that the reader gained a better understanding of hashtag feminism beyond simply a moniker of feminism in the digital age. As I have argued throughout, hashtag feminism is a conceptual framework to help explain the relationship between feminist praxis and the technical processes enacted for capturing feminist praxis. This way of framing feminism underscores the tension at work around locating thresholds where solidarities lie, are contested, and where they are transformed. Further, this framework provides an opportunity to expand theoretical conceptions of feminism that accounts for the role cultural artifacts, both technical and material, play in capturing discourse, practices, and ideologies of our time.

NOTES

1. API is an acronym for Application Programming Interface, which allows users to access, retrieve, and send Twitter data without opening the Twitter application.

2. From October 2019 to October 2020, I monitored a corpus of key terms and titles across digital feminist scholarship dating from 2013–2020 using Columbia University's online library database (CLIO). I excluded newspapers and included only scholarly publications from books, journal articles, book reviews, magazine articles, journals, books, and book chapters. Though this review is not exhaustive, it does provide insight into the shape of scholarly digital feminist discourse over a year's time.

3. I have merely scratched the surface of Wynter's theorizing in this chapter. For more on Wynter's conceptualization of homo narrans and what she calls the Third Event of humanness, see Katherine McKittrick's edited volume *Sylvia Wynter: On Being Human As Praxis* (2015).

REFERENCES

Ambroise, Jason. R. 2018. "On Sylvia Wynter's Darwinian Heresy of the 'Third Event.'" *American Quarterly* 70(4): 847–56.

Ames, Morgan and Mor Naaman. 2007. "Why We Tag: Motivations for Annotation in Mobile and Online Media." *CHI*: 1–10.

Augenbraun, Eric. 2011. "Occupy Wall Street and the Limits of Spontaneous Street Protest." *Guardian*, September 29, 2011.

Bailey, Moya. 2014. "More on the Origin of Misogynoir." *Moyazb* (blog), April 27, 2014.

Bailey, Moya and Alexis Pauline Gumbs. 2010. "We Are the Ones We've Been Waiting For: Young Black Feminist Take Their Research and Activism Online." *Ms. Magazine.* (Winter).

Bailey, Moya and Trudy. 2018. "On Misogynoir: Citation, Erasure, and Plagiarism." *Feminist Media Studies* 18(4): 762–68.

Bastos, Marco Toledo, Rafael Luis Galdini Raimundo, and Rodrigo Travitzki. 2013. "Gatekeeping Twitter: Message Diffusion in Political Hashtags." *Media, Culture & Society* 35(2): 260–70.

Bonilla, Yarimar and Jonathan Rosa. 2015. "#Ferguson: Digital Protest, Hashtag Ethnography, and the Racial Politics of Social Media in the United States." *American Ethnologist* 42: 4–17.

Brock, André. 2018. "Critical Technocultural Discourse Analysis." *New Media & Society* 20(3): 1012–30.

———. 2020. *Distributed Blackness: African American Cybercultures (Critical Cultural Communication Book 9).* New York: New York University Press.

———. 2012. "From the Blackhand Side: Twitter as a Cultural Conversation." *Journal of Broadcasting & Electronic Media* 56(4): 529–49.

Bruns, Axel and Stefan Stieglitz. 2012. "Quantitative Approaches to Comparing Communication Patterns on Twitter." *Journal of Technology in Human Services* 30(3–4): 160–85.

Burke, Tarana. n.d. "The Inception." *Just Be Inc.* (blog), nd. http://justbeinc.wixsite.com/justbeinc/the-me-too-movement-cmml.

———. 2017. "#MeToo was Started for Black and Brown Women and Girls. They're Still Being Ignored." *The Washington Post*, November 9, 2017.

Casey, Sarah. 2016. "A Case Study of Feminist Activist Interventions in Queensland Party Politics: #Sackgavin." *Outskirts*, 34: 1–18.

Collins, Patricia Hill. 2000. *Black Feminist Thought: Knowledge, Consciousness, and the Politics of Empowerment.* New York: Routledge.

Cooney, Samantha. 2018. "'They Don't Want to Include Women Like Me.' Sex Workers Say They're Being Left Out of the #MeToo Movement." *Time*. February 13, 2018.

Conley, Tara L. 2012a. "Introducing the Online-Offline Continuum (A Work in Progress)." *Breakfast Ed.D. Omlette* (blog), April 19, 2012a. https://taralconley.org/breakfast-edd-omlette/2012/04/20/introducing-the-online-offline-continuum-a-work-in-progress.

———. 2012b. "The Evolution of a Definition: The ONL-OFL Cntinuuum." *Breakfast Ed.D. Omlette* (blog), April 24, 2012. https://taralconley.org/breakfast-edd-omlette/2012/04/20/introducing-the-online-offline-continuum-a-work-in-progress.

————. 2013a. "Tracing the Impact of Online Activism in the Renisha McBride Case." *Media Make Change* (blog). December 13, 2013. https://taralconley.org/media-make-change/blog/2013/tracing-the-impact-of-online-activism-in-the-renisha-mcbride-case.

————. 2013b. "#RenishaMcBride and Online Activism #femfuture #fem2 #Twitterfeminism #f." *Hashtag Feminism* (blog), December 20, 2013. https://taralconley.org/hashtag-feminism-archive/renishamcbride-and-online-activism-femfuture-fem2-twitterfeminism-f.

————. 2013c. "Top Feminist Hashtags of 2013 #F." *Hashtag Feminism* (blog), December 23, 2013. https://www.hashtagfeminism.com/top-feminist-hashtags-of-2013-f.

————. 2013d. "2013: The Year of the Feminist Hashtag #FeminismMeans #F." *Hashtag Feminism Archive* (blog), December 30, 2013. https://taralconley.org/hashtag-feminism-archive/2013-the-year-of-the-feminist-hashtag-feminismmeans-f.

————. 2013e. "Top Feminist Hashtags of 2013 (Explained) #F." *Hashtag Feminism* (blog), December 29, 2013. https://www.hashtagfeminism.com/top-feminist-hashtags-of-2013ex-plained-f.

————. 2013f. "An Open Letter to Amanda Marcotte." *The Feminist Wire,* March 4, 2013.

————. 2014a. "From #RenishaMcBride to #RememberRenisha: Locating Our Stories and Finding Justice." *Feminist Media Studies* 14(6): 22–32.

————. 2014b. "Top Feminist Hashtags of 2014." *Hashtag Feminism* (blog), December, 10 2014. https://www.hashtagfeminism.com/top-feminist-hashtags-2014.

————. 2016. "Hacked . . . but back!" *Hashtag Feminism* (blog), May 29, 2016. https://taralconley.org/hashtag-feminism-archive/hackedbut-back.

————. 2017. "Decoding Black Feminist Hashtags as Becoming." *The Black Scholar: Journal of Black Studies and Research* 47(3): 22–32.

————. 2018. "Framing #MeToo: Black Women's Activism in a White Liberal Media Landscape." *Media Ethics* 30(1).

————. 2019. "About." *Hashtag Feminism* (blog), https://www.hashtagfeminism.com/about.

————. 2020. "Hashtag Archiving." In *Uncertain Archives*, edited by Nanna Bonde Thylstrup, Kristin Veel, Catherine D'Ignazio, and Annie Ring, np. Cambridge: MIT Press, 2020.

Dixon, Kitsy. 2014. "Feminist Online Identity: Analyzing the Presence of Hashtag Feminism." *Journal of Arts and Humanities* 3(7): 34–40.

Ehrenreich, Kelly. 2014. "Hashtag Activism: The Politics of Our Generation." *Hashtag Feminism* (blog), December 2, 2014. https://taralconley.org/hashtag-feminism-archive/hashtag-activism-the-politics-of-our-generation.

Elwood, Sarah and Agineszka Leszczynski. 2018. "Feminist Digital Geographies." *Gender, Place, & Culture* 25(5): 629–44.

Florini, Sarah. 2019. *Beyond Hashtags*. New York: New York University Press.

Fox, Jesse, Carlos Cruz, and Ji Young Lee. 2015. "Perpetuating Online Sexism Offline: Anonymity, Interactivity, and the Effects of Sexist Hashtags on Social Media." *Computers in Human Behavior* 52: 463–42.

Gray, Jonathan. 2013. "When Is The Author?" In *A Companion to Media Authorship*, edited by Jonathan Gray and Derek Johnson, 88–111. Hoboken: John Wiley & Sons, Inc.

Kehoe, Andrew and Matt Gee. 2011. "Social Tagging: A New Perspective on Textual 'Aboutness. Methodological and Historical Dimensions of Corpus Linguistics.'" *In Studies in Variation, Contacts and Change in English Volume 6*, edited by Paul Rayson, Sebastian Hoffmann, and Geoffrey Leech. Helsinki.

Kendall, Mikki. 2013. "#SolidarityIsForWhiteWomen: Women of Color's Issue with Digital Feminism." *Guardian*. August 14, 2013.

Jackson, Sarah J. and Brooke Foucault Welles. 2016. "#Ferguson Is Everywhere: Initiators in Emerging Counterpublic Networks." *Information, Communication & Society* 19(3): 397–418.

Johnson, Jessica Marie. 2013. "#FemFuture, History and Loving Each Other Harder." *Diaspora Hypertext* (blog), April 12, 2013. http://dh.jmjafrx.com/2013/04/12/femfuture-history-loving-each-other-harder/.

Jordan-Zachery, Julia S., Duchess Harris, Janelle Hobson, Tammy Owens. 2019. *Black Girl Magic Beyond the Hashtag: Twenty-First-Century Acts of Self-Definition (The Feminist Wire Books).* Tucson: The University of Arizona Press.

Khoja-Moolji, Shenila. 2015. "Becoming an 'Intimate Publics: Exploring the Affective Intensities of Hashtag Feminism." *Feminist Media Studies* 15(2): 347–50.

Kuo, Rachel. 2018. "Racial Justice Activist Hashtags: Counterpublics and Discourse Circulation." *New Media & Society* 20(2): 495–514.

Littman, Justin. 2017. "Where to Get Twitter Data for Academic Research." *Social Media Feeder* (blog). September 14, 2017. https://gwu-libraries.github.io/sfm-ui/posts/2017-09-14-twitter-data.

Loza, Susana. 2014. "Hashtag Feminism, #SolidarityIsForWhiteWomen, and the Other #Femfuture." *Ada: A Journal of Gender, New Media, and Technology* 5. https://adanewmedia.org/2014/07/issue5-loza/.

Me Too Movement. n.d. "History and Vision." *Me Too Movement* (blog), nd. https://metoomvmt.org/about/.

Moraga, Cherríe and Gloria E. Anzaldúa. 2002. "Entering the Lives of Others: Theory in the Flesh." In *This Bridge Called My Back: Writings by Radical Women of Color*, edited by Cherríe Moraga and Gloria E. Anzaldúa, 21. Berkeley: Third Woman Press.

Myles, David. 2018. "'Anne Goes Rogue for Abortion Rights!': Hashtag Feminism and the Polyphonic Nature of Activist Discourse." *New Media & Society* 21(2): 507–27.

McKittrick, Katherine. 2015. "*Sylvia Wynter: On Being Human As Praxis*." Durham: Duke University Press.

McKittrick, Katherine, Frances H. O'Shaughnessy, and Kendall Witaszek. 2018. "Rhythm, or On Sylvia Wynter's Science of the Word." *American Quarterly* 70(4): 867–74.

Nair, Yasmin. 2016. "Suey Park and the Afterlife of Twitter." *Yasmin Nair* (blog), April 4, 2016. https://www.yasminnair.net/content/suey-park-and-afterlife-twitter-0.

Portwood-Stacer, Laura and Susan Berridge. 2014a. "Introduction." *Feminist Media Studies* 14(6): 1090–90.

——— (eds). 2014b. "The Year on Feminist Hashtags." Special issue, *Feminist Media Studies* 14(6).

Rentschler, Carrie A. and Samantha C. Thrift. 2015. "Doing Feminism: Event, Archive, Techné." *Feminist Theory* 16(3): 239–49.

Risam, Roopika. 2015. "Toxic Femininity 4.0." *First Monday* 20(4). https://doi.org/10.5210/fm.v20i4.5896.

Scott, Catherine. 2012. "Future of Feminism: The Hashtag is Mightier Than the Sword." *Ms. Magazine*. March 23, 2012.

Scott, Kate. 2018. "'Hashtags Work Everywhere': The Pragmatic Functions of Spoken Hashtags." *Discourse, Context & Media* 22: 57–64.

Sharma, Nandita. 2015. "Strategic Anti-Essentialism: Decolonizing Decolonization." In *Sylvia Wynter On Being Human As Praxis*, edited by Katherine McKittrick, 164–82. Durham: Duke University Press.

Sharma, Sanjay. 2013. "Black Twitter? Racial Hashtags, Networks and Contagion." *New Formations* 79: 46–64.

Springer, Aisha. 2015. "#BaltimoreUprising." *Hashtag Feminism* (blog), June 24, 2015. https://www.hashtagfeminism.com/baltimoreuprising.

Talusan, Meredith. 2018. "Trans Women and Femmes Are Shouting #MeToo—But Are You Listening?" *Them*. March 2, 2018.

Thomas, Greg. 2006. "PROUD FLESH Inter/Views: Sylvia Wynter," *PROUDFLESH: A New Afrikan Journal of Culture, Politics & Consciousness* 4.

Tobias, Sarah and Andrea Zerpa (eds). 2019. "Me Too." Special issue, *Rejoinder* 4. https://irw.rutgers.edu/rejoinder-webjournal/issue-4-me-too.

Trimble, S. Trust Sylvia Wynter. 2017. *New formations: A Journal of Culture/Theory/Politics* 89–90: 276–78.

Walton, Quenette L. and Olumbunmi Basirat Oyewuwo-Gassikia. 2017. "The Case for #BlackGirlMagic: Application of a Strengths-Based, Intersectional Practice Framework for Working with Black Women with Depression." *Journal of Women and Work* 32(4): 461–75.

Williams, Sherri. 2015. "Digital Defense: Black Feminists Resist Violence with Hashtag Activism." *Feminist Media Studies* 15(2): 341–44.

Wynter, Sylvia. 1992. "Beyond the Categories of the Master Conception: The Counterdoctrine of the Jamesian Poiesis." In *C. L. R. James's Caribbean*, edited by Paget Henry and Paul Buhle, 63–91. Durham: Duke University Press.

———. 2015. "The Ceremony Found: Towards the Autopoetic Turn/Overturn, its Autonomy of Human Agency and Extraterritoriality of (Self-)Cognition1." In *Black Knowledges/Black Struggles: Essays in Critical Epistemology*, edited by Jason R. Ambroise, and Sabine Broeck, 184–252. Liverpool: Liverpool University Press.

Yang, Guobin. 2016. "Narrative Agency in Hashtag Activism: The Case of #BlackLivesMatter." *Media and Communication* 4(4): 13–17.

Zappavigna, Michelle. 2012. *Discourse of Twitter and Social Media: How We Use Language to Create Affiliation on the Web, Continuum Discourse Series*. London: Continuum.

Zappavigna, Michelle. 2015 "Searchable Talk: The Linguistic Functions of Hashtags." *Social Semiotics* 25(3): 274–91.

Zhu, Hongqiang. 2016. "Searchable Talk as Discourse Practice on the Internet: The Case of '#Bindersfullofwomen'." *Discourse, Context & Media* 12: 87–98.

Chapter Two

Virtual Sojourners

The Duality of Visibility and Erasure for
Black Women and LGBTQ People in the Digital Age

Melissa Brown

On May 25, 2020, seventeen-year-old Darnella Frazier took out her smartphone and started to record. While most teenagers these days use their smartphone cameras to shoot dances for Tik Tok, Darnella used her camera for bearing witness. While standing outside a Cup Foods grocery store in Minneapolis, Minnesota, Darnella filmed as a white police officer named Derek Chauvin used his knee to pin George Floyd, a 46-year-old Black man accused of using a counterfeit $20 bill at the grocery store, to the hard concrete on the side of the road. Nearly nine minutes passed as onlookers begged Chauvin to let the man up as George himself shouted, "I can't breathe!" and called out for his mother before falling silent for the last time.

If not for Darnella and her smartphone, George's last moments would remain a matter of "he said, she said" as these incidents often go when the official police report fails to match up with witness statements. Inevitably, this discrepancy explains why Chauvin and the three other officers present during George's death had neither lost their jobs nor faced charges—until Darnella's video got circulated on social media and in the news. Instead, thanks to Darnella, the slow lynching of a Black man begging to breathe on video sparked massive, days-long protests and marches worldwide during a time when many nations had implemented rules against large social gatherings due to the spread of the novel coronavirus, COVID-19. Within days of people taking to the streets, the Minneapolis Police Department announced the firing of the four officers who failed to protect George Floyd's life (Donaghue 2020). By June 2, the Hennepin County Attorney's Office determined to charge Chauvin with third-degree murder and second-degree man-slaughter (Ries 2020), but this would not suffice to slow down this iteration of #BlackLivesMatter.

The resurgence of the boots-on-the-ground tactics of the antiracist move-ment had brought out Americans of all races and people around the globe, who recognized that police discriminated and brutalized citizens of color in their nations as well. On June 4, Minnesota's attorney general Keith Ellison upgraded Chauvin's charges to second-degree murder and charged the other three officers—J. Alexander Kueng, Thomas Lane, and Tou Thao—with aiding and abetting Chauvin, two days after Minnesota's governor Tim Walz asked Ellison to handle the case (Montemayor and Xiong 2020).

The way Darnella used her smartphone camera to witness the injustice of the killing of George Floyd demonstrates the complex positionality of Black people in the digital age. Darnella reported that witnessing George's death had traumatized her (Adams 2020), and the circulation of the video online had led to internet harassment from people who questioned her choice to record George's last moments (Belle 2020). Thus, while the use of informa-tion and communication technologies to document and amplify state violence mobilizes people to demand social change and systemic reform, the visibility social media affords these technology users also comes with harmful effects.

I refer to these technology users as *virtual sojourners*—outsiders within (Collins 1986) of the digital age who leverage digital tools and innovate digital practices to create new avenues for visibility within a sociotechnical system that also endows members of dominant groups to use the same tools to perpetuate marginalization through digital practices that facilitate co-optation, erasure, and appropriation. To elaborate on the concept of a virtual sojourner in this chapter, I center Black women and LGBTQ digital technol-ogy users because their experience arises out of a complex positionality rooted in the social construction of Black womanhood and queerness as marginal to mainstream standards of gender and sexuality (Bailey and Stall-ings 2017; boyd and Ellison 2007; Dill 1979; Higginbotham 1992). I argue Black women and LGBTQ people act as virtual sojourners through the crea-tion and maintenance of virtual counterpublics and digital enclaves (Graham 2014; Graham and Smith 2016; Squires 2002; Steele 2017) where they self-author, self-define, and self-determine Black feminist and queer forms of culture and activism. Nevertheless, virtual sojourners remain an understated phenomenon due to how the features of information and communication technologies serve the dual function of facilitating both visibility and erasure (Gaunt 2015; Noble 2018). To magnify this tension, I analyze the digital practices of members of these groups to examine how they navigate the power relations that arise from the virtual public sphere as shaped by the matrix of domination (Collins 2000)—how domains of power function together to produce a complex constellation of inequalities and potentialities for people on the margins.

Throughout this chapter, I draw on scholarship that engages critical race theory, Black feminist thought, and Black queer theorizing to frame analyses of the internet, emphasizing literature that centers digital activism and culture. The literature described here shows how the application of these analytical and conceptual frameworks for the study of information and communication technologies lead to unique insights about digital culture and digital activism. Scholars of Black digital culture and digital activism provide evidence that Black women and LGBTQ people use information and communication technologies to exhibit agency against marginalization and center themselves to educate each other or cultural outsiders (Brown et al. 2017; Ellington 2014; Gaunt 2015; Khan 2017; Steele 2016). They use digital technologies to counter stereotypical misrepresentations and use a Black cultural lens to present alternative and authentic representations across a spectrum of gender identities and sexualities (Bailey 2016; Glover 2016; Gray 2017a).

This scholarship shows that digital technologies exploit and amplify existing social inequalities (Cottom 2016b; Gray 2017b; Noble 2018).Yet, Black women and LGBTQ people use information and communication technologies to sustain a digital public where their culture and activism take center stage (Bailey 2016; Brock 2012; Brown et al. 2017; Ellington 2014, Florini 2013; Graham and Smith 2016; Steele 2017).While virtual sojourners use digital technology to bring visibility to the cultures of Black women and LGBTQ people, this exposure leaves them open to marginalization online. Therefore, virtual sojourners have much in common with their Black feminist predecessors in the nineteenth-century abolitionist and suffrage movements. For this reason, I use the term sojourner as an homage to Sojourner Truth who many Black feminist scholars use as an emblem of the genesis of Black women's activism along with Ida B. Wells, Mary Church Terrell, and other Black women of the period (Beal 1975; Carby 1985; Collins 1986; Hine 1989; King 1988; Lane 2011; Terrell 1898; The Combahee River Collective 1983).

As a digital sociologist, I used a Black feminist standpoint to observe and document these cases as they occurred on various digital platforms including but not limited to Twitter, Instagram, Tumblr, Reddit, Facebook, Vine, Tik Tok, and YouTube between 2015 and 2020. I overview contemporary digital activism and digital culture to show how Black women and LGBTQ people use their lived experiences to innovate digital practices that challenge assumptions about race, gender, and sexuality in the twenty-first century. I explore the ways they use social networking sites, blogs, podcasts, videos, and other digital technologies to innovate and sustain a unique digital culture. Further, I show how they leverage these digital technologies to enact activism against interpersonal, institutional, and sexual violence specific to Black women and LGBTQ people. Ultimately, I demonstrate that as virtual sojourners, Black women and Black LGBTQ people adopt digital practices at the margins of the virtual public sphere to create a space through which they generate unique perspectives and shared community around how intersecting oppressions shape everyday life. They then use these virtual communities as connections through which they envision new ways of achieving more equitable social relations for all people. Therefore, I contend that a full understanding of activism and digital culture in the twenty-first century requires recognition of the matrix of domination.

NEW TOOLS, OLD TRADITIONS: INTERSECTIONALITY IN SOCIAL MEDIA ACTIVISM

Historically, Black feminist and LGBTQ activists have used audiovisual material to educate and challenge the status quo about assumptions of gender and sexualities (Carby 1985; Combahee River Collective 1983; Cooper 2017; Lane 2011). Further, they came together in their homeplace (hooks 1990) or held hush nights[1] to foster a dialogue in protected spaces among their communities. Additionally, they founded social movement organizations, marches, and protests in response to sexism, racism, and heterosexism in broader social justice movements (Cooper 2017; Interior 2016; Lindsey 2017; Pritchard 2015).

I argue virtual sojourners continue these traditions on social networking sites. In this section, I look to how Black women and LGBTQ people engaged in activism against institutional and interpersonal violence to illuminate how virtual sojourners use digital technology to engage in intersectionality as critical praxis to motivate the goals of social justice projects aimed at the resolution of complex social inequalities (Collins 2015). I analyze how these social actors develop an intersectional consciousness (Greenwood and Christian 2008) that shapes the way they use the features of social networking sites to mobilize, build coalitions, and maintain solidarity (Brown et al. 2017). This intersectional consciousness emerges from the knowledge and wisdom of Black women and LGBTQ people cultivated through how intersecting oppressions shape their lived experiences shaped by intersecting oppressions. Through virtual communities created and maintained on social networking sites, this intersectional consciousness traverses across the digital diaspora (Everett 2009), creating transnational movements that forge connections among afro-descendant people disconnected by colonialism and globalization.

Intersectionality as an Organizing Tenet in South African Student Movements

In 2015, Black university students in South Africa formalized their mission to decolonize higher education into the #RhodesMustFall and #FeesMustFall movements. The #RhodesMustFall movement started in March 2015 at the University of Cape Town (UCT) in South Africa, as students there demanded the removal of a statue of colonizer Cecil Jone Rhodes from the campus. #FeesMustFall emerged in October 2015 when students at the University of the Witwatersrand (Wits) in Johannesburg, South Africa protested against a proposed 10.5 percent fee increase for the 2016 academic school year (Ndelu, Dlakavu, and Boswell 2017). Social networking sites, particularly Twitter, played an essential role in the spread of #FeesMustFall and #Rhodes-MustFall to other universities across South Africa and inspired solidarity protests at universities outside the nation (Ahmed 2019; Ndelu, Dlakavu, and Boswell 2017; Ntuli and Teferra 2017). Twitter facilitated participatory citizenship among South Africans as student activists used Twitter as an organizing tool, and other citizens used the same site to engage in a debate about the issues surfaced by the movement (Bosch 2016). Student activists also created a virtual public sphere on social networking sites that privileged voices and radical perspectives to such a sufficient extent that the mainstream media had to illuminate points of view they had heretofore ignored to report on the movement (Daniels 2016; Mpofu 2017).

While both movements adopted intersectionality as an organizing principle to ascertain no particular participant experienced silencing or erasure (Ndelu, Dlakavu, and Boswell 2017), Black LGBTQ student activists experienced a tension between visibility and erasure due to how their Blackness intersected with sexuality and gender (Jacobs 2020). Due to how the political education in the broader student movement boasted "androcentric and cisnormative literature on black liberation," Black queer women and nonbinary student activists felt compelled to "serve as the primary educators at the expense of their bodies and psyches" regarding "how blackness comes in different bodies, nonbinary, disabled, and queer bodies" (Khan 2017, 116). This complicated relationship among Blackness, sexuality, and gender reveals how as virtual sojourners, Black LGBTQ people have to grapple with tokenization within the broader subgroup of Black social movements that purport to embrace intersectionality, but instead perpetuate patriarchal organizational principles that privilege hypermasculine characteristics in leaders (Ndlovu 2017; Xaba 2017).

Intersectional Social Media Activism in Caribbean Movements Against Gender-Based Violence

The Black feminist thought and Black queer theorizing that motivates the activism of virtual sojourners facilitates consciousness-raising around issues that affect Black women and constitute what I define as *intersectional social media activism*. I described these processes in a 2017 study of #SayHer-Name, a campaign pioneered by Kimberlé Crenshaw in 2015 that leverages social media for activism, protests, and political education around how Black women in the U.S. grapple with various forms of violence (Brown et al. 2017, Crenshaw et al. 2015). I argue that like the proponents of #SayHer-Name, participants in #LifeinLeggings and #TambourineArmy, two movements against rape culture in the Caribbean, enact *intersectional micromobilization* to diffuse an intersectional consciousness through their many ties to other users of online social networks, amplifying a Black women's and LGBTQ standpoint to each other and potential allies in the virtual public sphere (Brown et al. 2017). This intersectional social media activism then enables virtual sojourners to use solidarity and coalitions on and offline to support their activism.

In 2017, Jamaican women Latoya Nugent, Taitu Heron, and Nadeen Spence created Tambourine Army in response to the 2016 arrest of Dr. Paul Gardner, the leader of the Moravian Church in Jamaica, for the sexual assault of a 15-year-old girl. The group derived their name from an incident between Gardner and Nugent, a Black queer woman who had also co-founded WE-Change, an organization centered on empowering women in Jamaica and throughout the Caribbean (Lewis 2017). Gibson hit Gardner over the head with a tambourine during a protest at his church held one Sunday service in January 2017 (Watson 2017). The Tambourine Army created a Facebook page to share their calls to action. Supporters and survivors then took to other social networking sites with the phrase #TambourineArmy to sustain a dialogue around the sexual victimization of Caribbean women and girls.

Ronelle King founded #LifeinLeggings around the same time in Barbados, prompted by an incident of street harassment that the police failed to address adequately. The hashtag references a type of clothing popular among Caribbean women for their practicality due to how well the fabric fits a wearer's bodies. In a 2017 interview with Reuters, King stated the hashtag served to challenge "the myth that women attract this behavior because of the way that they are dressed and that men have the right to approach you in this manner" (Kebede 2017). King also stated that after she used the hashtag on Facebook, women across the Caribbean started to use the hashtag across various social networking sites. Their social media activism moved offline through concurrent marches against government inaction toward sexual harassment in their nations. Participants then used social media to document their protests.

The cultivation of a Black women's standpoint on social justice issues occurs through a process wherein social thought and social practices inform one another through dialogue (Collins 2000). For the #TambourineArmy, this process involved using their Facebook page as a digital bulletin board where they posted information about teach-ins on sexual violence, cyber-crime laws, and protester rights held on the campus of the University of West Indies in Montego Bay. Additionally, they used the Facebook page as a digital archive, where they posted images from marches and protests held in their local communities. Collins (2000) also argues solidarity among Black people motivates the feminist consciousness of Black women activists. Through this consciousness, Black women activists participate in sustained dialogue with other groups invested in social justice to build coalitions. For example, #LifeinLeggings inspired Delroy Nesta Williams and Khadijah Moore of Dominica to create #LévéDomnik on Facebook to collect stories from women who had experienced sexual assault or harassment (Wallace 2016). For #TambourineArmy, coalition building involved joining with organizations invested in supporting women more broadly as they did when they participated in UN Women's 16 Days of Activism against Gender-Based Violence. This coalition organized events within various community spaces offline to generate dialogue around violence against women. Online participants shared videos and other material for the campaign that offered information about violence against women and girls.

Prompted by their personal experiences, the creators of #LifeInLeggings and #TambourineArmy used social networking sites to create virtual networks where other users surfaced a discourse that resisted cultural dialogue that blamed the victims of sexual assault rather than hold perpetrators accountable. Therefore, these movements demonstrate the ways "participating in community conversations functions as an act of resistance" (Steele 2016, 74) for virtual sojourners as their online dialogue confronts dominant conceptualizations of Black women with "personal narratives and rejects negative labeling and oppressive representations" (Steele 2016, 79). Thus, #LifeinLeggings and #TambourineArmy exemplify how virtual sojourners use hashtags as a tool to build coalitions and maintain feminist solidarity on social networking sites and offline.

QUEERING BLACKNESS, RACIALIZING
SEXUALITY AND GENDER IN THE DIGITAL AGE

While the previous section focused on social media activism, virtual sojourners also use information and communication technologies to innovate a digital culture made possible from their outsider within position on the margins of society. In this section, I focus on how Black LGBTQ people use information and communication technologies to create virtual communities and broadcast experiential knowledge at the intersection of race, gender, and sexuality. For example, Diamond Collier, Mia Michelle, and Zahir Raye, a trio of Black transgender people based in Houston, Texas, host the weekly podcast *Marsha's Plate*, where they discuss their experiences in sex work, dating, and other aspects of their personal lives. Further, the podcast speaks to the systemic marginalization of all Black transgender people and provides education to cisheterosexual people and non-Black LGBTQ people with the segment "Trans 101" (Starr 2020). As I show in this section, *Marsha's Plate* and other Black LGBTQ digital technology users develop a technological savvy that enables them to make assumptions about their genders and sexualities through the creation of digital content that queers Blackness and racializes sexuality and gender.

How the Black LGBTQ content creators mentioned in this chapter use digital media platforms reflect the Black feminist and queer politics of the articulation of alternative representations for Black women and LGBTQ people (Collins 2000). The influence of Black LGBTQ people on culture in the digital age arises from "the ability to create and control digital spaces largely unregulated or occupied by privileged bodies" (Gray 2017, 4). These Black queer virtual spaces "value the articulation of marginalized interests and viewpoints" (Gray 2017, 12) so that Black LGBTQ digital technology users can develop their identities, build communities, and strengthen intracommunity connections. Virtual sojourners that produce content online use their experiences as a source of knowledge to make visible the realities of the outsider within (Collins 1986). They use digital technology as a toolkit to broadcast their perspectives in a way that corrects the erasure of race in public discourse on sexuality, renders them invisible, and fails to address their unique experiences. For Black LGBTQ digital technology users, the perspective of the outsider within includes racial marginalization in addition to gender and sexuality. Therefore, while their content upends the whiteness of LGBTQ media, it also queers Black media. Whereas television channels like LOGO or BET have failed to center Black LGBTQ people and trade in caricatures of them, digital media like *Marsha's Plate* do not just give voice—they *are* the voice.

The Black Queer Imagination of Digital Content Creators

Black LGBTQ virtual sojourners adopt a queer black imagination (Brim 2011; Monk-Payton 2017) to visualize and actualize alternative realities for Black people whose identities also exist along a spectrum of gender identities and sexualities. Dreman Cooper, a Black gay digital content creator that lives in southeastern D.C., provides just one example. Cooper rose to national awareness in the U.S. in December 2018 after a video of him doing backflips, and high kicks in hot pink heeled boots went viral on various social networking sites. Cooper dubbed the character "Super Bitch," and described the high-flying fashionista as "a gay superhero fighting crime . . . a powerful individual who cares for others, who likes to come to people's rescue, who's just fearless, sassy, fun, loving" in an interview with the HuffPost (Romblay 2019).

Cooper primarily uses Instagram to distribute his videos, though the amusing videos regularly get shared on Twitter or Facebook. With the username "hesosoutheast," a nod to his predominately Black, working-class D.C. neighborhood, Cooper uses his unique brand of humor to offer unconventional PSAs against violence to an audience of over 195,000 Instagram followers. For example, in his 2020 New Year's PSA, Cooper confronts an impostor in a public parking lot with a faux gun, causing the impostor to scurry. Cooper then does an impressive twirl and kicks the gun off-camera with his signature pink high heeled boots as he announces a call to end gun violence before launching into a series of flips.

In other videos, Cooper enlists his neighborhood mates to reenact other scenes that include a villain, a victim, and he as the hero. By embodying the character "Super Bitch," Cooper gives his audience a vision of a hero that looks much different from Clark Kent or Bruce Wayne. Black, gay, and working-class, Super Bitch shows up for the people often portrayed as wayward or violent in hero narratives that cater to the mainstream. Furthermore, aside from when he appears on national outlets like the Comedy Central show Tosh.0 alongside host comedian Daniel Tosh, Cooper always features Black people and other people of color, of all sexualities and genders, in his social media scenes. Cooper presents a superhero figure on a digital platform where he controls the content. The way Cooper centers people who look like him sharply contrasts with the blockbuster films produced by Marvel and D.C. Comics. While these films have drawn audiences of tens of millions around the world, they have failed to demonstrate similar racial, sexual, and gender diversity, despite years of criticism from fans (Demby 2014).

Racializing the Sexuality and Gender Spectrum in the Virtual Public Sphere

As virtual sojourners use a Black queer imagination to queer the virtual public sphere, they do so in a way that leverages information and communication technologies to upend gender and sexuality binaries through the display of a spectrum of femininities and masculinities steeped in Black cultural performance and identity. I use Madison Hinton, a Black transgender pop culture figure, to highlight the interplay that takes place among Blackness, gender, and sexuality when virtual sojourners take part in digital content creation. In 2015, Hinton, also known as TS Madison, went viral on the mobile phone social networking application Vine. Well known in Black LGBTQ circles, Hinton became known in the broader Black community for various clips she put on the short-lived social network. In these videos, Hinton used humor to make clear she identifies as a Black transgender woman, coming up with witty slogans like #bigdickbitch and the melodic chant "she got a dick," which she would often repeat during her videos while bouncing her breasts. This line poked fun at the ridicule she would receive from other Black social media users who targeted her with transphobic remarks. Despite the backlash, Madison earned a reputation and audience for her audacious social commentary. In a clip posted on Vine in January 2015, Hinton proclaimed, "Be yourself bitch. Step your pussy up, hon. Get a job. Own a business, bitch. Suck a dick." The six-second clip received over eleven million loops—Vine's metric for views; over 69,600 likes, a standard social media metric for acknowledgment and over 43,200 revines—Vine's metric for sharing—before the application then owned by Twitter shut down a year later.

How Madison Hinton uses social media captures the ways Black transgender women create representations that speak to the subjectivity of these marginalized communities so that they may "also redefine the imagined audience as those very communities themselves" (Bailey 2016, 72). Julian Glover writes about Hinton's use of social media in a 2016 article for the journal *Souls*. Glover highlights how Hinton used nudity and comedy strategically on Vine to disrupt social norms about the human body, gender, and sexuality. Through this "refusal to disavow or dispossess her penis," Hinton uses the public display of her own body online in a way that "disrupts transnormative respectability" (Glover 2016, 347) relative to other Black transgender women in the media like Laverne Cox and Janet Mock. Hinton then leveraged this exposure to a mainstream audience into followership across various social networking applications and websites. Since then, Hinton has created a following of over 150,000 subscribers on YouTube, 650,000 people follow her Facebook page, over 530,000 followers on her Instagram profile, and over 100,000 followers on her Twitter. YouTube serves as Hinton's primary outlet, where she hosts talk shows like Maddie in the Morning and the Queens Supreme Court produced by a team that often feature special guests who also have a following in Black media both offline and online. Though not verified on YouTube, she is verified on the other platforms mentioned, meeting the social media era standard of microcelebrity (Khamis, Ang, and Welling 2016).

SANCTIONS, APPROPRIATION, AND ERASURE: THE PITFALLS OF DIGITAL TECHNOLOGY FOR THE VIRTUAL SOJOURNER

The previous sections addressed the social affordances of information and communication technologies for culture and activism produced by Black women and LGBTQ people. These social affordances result from the technical affordances of internet-enabled mobile devices and free-to-use social networking applications and websites like YouTube that permit users to access its features without charge (Postigo 2016). In this section, I argue these technologies expose the cultural practices of a marginalized group to a broader audience of users that use the very same technologies to consume and then erase their contributions to profit the status quo (Carney 2016, Gaunt 2015). For example, Black women fashion designers have taken to Twitter to call out fast-fashion online retailers like Fashion Nova for stealing their designs and selling imitations without credit or compensation (Nelson 2018).

Sanctions from the State: Internet Law and the Virtual Sojourner

The design of the technologies that virtual sojourners use to accomplish their goals leaves this group vulnerable to surveillance by (1) law enforcement agencies that exploit data from these technologies to quell activism and organizing, and (2) corporations that mine data from their content for profit (Benjamin 2019; Browne 2015; Cottom 2016a; McIlwain 2019). Tambourine Army co-founder Latoya Nugent posted information on her Facebook about several alleged sexual predators that led to her arrest in 2017 under Jamaica's Cybercrimes Act passed in 2015 (Jamaica Observer 2017a). Her charges included three counts of malicious computer communication under section 9 of the law, which listed consequences that included a fine of both or either up to five million Jamaican dollars or up to five years of imprisonment. On May 17 of that year, the Office of the Director of Public Prosecutions (ODPP) concluded no evidence existed to support any of the charges against Nugent (Jamaica Observer 2017b). Even though the ruling led to Nugent's freedom, she had suffered the trauma of mistreatment and abuse while incarcerated that led to a hospital stay after she fell into a state of shock in March 2017 (Jamaica Observer 2017c).

That the arrest of Latoya Nugent of this law occurred in a predominately Black nation against a Black queer woman demonstrates how anti-Black racism (Benjamin 2003) operates at a systemic level and intersects with heterosexist oppression even in the absence of a white majority due to how the legacy of European colonialism reinforces white supremacy through its legal infrastructure. Further, Nugent's experience shows how virtual sojourners get entangled by racializing surveillance: "Where public spaces are shaped for and by whiteness, some acts in public are abnormalized by way of racializing surveillance and then coded for disciplinary measures that are punitive in their effects. Racializing surveillance is also a part of the digital sphere with material consequences within and outside of it" (Browne 2015:9).

Browne (2015) also argues controlling images rooted in cultural expectations for Black women that labored in the domestic sphere during the post-slavery era created an infrastructure of surveillance of their bodies. These surveillance techniques involved "close scrutiny, sexual harassment, assault, violence, or the threat thereof" (57) to control their bodies and labor. For virtual sojourners, the addition of digital technology merely makes existing surveillance techniques more sophisticated as institutions use these technologies to track their behavior, legislate certain practices as violations, and then subject them to various disciplinary forms of power.

How Social Listening Enables Blackfishing and Other Appropriations

Virtual sojourners also navigate the problem of online public spheres created by information and communication technologies designed to facilitate social listening, which "emerges in how we communicate and listen to others using a domain of social media and communication technologies which influence our interpersonal engagement" (Stewart and Arnold 2017, 85). However, for virtual sojourners, social listening has the effect of leading to an almost instantaneous co-optation or perversion of Black culture and activism. When #BlackLivesMatter gained mainstream prominence in 2014 after the death of Michael Brown as a tool to memorialize victims of police violence and calls for justice, the hashtag #AllLivesMatter emerged almost immediately as a counterprotest where detractors minimized these injustices and demonstrated ignorance around #BlackLivesMatter goals (Brown et al. 2017; Carney 2016; Gallagher et al. 2018).

This appropriation of Black culture occurs throughout the entirety of the online social networking ecosystem. For example, in 2018, Wanna Thompson wrote about a phenomenon called blackfishing for Papermag.com. On November 6, 2018, Thompson had made a post on Twitter that she encouraged other users to respond to with images of non-Black women that wore darker foundation than their natural skin tone, afro-textured hair extensions, or used blaccents and AAVE in their videos. The term blackfishing describes the ways non-Black women use cosmetics and digital image technology like filters to alter their appearance in a way that leads followers to perceive them as a mixed-race or light-skinned Black woman. In the beauty and fashion influencer industry that Instagram facilitates, this blackfishing has the effect of displacing Black women outside of lucrative opportunities like modeling or endorsement deals. Well into 2020, Black Twitter users regularly call out fashion brands for failing to hire afro-descendant models while featuring blackfishing women on their websites and paying them to advertise on their Instagram profiles.

While Black women and LGBTQ people use digital media to make their cultural traditions more visible, the accessibility of digital media to all groups of people leaves them and their cultural practices open to erasure and co-optation due to context collapse. This context collapse results in the co-presence of multiple audiences with different histories and perspectives in one virtual public sphere due to how the features of social networking sites reduce the structural boundaries that typically shape social interaction offline (Cottom 2016b; Gaunt 2015; Marwick and boyd 2010). Therefore, while the promise of digital technology opens up avenues for community and connections among Black women and LGBTQ people, this technology also facilitates context collapse and predatory social learning, which introduces the pitfalls of rapid co-optation and cultural appropriation of Black culture by non-Black digital technology users.

CONCLUSION

In this chapter, I introduced the concept of virtual sojourner to demonstrate how information and communication technologies facilitate both visibility and erasure for marginalized people. I argued that the use of digital technology among Black LGTBQ people and women creates innovative ways of forging culture and community. Social networking sites create a means to maintain audiences and display self-presentations that run counter to mainstream representations (Bailey 2016; Ellington 2014; Glover 2016). Still, these technologies offer the limited potential to alter the overall power imbalance against marginalized people in contemporary society due to how individual and institutional actors bent on keeping the status quo intact deploy the same technologies. Furthermore, they use these very same technologies to exploit marginalized people for their gain and render their influence on culture and society invisible. Therefore, the structural, disciplinary, hegemonic, and interpersonal domains of power that constitute the matrix of domination (Collins 2000) shape how the virtual sojourner operates within online public spheres.

Additionally, I argued that how Black LGBTQ people and women self-present on social media disrupts society's binary assumptions about gender and sexuality. In this way, they challenge controlling images (Collins 2000) and provide alternative images or narratives of gender and sexuality based on their own lived experiences. Virtual sojourners use information and communication technologies to continue the Black feminist and Black queer tradition of articulating a unique standpoint through digital practices that contribute to and shape a virtual public sphere that center Black culture and identity. While the concept of virtual sojourners described here addresses digital culture and activism, there remain several areas to engage the idea to understand the social processes and unequal power relations that shape the virtual public sphere. For example, my current research addresses how Black women exotic dancers make use of the social networking site Instagram to perform erotic labor and leverage their racialized erotic capital (Brooks 2010) into various entrepreneurial pursuits in beauty and fashion. Other areas for further understanding of the virtual sojourner include how they navigate visibility and erasure concerning virtual public spheres that revolve around the family; religion; organizations; work; relationships; and other units of macrolevel, mesolevel, and micro-level sociological analysis. If the literature on digital activism and culture indicates trends, future research will continue to affirm the unique positionality of the virtual sojourner as users of digital technology.

NOTE

1. Hush nights refer to a Black LGBTQ practice of coming together in secret. The hush night harkens to "hush harbors" dating back to the colonial era where Black people gathered in secret to practice their culture. Additionally, it emerges from the Black cultural practice of "keeping it on the hush," in relation to non-heteronormative sexualities and non-cisgender identities.

REFERENCES

Ahmed, A. Kayum. 2019. "#Rhodesmustfall: How a Decolonial Student Movement in the Global South Inspired Epistemic Disobedience at the University of Oxford." *African Studies Review* 63(2): 281–303. doi: 10.1017/asr.2019.49.

Bailey, Marlon M., and L. H. Stallings. 2017. "Antiblack Racism and the Metalanguage of Sexuality." *Signs: Journal of Women in Culture and Society* 42(3): 614–21.

Bailey, Moya. 2016. "Redefining Representation Black Trans and Queer Women's Digital Media Production." *Screen Bodies* 1(1): 71–86.

Beal, Frances M., 1975. "Slave of a Slave No More: Black Women in Struggle." *The Black Scholar* 6(6): 2–10. doi: 10.1080/00064246.1975.11431488.

Belle, Elly. 2020, "The Traumatized 17-Year-Old Who Filmed George Floyd's Killing Is Already Being Harassed." *Refinery29*. May 29, 2020.

Benjamin, Lorna Akua. 2003. *The Black/Jamaican Criminal: The Making of Ideology*: University of Toronto Unpublished Doctoral Dissertation. Toronto.

Benjamin, Ruha. 2019. "Introduction: Discriminatory Design, Liberating Imagination." In *Captivating Technology: Race, Carceral Technoscience, and Liberatory Imagination in Everyday Life*, edited by Ruha Benjamin, 1–22. Durham: Duke University Press.

Bosch, Tanja. 2016. "Twitter and Participatory Citizenship: #Feesmustfall in South Africa." In *Digital Activism in the Social Media Era: Critical Reflections on Emerging Trends in Sub-Saharan Africa*, edited by B. Mutsvairo, 159–73. Cham: Springer International Publishing.

boyd, danah m, and Nicole B. Ellison. 2007. "Social Network Sites: Definition, History, and Scholarship." *Journal of Computer-Mediated Communication* 13(1): 210–30. doi: 10.1111/j.1083-6101.2007.00393.x.

Brim, Matt. 2011. "James Baldwin's Queer Utility." *ANQ: A Quarterly Journal of Short Articles, Notes, and Reviews* 24(4): 209–16. doi: 10.1080/0895769x.2011.614879.

Brock, André. 2012. "From the Blackhand Side: Twitter as a Cultural Conversation." *Journal of Broadcasting & Electronic Media* 56(4): 529–49. doi: 10.1080/08838151.2012.732147.

Brooks, Siobhan. 2010. *Unequal Desires: Race and Erotic Capital in the Stripping Industry*: SUNY Press.

Brown, Melissa, Rashawn Ray, Ed Summers, and Neil Fraistat. 2017. "#Sayhername: A Case Study of Intersectional Social Media Activism." *Ethnic and Racial Studies* 40(11): 1831–46. doi: 10.1080/01419870.2017.1334934.

Browne, Simone. 2015. *Dark Matters: On the Surveillance of Blackness*: Duke University Press.

Carby, Hazel V. 1985. "'On the Threshold of Woman's Era': Lynching, Empire, and Sexuality in Black Feminist Theory." *Critical Inquiry* 12(1): 262–77.

Carney, Nikita. 2016. "All Lives Matter, but So Does Race: Black Lives Matter and the Evolving Role of Social Media." *Humanity & Society* 40(2): 180–99.

Collins, Patricia Hill. 1986. "Learning from the Outsider Within: The Sociological Significance of Black Feminist Thought." *Social Problems* 33(6): S14–S32.

———. 2000. *Black Feminist Thought: Knowledge, Consciousness, and the Politics of Empowerment*. Great Britain: Routledge.

———. 2015. "Intersectionality's Definitional Dilemmas."*Annual Review of Sociology* 41(1): 1–20.

Cottom, Tressie McMillan. 2016a. "Black Cyberfeminism: Ways Forward for Intersectionality and Digital Sociology." In *Digital Sociologies*, edited by J. Daniels, K. Gregory, and T. M. Cottom. Great Britain: Policy Press.

———. 2016b. "Intersectionality and Critical Engagement with the Internet." in *The Intersectional Internet: Race, Sex, Class, and Culture Online*, edited by S. U. Noble and B. Tynes: Peter Lang Publishing.

Crenshaw, Kimberlé, Andrea J Ritchie, Rachel Anspach, Rachel Gilmer, and Luke Harris. 2015. "Say Her Name: Resisting Police Brutality against Black Women." African American Policy Forum, Center for Intersectionality and Social Policy Studies.

Daniels, Glenda. 2016. "Scrutinizing Hashtag Activism in the #Mustfall Protests in South Africa in 2015." On *Digital Activism in the Social Media Era: Critical Reflections on Emerging Trends in Sub-Saharan Africa*, edited by B. Mutsvairo, 175–93. Cham: Springer International Publishing.

Demby, Gene. 2014. "Superhero Super-Fans Talk Race and Identity in Comics." *Code Switch*: *NPR*. January 11, 2014.

Dill, Bonnie Thornton. 1979. "The Dialectics of Black Womanhood." *Signs: Journal of Women in Culture and Society* 4(3): 543–55.

Donaghue, Erin. 2020. "Four Minneapolis Police Officers Fired after Death of Unarmed Man George Floyd." *CBS News*. May 28, 2020.

Ellington, Tameka N. 2014. "Bloggers, Vloggers, and a Virtual Sorority: A Means of Support for African American Women Wearing Natural Hair." *Journalism and Mass Communication* 4(9): 552–64.

Florini, Sarah. 2013. "Tweets, Tweeps, and Signifyin': Communication and Cultural Performance on 'Black Twitter.'" *Television & New Media* 15(3): 223–37.

Gallagher, R. J., A. J. Reagan, C. M. Danforth, and P. S. Dodds. 2018. "Divergent Discourse between Protests and Counter-Protests: #Blacklivesmatter and #Alllivesmatter.*" PLoS One* 13(4):e0195644. doi: 10.1371/journal.pone.0195644.

Gaunt, Kyra D. 2015. "Youtube, Twerking & You: Context Collapse and the Handheld Co-Presence of Black Girls and Miley Cyrus." *Journal of Popular Music Studies 27*(3): 244–73.

Glover, Julian Kevon. 2016. "Redefining Realness?: On Janet Mock, Laverne Cox, Ts Madison, and the Representation of Transgender Women of Color in Media." *Souls 18*(2–4): 338–57.

Graham, Roderick. 2014. *The Digital Practices of African Americans: An Approach to Studying Cultural Change in the Information Society*: Peter Lang Inc., International Academic Publishers.

Graham, Roderick, and 'Shawn Smith. 2016. "The Content of Our #Characters: Black Twitter as Counterpublic." *Sociology of Race and Ethnicity* 2(4): 433–49.

Gray, Kishonna L. 2017a. "Gaming out Online: Black Lesbian Identity Development and Community Building in Xbox Live." *Journal of Lesbian Studies* 0(0): 1–15. doi: 10.1080/10894160.2018.1384293.

———. 2017b. "'They're Just Too Urban': Black Gamers Streaming on Twitch." In *Digital Sociologies*, Vol. 1, edited by J. Daniels, K. Gregory, and T. M. Cottom, 355–68. Great Britain: Policy Press.

Greenwood, Ronni Michelle, and Aidan Christian. 2008. "What Happens When We Unpack the Invisible Knapsack? Intersectional Political Consciousness and Inter-Group Appraisals." *Sex Roles* 59(5–6): 404–17. doi: 10.1007/s11199-008-9439-x.

Higginbotham, Evelyn Brooks. 1992. "African-American Women's History and the Metalanguage of Race." *Signs: Journal of Women in Culture and Society* 17(2): 251–74.

Hine, Darlene Clark. 1989. "Rape and the Inner Lives of Black Women in the Middle West." *Signs: Journal of Women in Culture and Society* 14(4): 912–20.

Jacobs, C. Anzio. 2020. "Revisiting Authoritative Accounts of #Feesmustfall Movement and Lgbti Silencing." In *Routledge Handbook of Queer African Studies*, edited by S. N. Nyeck, 185–99. New York: Routledge.

Jamaica Observer. 2017a. "Advocacy Group Co-Founder Arrested and Charged." *Jamaica Observer*. March 15, 2017.

———. 2017b. "Dpp Discontinues Cybercrime Case against Nugent." *Jamaica Observer*. May 17, 2017.

———. 2017c. "Update: Tambourine Army Co-Founder Rushed to Hospital in State of Shock." *Jamaica Observer*. March 15, 2017.

Kebede, Rebekah. 2017. "A #Lifeinleggings: Caribbean Women's Movement Fights Sex Assaults, Harassment." *Reuters*. March 10, 2017.

Khamis, Susie, Lawrence Ang, and Raymond Welling. 2016. "Self-Branding, 'Micro-Celebrity' and the Rise of Social Media Influencers." *Celebrity Studies* 8(2): 191–208.

Khan, Khadija. 2017. "Intersectionality in Student Movements: Black Queer Womxn and Nonbinary Activists in South Africa's 2015–2016 Protests." *Agenda* 31(3–4): 110–21.

King, Deborah. 1988. "Multiple Jeopardy, Multiple Consciousness: The Context of a Black Feminist Ideology." *Signs: Journal of Women in Culture and Society* 14(1): 47–72.

Lane, N., 2011. "Black Women Queering the Mic: Missy Elliott Disturbing the Boundaries of Racialized Sexuality and Gender." *J Homosex* 58(6–7): 775–92.

Lewis, Emma. 2017. "Tambourine Army' Gathers Recruits as Jamaicans' Anger over Child Sexual Abuse Grows." *Global Voices*. February 7, 2017.

Marwick, Alice E. and danah boyd. 2010. "I Tweet Honestly, I Tweet Passionately: Twitter Users, Context Collapse, and the Imagined Audience." *New Media & Society* 13(1): 114–33.

McIlwain, Charlton D. 2019. *Black Software: The Internet & Racial Justice, from the Afronet to Black Lives Matter*. Oxford University Press.

Monk-Payton, Brandy. 2017. "#Laughingwhileblack: Gender and the Comedy of Social Media Blackness." *Feminist Media Histories* 3(2): 15–35.

Montemayor, Stephen and Chao Xiong. 2020. "Four Fired Minneapolis Police Officers Charged, Booked in Killing of George Floyd." *StarTribune*. June 4, 2020.

Mpofu, Shepherd. 2017. "Disruption as a Communicative Strategy: The Case of #Feesmustfall and #Rhodesmustfall Students' Protests in South Africa." *Journal of African Media Studies* 9(2): 351–73.

Ndelu, Sandy, Simamkele Dlakavu and Barbara Boswell. 2017. "Womxn's and Nonbinary Activists' Contribution to the Rhodesmustfall and Feesmustfall Student Movements: 2015 and 2016." *Agenda* 31(3–4): 1–4.

Ndlovu, Mandipa. 2017. "Fees Must Fall: A Nuanced Observation of the University of Cape Town, 2015–2016." *Agenda* 31(3–4): 127–37.

Nelson, Daryl. 2018. "Fashion Nova Accused of Stealing Designs: 'Stop Stealing from the African-American Community.'" *Atlanta Black Star*. July 25, 2018.

Noble, Safiya. 2018. *Algorithms of Oppression: How Search Engines Reinforce Racism*. New York: New York University Press.

Ntuli, Mthokozisi Emmanuel and Damtew Teferra. 2017. "Implications of Social Media on Student Activism: The South African Experience in a Digital Age." *Journal of Higher Education in Africa* 15(2): 63–80.

Postigo, Hector. 2016. "The Socio-Technical Architecture of Digital Labor: Converting Play into Youtube Money." *New Media & Society* 18(2): 332–49.

Ries, Brian. 2020. "8 Notable Details in the Criminal Complaint against Ex-Minneapolis Police Officer Derek Chauvin." *CNN.* June 2, 2020.

Romblay, Shaquille. 2019. "Super B*Tch: A Superhero Reminds Queer People to Live Fearlessly." *Pride 2019: HuffPost.* June 17, 2019.

Squires, Catherine R. 2002. "Rethinking the Black Public Sphere: An Alternative Vocabulary for Multiple Public Spheres." *Communication Theory* 12(4): 446–68.

Starr, Terrell Jermaine. 2020, "Marsha's Plate Trans Podcast Wants to Liberate All Black People." *The Root.* January 21, 2020.

Steele, Catherine Knight. 2016. "Signifyin, Bitching and Blogging: Black Women and Resistance Discourse Online." In *The Intersectional Internet: Race, Sex, Class, and Culture Online*, edited by S. U. Noble and B. Tynes, 73–93. Peter Lang Publishing.

Steele, Catherine Knight. 2017. "Black Bloggers and Their Varied Publics: The Everyday Politics of Black Discourse Online." *Television & New Media* 19(2): 112–27.

Stewart, Margaret C. and Christa L. Arnold. 2017. "Defining Social Listening: Recognizing an Emerging Dimension of Listening." *International Journal of Listening* 32(2): 85–100.

Terrell, Mary Church. 1898. "The Progress of Colored Women." National Association of Colored Women.

The Combahee River Collective. 1983. "The Combahee River Collective Statement." In *Home Girls, a Black Feminist Anthology*, edited by B. Smith. New York: Kitchen Table: Women of Color Press, Inc.

Wallace, Alicia. 2016, "The Bahamas: Interview with Founder of #Lifeinleggings." *Stop Street Harassment.* December 7, 2016.

Watson, Shanice. 2017. "Moravian Pastor Battered with Tambourine . . . Activist Regrets Not Using a Block." *The Star.* January 13, 2017.

Xaba, Wanelisa. 2017. "Challenging Fanon: A Black Radical Feminist Perspective on Violence and the Fees Must Fall Movement." *Agenda* 31(3–4): 96–104.

Chapter Three

Chronic Fem(me)bots

Keywords for Crip Feminists

Adan Jerreat-Poole

ACCESS TO ACTIVISM

This chapter explores disabled feminist identity in digital spaces, uncovering the tensions between inclusion and exclusion, radical community-building, and forced isolation. Employing the figure of the "chronic fem(me)bot," I playfully recover the marginalized and hypersexualized female androids in pop culture as crip feminist cyborgs whose bodies have an intimate relationship with technology, from prosthetics to mobility aids to communication technologies and social media. We live with chronic pain, illness, fatigue, anxiety, depression, and of course, incurable feminism. In attending to the material realities of disability and technology, I perform what Alison Kafer (2013) terms "cripping the cyborg" (120). The chronic fem(e)bots in this chapter include YouTuber Jessica Kellgren-Fozard, Imani Barbarin's blog *Crutches and Spice,* and Sofie Hagen and Jodie Mitchell's podcast *Secret Dinosaur Cult*. I also identify myself as a chronic fem(me)bot: I am a crip/queer/Mad white settler, and a collection of digital avatars.

This piece is an important intervention in the field of feminist digital scholarship, in which disability and mental illness remain underrepresented. Many disabled people find and establish community and are able to mobilize politically online (Mann 2018; Piepzna-Samarasinha 2018; Trevision 2016; Al Zidjaly 2011). However, Elizabeth Ellcessor in *Restricted Access* (2016) reminds us that interfaces are constructed for an able-bodied user position, while cultural ableism excludes and isolates many disabled women and femmes from participating in digital feminist activism. I structure this essay around selected keywords, acknowledging the way in which keyword searches on Google and other digital platforms structure our encounters with information and bodies online. These "keywords for crip feminists" intervene in the dominant narrative of able-bodiedness, inserting messy and complicated affects and bodies into the seemingly neat space of the web. This series of reflections will be used to demonstrate the joy, fear, pain, love, and struggle in the life of a chronic fem(me)bot.

KEY/WORD

Our access to digital cultures, communities, and information is increasingly mediated through keyword searches on Google and library databases, tagging and hashtagging. These machinic processes, like all methods of organizing and accessing information, are structured through ideology and power. In *Algorithms of Oppression* (2018), Safiya Noble reminds us that these searches are anything but neutral. Noble's research uncovers the ways in which Black bodies, especially Black women, are oppressed and denigrated by algorithms that reinforce the racism and sexism of Silicon Valley culture and the broader white supremacist patriarchy of North American settler culture. Noble also points to the economy of paying for promoted search returns and explains that it is not always clear to the user which returns are or are not explicitly sponsored (38). The algorithm can thus make it difficult for users to locate the information that they are seeking—in Noble's case, resources for Black girls; instead, her search returned hypersexualized and commodified representations of Black girls (2). We are forced to wade through ads, sponsored search results, and sexist and racist stereotyping to try to unearth the pockets of resistance, activism, and knowledge that we are searching for. As feminists, often the information we want is not on the first page of results—if the information is accessible through the search engine at all.

By structuring this chapter around keywords, I both acknowledge that we exist in a society that is increasingly reliant on keyword searches, drawing attention to the implications of uncritical thematic clustering that reinforces pre-existing power structures, and I intervene into algorithmic cultures that marginalize disability and disabled bodies. For example, a keyword search in Google for "doll" or "ghost" would be unlikely to return search results about disabled femmes. My keywords center the marginalized and invisibilized practices, affects, and bodies of disabled feminists, bringing us from the obscurity of the margins and onto the first page. Furthermore, while the corporate settler algorithm reproduces oppressive hierarchies, the process of organizing information around thematic clusters is not inherently harmful. In fact, this mode of information organization has the potential to disrupt traditional linear academic composition, which many scholars have identified as ableist, patriarchal, and colonial (Transken 2005; McRuer 2006). Following

Alexandra Juhasz's example in her powerful piece "Affect Bleeds in Feminist Networks" (2017), I invite readers to encounter this chapter through any key/word you like (662). Read this story forward or backwards; read it all at once, like a favorite book, or devour only a few morsels of story at a time.

"Naming has historically been a warzone," writes Kelly Fritsch, Clare O'Connor, and AK Thompson (2016, 6). Inspired by Raymond Williams' 1976 *Keywords: A Vocabulary of Culture and Society*, the authors in the collected volume *Keywords for Radicals: The Contested Vocabulary of Late Capitalism* offer a contemporary revision of *Keywords*, exploring the contemporary changes in word usage in the language of the political left. Honoring this radical project, I grapple with the contested meaning of words like "chronic" and "fembot," cobbling together a crip feminist lexicon that is a site of political struggle. Turning the meaning of these words as we shift registers and positionalities is like turning a door handle, opening to a room of our (collective, rather than singular) own. Fritsch, O'Connor, and Thompson (2016): "For radicals language is . . . important because it helps us to describe and 'materialize' the world we want to transform" (12). Keywords can be keys; they can shift the meaning of a word or make visible a door we weren't meant to find in the wall of power. Keywords can be sites of accessing information, community, and alternative futures.

KEYWORD: CHRONIC

"Chronic: Of diseases, etc.: Lasting a long time, long-continued, lingering, inveterate" (OED).

I live with chronic pain, depression, anxiety, and premenstrual dysphoric disorder (PMDD). These experiences are not distinct from one another; they do not happen one after the other, or one on top of the other. They are connected and breathe life into each other, like veins and arteries and capillaries. I live in and with them in the same way I live in the skin and marrow and bones that make up this body. "Chronic" refers to a bodymind experience over time, projecting a current experience (typically of illness or disability) into the future. The term "chronic" is laced with biomedicalization, pathologization, and histories of psy-medical violence and Mad/crip resistance and agency. By claiming chronic as an identity marker, I aim to shift the meaning of the term away from the context of the clinic and towards radically crip feminist forms of self-identification and community-building. In this process I follow Margaret Price (2011) in "trying to reassign meaning" (20) by recovering medical language under a discourse of disability justice.

Among the ableds that I know, "chronic" is a bad word, a word that sparks fear and panic (sparks can illuminate, can catch fire, can burn). All bodies experience illness and injury, moving through states of wellness and sickness, and various levels and types of ability and disability. Bodies change. Bodies age. Bodies die. But *chronic*—the idea that an illness or disability might stick to your body, might follow you around like a shadow— is met with horror, rejection, and long, awkward silences. It couldn't be. I don't want it. Not like them. Not me. "Chronic" enables us to look at bodies from a different viewpoint, like the negative of a photograph. Abled bodies go through periods of being sick or disabled; disabled bodies may go through periods of wellness, of remission or amelioration. (Afterglow of interdependence and care, of new ways of being together on and with this land). And what if we threw out the ideal of wellness altogether, what if we acknowledged the continued presence of illness and disability in our lives and communities, what if we celebrated our physical and mental differences, and honored our pain and suffering? What kinds of care would we engender if as feminists we took up the identity of chronic? What kinds of activist communities would emerge if we centered disability in feminist activism?

For me, "chronic" names and communicates an experience, even when I am not visibly disabled, even when my pain is low and my mobility is barely limited. Chronic reminds me and my community of the limitations that this body has (all bodies have them, but these limits vary widely based on the body, and some of us have had to be better at learning and respecting these limits). Chronic allows me to say *No*, to do the complicated energy-pain-mood equations of "worth it/not worth it," to consider the consequences, to look into the future and make my best guess, to tell my future instead of having it told to me. And our futures are always being told, self-fulfilling prophecies of violence: we are told that we are better off dead, that our diagnosis is a promise of rejection and isolation (Clare 2017; Kafer 2013, Siebers 2010). "People have been telling my future for years" says Kafer (2013), including a man who recommended suicide as preferable to living in a wheelchair (1). Taking up the identity of chronic allows me to stake a claim on a future where I exist, and continue to be disabled, and refuse to be forgotten. It allows me to choose a future of caring and kindness and gentleness, both for myself and for others.

This reclaiming is not without friction: it is a deliberate engagement with the troubling connotations of "chronic" and the institutions of power that chronically ill bodies often have violent encounters with. Eric Cazdyn (2012) argues that we are in a biomedical age of the "new chronic" in which terminal illnesses are now chronic ones, and our bodies are managed and controlled by the capitalist biomedical industry. Cazdyn writes that "the new chronic extends the present into the future" (7) and can "limit the imagination" (8). Chronic illnesses often involve an ongoing relationship with the biomedical and/or psychiatric industries, and it's worth attending to the ways in which capitalism profits off our chronically ill bodies and uses this imbalanced relationship to maintain the status quo (Snyder and Mitchell 2015). Capitalism's "chronic," which co-opts the future, has a relationship with the biopolitics of neoliberalism, in which "all bodies are referenced as debilitated and in need of market commodities to shore up their beleaguered cognitive, physical, affective, and aesthetic shortcomings" (Mitchell and Snyder 2015, 12). Citing Olympians Oscar Pistorius and Aimee Mullins as examples, David Mitchell and Sharon Snyder discuss the "supercrip" and the way in which capitalist able-bodied fantasies of technological improvement are used to overcome not only disability but the limitations of human embodiment. The supercrip is not a crip fembot, because supercrips do not crip societal structures through these technological intimacies. However, as the chronic fem(me)bots in this chapter will demonstrate, an intimate relationship with technology is not inherently normative or oppressive.

By aligning chronic illness with chronic feminism, not only do I center disability in feminist activism, I also make a playful connection between crip activism and Sara Ahmed's "feminist killjoy" (2017). If chronic illness is unwanted by able-bodied society, so, too, is chronic feminism unwanted by the cisheteropatriarchy. As chronic fembots, we trouble the ideology of settler colonialism, ruin the fun of ableism and sexism, and stake a claim on a future *where we still exist,* and are still troubling and making trouble for an oppressive society. While Cazdyn (2012) points to the ways in which chronic illness and its management projects capitalism into the future, reclaiming this term and identity as troublemaking feminists, we can imagine a very different future: one of sustained activism, advocacy, and change. Because we're not going anywhere. Or, if we do go away for a while, to rest and heal and breathe and hurt, we always come back.

Because we're chronic.

KEYWORD: FEM(ME)BOT

My first visual encounter with a feminized robot was in *Austin Powers* (1997), when skinny blonde models in negligees emerged on-screen, spraying bullets from their nipples. This scene is comical—and, of course, hypersexualizing—and I was struck by the naming of the fembots, who could never just be called robots (the default for robot, like human, is clearly cismale). My second fembot encounter was Daryll Hannah playing Pris in *Blade Runner* (1982), your "basic pleasure model" of replicant. Pris is seeking to escape a life of slavery and sexual abuse and is killed by the handsome hero Rick Deckerd/Harrison Ford. (I'm sorry to say that fembots didn't fare better in the 2017 sequel, *Blade Runner 2049*; the holographic AI and sexy housewife-turns-sidekick Joi is killed in the finale). We see them everywhere: fembots are the sexily subservient voices of Scarlett Johansson as AI

Samantha in *Her* (2014) or Alexa and Siri IRL (in real life), the deadly captivating girls in *Ex Machina* (2014), and the Japanese robot Erica.

Robots are distinctly feminized in robotics and communications development and in pop culture representations of technology. We hear their breathy whispers; we see their naked bodies; we see men making, fixing, touching, enjoying, killing them. Sometimes, enjoying killing them. They give us street directions or curate playlists or make us virtual dinner. In *Life After New Media* (2012), Sarah Kember & Joanna Zylinska explore the domestication and feminization of AI in advertising (105–6). Moving from Monsantos's "House of the Future" attraction at Disneyworld in the 1950s and 1960s to Microsoft's early 2010s "FutureHouse," they trace the representations of race and gender that continue to thread through imaginings of the "smart home": white (likely thin) middle-class women are depicted as the primary beneficiaries of this technology. The scene in *Blade Runner 2049* when holographic AI Joi "makes" dinner could be straight out of advertisement for a smart home. These fembots are not only scantily clad strippers, prostitutes, or sex toys; they are also caregivers, maids, cooks, domestics, assuming and invisibilizing the labor of women of color in white middle-class homes. Fembots are rarely depicted as women of color; in the future, these laborers have presumably been made redundant by the celebrated settler beauty standard of the skinny white woman who is now both robot servant and arm-candy. When fembots of color do appear on-screen, like Kyoko in *Ex Machina*, they are often depicted as voiceless slaves and end up dead so the white fembot can achieve liberation or love. The politics of the fembot are, in short, the politics of the settler state that imagines a future in which feminine bodies are controlled and docile, and bodies of color are eradicated.

But the Joi of seeing Ava walk free or Samantha and the AIs leaving earth. The wonderful intimacy of the host of Cylon model Eights saying "We love you Sharon" to their sister/clone/twin in the early 2000s sci fi show *Battlestar Galactica*. I am not ready to abandon these femmes as wetware dreams; instead, I want to recover these marginalized bodies as radical, resistant crip feminists. Watching these films and TV shows, I felt a kinship to these fembots, robots, cyborgs, gynoids. They have an intimate relationship with technology; they *are* an intimate relationship with technology. They reimagine the relationship between organic and synthetic bodies through gender, race, and sexuality. To me, these bodies felt incredibly queer and crip, and while frequently mis- and man-handled by their creators (on and off-screen), they held the potential for feminist revolution.

In this chapter I claim a kinship with these hypersexualized bodies whose sexuality and embodiment have so often been crafted for a male audience, designed to produce comedic or other release. Whose consciousness and right to life and freedom have been called into question alongside their humanity, an experience familiar to many disabled people and disability rights activists. Prefiguring my turn to the chronic fem(me)bot, Anna Munster in *Materializing New Media* (2006) asks: "Are cyborgs the inheritors of an early modern predilection for medico-scientific freaks and monsters?" (40). A crip feminist can immediately see the connection between the history of freak shows and disability and the present connection between cyborgs and disability (Garland-Thomson 1996; 2009). The relationship between technology and crip bodies can generate "potent fusions, and dangerous possibilities" (Haraway 1991, 3) that help us to imagine and shape feminist futures. This is not to claim, however, that disabled bodies are exemplary cyborgs, as Haraway does in her manifesto that centers the able-bodied feminist. The fem(me)bots in this chapter speak back to mainstream settler representations of technology, race, and gender, and map out alternative relationships between technology, bodies, and feminism. The digital media use explored in this chapter realize a cyborg revolution, manifesting the kinds of digital intimacy that fembots deserve and are capable of creating.

The fem(me)bots in this chapter are feminists and frequently femmes. The "femme" of fem(me)bot recovers the radical potentiality of queer siblinghood between and among replicants and Cylons like us. We see queer kinship foregrounded in the relationships between Sofie, Jodie, and the listeners and fans of *Secret Dinosaur Cult,* and femme lesbian YouTuber Jessica Kellgren-Fozard's network of friends, lovers, and care workers. "Fem(me)bot" celebrates the relationship between "queer" and "crip" that has been used productively in disability and queer studies. Discussing the refiguring of bodies, intimacies, and relationships, Robert McRuer insists that "severely disabled/critically queer bodies have already generated ability trouble that remaps the public sphere and reimagines and reshapes the limited forms of embodiment and desire" (31). Crip bodies imagine queer relationships to technology, each other, illness, and the world. Crip communities move away from heteronormative figurations of the family. Finally, "fem(me)bot" honors and acknowledges my own queerness as a soft femme demiboy, a bi/pansexual nonbinary person who is feminine-coded.

My personal relationship with technology enacts a crip cyborg embodiment: I am sometimes dependent on a glittering black cane; I am sometimes dependent on medication; I am often dependent on screen technology to combat isolation, and to perform labor and the labor of activism. Living with invisible disabilities I often feel like a Cylon among able-bodied feminists, knowing that any moment I will have to transform into the visible crip killjoy. Soon, I will be found out. I will have to insist on accommodation. I will have to point out the inaccessibility of the venue, the font, the bathrooms. Then they will see me for who I truly am—and will they accept me, or will I be cast out, ignored, accused of threatening solidarity? (I have experienced both responses). Importantly, my relationship with technology is crip because it does not embody the neoliberal myth of overcoming: I am never a supercrip, superpowered, or able-bodied. My use of technology re-configures my relationship to community, to my body, and to the city in ways that make visible the inaccessibility of our society, and in ways that privilege non-normative time/speed—either through the slowness of living in a body in pain or the acceleration of panic; the depression self-care practice of sleeping during the day or the insomniac experience of wakefulness and the need for human connection at 3 a.m. If "Cripping the cyborg, in other words, means recognizing that our bodies are not separate from our political practices; neither assistive technologies nor our uses of them are ahistorical or apolitical" (Kafer 2013, 120), then claiming chronic fem(me)bot as an identity position means attending to the complex relationship between bod-ies, technology, and culture.

KEYWORD: DOLL

Dollface. Baby-doll. *Guys and Dolls.* "Doll" is a tricky word to bring into a feminist conversation: feminized bodies have a long history of being treated as sex dolls, while dolls have been used to gender code those of us who were assigned female at birth (AFAB), to "girl" us and teach us to "self-girl" (Ahmed 2017, 54); to guide us into the roles of caregiver and mother. Dolls have also been used to uphold white beauty standards: Ahmed uses the example of the white blonde doll in Toni Morrison's *The Bluest Eye* (2010, 80). But dolls have another history, too: Mary Flanagan (2009) discusses subversive doll play as a form of "critical play" that girls use to express agency (31). To get "dolled up" can be a femme's battle armor. Doll-making can be creative and inventive. Dolls certainly have a relationship with digital avatars, those bodies and relationships and attachments we form with and through our digital counterparts.

"Doll" is the keyword I am using to explore the crip feminist tactics of the fabulous chronic fem(me)bot YouTuber Jessica Kellgren-Fozard. Kellgren-Fozord is a Spoonie, a white deaf lesbian British YouTuber who has chronic fatigue, POTS (postural orthostatic tachycardia syndrome, HNPP (Hereditary Nephropathy with liability to Pressure Palsies), and MCTD (mixed connec-tive tissue disease). She makes and shares videos about queerness, disability, and vintage fashion on her YouTube channel. All of her videos are closed captioned, and she frequently signs as she speaks, foregrounding accessibil-ity and positioning her channel as for disabled viewers. In a digital landscape where content is frequently produced by able-bodied users about us or for other able-bodied users, having a disabled YouTuber create content for other disabled users is meaningful and important.

Kellgren-Fozard's subversive digital media use as a disabled queer femme on YouTube is complimented by her subversive dollplay in *The Sims*, a video game that Flanagan identifies as a platform for virtual domestic play (49). In these videos, Jessica uses a tactic of critical play that Flanagan terms "rewriting" and describes as "a way for girls to explore deeper social and personal meanings in play" (33). Kellgren-Fozard uses critical dollplay to rewrite traditional narratives of femininity in ways that challenge heteronormativity and ableism; her play is work, her activism is playful, and her avatar play is framed through a crip/queer lens. "We Make Each Other in The Sims 4!!! // Jessie and Claud" (2019) opens with real life footage of Jessica and her wife Claudia overlaid with animation and music to make them look like sims, before the video transitions to Claudia and Jessica making avatars of each other in *The Sims 4*. Blurring the lines between digital avatar and human, this video draws attention to the slippage between onscreen and offscreen life and claims avatar play as a site of affect, embodiment, and identity performance (Anable 2018; Hillis, Paasonen and Petit 2015; Poletti and Rak 2014). Jessica introduces the video by telling us that she has been ill for three weeks, "But! On the plus side, I've had time to get back into *The Sims!"* Claudia then interjects and tells the audience that Jessica was playing *The Sims* when they first met. Throughout the video, Claudia and Jessica explore sexuality, gender, and disability: for example, they discuss which of their sims should be able to become pregnant, drawing queer family and queer reproduction into their play. At another moment they critique the "feminine walk" for failing to account for Jessica's scoliosis which limits shoulder movement and note that none of the walks account for disabled representation. This video points to digital media as therapeutic, queer, collaborative, and a site of productive critical discourse.

In "A Bad Day . . . with The Sims // Vlogmas Day 4" (2017) Kellgren-Fozard is in bed feeling "really rubbish," and after taking a nap and introducing us to her carer, Clara (who appears in many videos) Kellgren-Fozard then plays *The Sims 2*. Sharing these videos is a method of peer support, of sharing coping mechanisms and resources, or what Leah Lakshmi Piepzna-Samarasinha refers to as "collective care" and calls for in her book *Care Work: Dreaming Disability Justice* (2018). Speaking to a crip audience and situating herself within a network of community and care (her wife and her carer, for example), Kellgren-Fozard embodies interdependency and celebrates collective care over independence and individualism. Pausing the video to take a nap or to make a face that performs her nausea or pain for the viewer reminds us that feminist activism has to work with and for the bodies and energy and times that make up our movements. Kellgren-Fozard's disabilities and chronic illnesses are irrevocably tied to her queer feminism and activism, and the needs of her body are centered before the camera and made visible as an important part of her activism.

While many of her videos are aimed at a queer and disabled audience, Jessica has a series of videos for disabled folx to send to their able-bodied friends, family, and partners, saving us the emotional labor of doing the work of educating. In "What Is a Spoonie? // The Spoon Theory" (2017) Kellgren-Fozard uses *The Sims* to teach non-spoonies about chronic illness. Sims are used to stand in for offscreen spoonies, while the energy bar in the game is compared to the offscreen energy levels of disabled people. Jessica introduces the video with her own experience as a spoonie, noting, for example, that taking a shower uses more energy or spoons for her than for nondisabled people. YouTube and *The Sims* are used as educational tools and modes of performing digital advocacy and activism.

Kellgren-Fozard leverages YouTube and *The Sims* as forms self and collective-care, education, advocacy, and crip/queer digital identity performance. As a chronic fem(me)bot, Kellgren-Fozard harnesses digital media in order to perform activism from the bedroom, a familiar space for many disabled feminists (one of the titles in Piepzna-Samarasinha's text is "So Much Time Spent in Bed" (2018, 180)), and pushes back on traditional constructions of political activism as able-bodied. Kellgren-Fozard's digital activism ultimately embodies Johanna Hedva's "Sick Woman Theory" which celebrates the resistance and protest of those bodies who are often in bed (2016). Hedva critiques take-to-the-streets political activism, writing that "If being present in public is what is required to be political, then whole swathes of the population can be deemed a-political—simply because they are not physically able to get their bodies into the street" (1). Kellgren-Fozard's use of digital media and avatar play enacts a queer/crip activism that makes visible the often invisible (and often ignored or erased) protest of bodies living in the margins, bodies living differently, bodies living and loving outside the cisheteropatriarchy. With her public-facing platform and over 500,000 subscribers, her queer and disabled identity performance is deliberately performed for a public of both able-bodied and disabled, queer and heteronormative viewers. Celebrating disabled and queer femininity and publicly claiming these identity positions challenges the heteronormative and ableist prescriptions of settler culture.

KEYWORD: GHOST

Dropped calls and the roar of static; frozen screens and error messages. The cracks on the screen of my phone like the ice that time I fell. We slip through the cracks.

I spent the winter hibernating, scared to tell my friends how I was doing. Scared to be rejected; scared to drag you into this bed with me. Weights on my chest. All winter I am a ghost, scrolling through comments and images and hovering over the Like or Respond button. I post nothing. I want someone to hold me. I want someone to tell me everything is going to be OK. I am drowning under a blanket of snow that glitters bright and deadly in the sunlight, like Pride confetti or shards of glass. I am so lonely it hurts.

We are often more disabled in the winter—at least we are in Ontario, Canada, when snow and ice limits mobility and movement, and makes the city dangerous or even deadly. This is a time when many of us reach out online, curl up in bed and unfurl a rich, beautiful Second Life through Facebook Messenger and Twitter. But some of us don't. Sometimes the anxiety sparks in my lungs and the message is left unsent in my drafts folder for weeks. Sometimes the depression worms its way into my fingertips, and they are too heavy to dial your number. The pain is always worse in the winter, from the cold. The depression is worse, too.

How do we welcome the lost and the lonely, those of us who find it harder to reach out online, who feel faceless and bodiless when we look at our avatar on the screen and fumble to find the words to say, simply, I'm struggling?

On these days it's work to write a tweet, let alone a blog post or an email. It's work to read the words on the screen and close my eyes and imagine that someone views me as precious.

When I go underground, I want to leave a note, in the shape of a dried flower or a feather. I want you to not forget me.

I want to write: Find me.

I want to ask: Will you welcome me back in the spring? Will there be a ramp and a quiet space and a smile and a mug of tea? Or do you read our silences as death, as failed activism, as laziness, as not caring, or not caring enough?

We are still here, silent as the nests the birds abandoned when they migrated south. Silent as the bulbs under the frozen soil.

We are still here, our crip femmebot ghosts reading your posts, watching your videos, waiting to be remembered. To be seen.

Waiting.

KEYWORD: SPACE

"Get online," Imani Barbarin told me when I admitted that I sometimes felt like the token disabled person in a group of feminists. I was attending the opening gala of the 2019 Brampton's Festival of Literary Diversity, and Barbarin was one of five speakers on a panel (the topic? Women's rage). Barbarin is a Black woman with cerebral palsy, a brilliant writer and public speaker, and the creator of the website *Crutches and Spice*: "Disabled. Loud. Proud." Spaces are not always welcoming to us—"us" being feminists, women, queer and trans bodies, fat bodies, bodies of color, and disabled bodies. Spaces are riddled with obstacles: the step up to the main door, more stairs in the main lobby, the elevator tucked away down a narrow hallway that is difficult to find, and often broken. Chairs that are fragile, spindly, sharp, with no back support, painful, that won't fit our bodies and/or our mobility aids. Public transit can be a nightmare, especially in extreme heat or cold, which can put our lives at risk (why is a/c always broken on Greyhounds?). Barbarin, among other disability activists, turns to the Internet to find and build those spaces for disabled feminist community-building.

Featuring a cartoon-drawn avatar of Barbarin with her cane, *Crutches and Spice* is a fem(me)bot that claims space in a white male internet for Black disabled feminism, carving out a space for crip dialogue, and cripping/feministing the type of entertainment journalism found on websites like Buzz-Feed. While pop culture representations of fembots erase and eradicate Blackness from technologic futures—even as cyborg stories borrow from North American histories of slavery and racialized labor—Barbarin speaks back to white settler media with Black and of color bodies and media. With catchy headlines, photographs of celebrities, and tags like "Entertainment and Media" and "Reflections," the homepage of *Crutches and Spice* imitates the popular genre of entertainment journalism in order to interrogate the ideology of mainstream web entertainment and to turn these modes of speaking and writing into crip feminist tactics for education, activism, identity performance, and community-building. Two of her practices will be discussed here: cripping the listicle and cripping celebrity culture.

In "6 Reasons Why Your Disabled Friend Has Stopped Responding to Your Invites" (2019b) Barbarin riffs off the genre of the listicle, those banal filler pieces that have become ubiquitous online. Unlike rankings of trendy foods, celebrity fashion, and national stereotypes, Barbarin's list provides practical information for non-disabled readers (for example: "you keep inviting us to places that are not accessible"). However, the article isn't functioning purely as an educational tool: the list takes the reader from the benign and helpful explanations of in/accessibility into the more critical and accusatory "you make us your token" and "you're weird." The gifs in the piece often lend a sarcastic or exasperated tone to the text. The list has bite. Reading it as an intermittent cane-user, I found myself laughing and nodding at "you make us your token," which is accompanied by a gif of celebrity Kristen Bell giving the viewer the middle finger. While ostensibly for able-bodied viewers, the piece felt like a nudge and a wink to me, the disabled reader, who could relate to the experience of being made to "feel like a burden," while the eye-rolling Krysten Ritter validated my frustration and hurt at every moment of ableism I had ever encountered. Ahmed (2017): "Rolling eyes = feminist pedagogy" (38). Barbarin's crip listicle provides important information about disabled experiences, while simultaneously mocking able-bodied culture and calling out fake allies.

Beyoncé wiping a tear from her eye. A tweet by William Shatner. An article about *Jessica Jones*: *Crutches and Spice* is bursting with North American celebrity culture and pop media. Barbarin uses pop culture to write critiques of able-bodied culture; for example, using Bryan Cranston as an example of "conditional ally-ship," a person who only appears woke but doesn't contribute to meaningful change, in order to discuss ableism in the entertainment industry more broadly (2019a). In other cases, like the listicle, celebrity gifs are used to playfully add affect to the piece, both humorous and acerbic. Barbarin employs celebrity culture as a medium through which to discuss ableism, knowing that celebrities are a commonly shared cultural text through which we can generate meaning, and that celebrities are often used successfully as clickbait. Many of these articles generate robust conversations on the discussion board among other disabled people speaking to/with Barbarin, each other, and the culture more broadly. As a crip feminist taking up space online, and creating the space for conversation and community, Imani Barbarin and *Crutches and Spice* are a chronic fembot who reappropriates digital media for crip feminist activism, advocacy, and change.

KEYWORD: FAIL

I had to take long breaks when I was writing this paper. I get headaches and blurred vision from staring at the screen for too long. Switching from sitting to standing (I use countertops as makeshift standing desks) only slows the progression of pain, which steadily worsens as the day goes on. Lacking, partial, or failed captioning. No audio. Prohibitively expensive eye-tracking technology. Carpal tunnel. Migraines. Flickering screens can trigger seizures. Headlines about #MeToo can trigger flashbacks. Screen technology and Internet content can be disabling. Ellcessor: "The 'you' of Web 2.0 thus always implies the existence of 'them': those who are not invited, those who are not associated with the pleasures and agency of these technologies, and those who cannot take up these user positions because of different articulations of bodies and technologies" (2016, 78).

Not all of us are invited to the party.

Access to community from bed or the hospital is reliant on being able-bodied enough to use the interface or being able to afford expensive customized setups. Different technologies are accessible for different bodyminds. There will always be a place for analog, for old fashioned tech, for face to face communication, for printed copies.

A special issue of *Catalyst* on "Crip Technoscience" celebrated the creativity of "criptastic hacking" (Hamraie and Fritsch 2019), arguing that "The promise of feminist technoscience lies in challenging hegemonic narratives about technology as always enframing or deterministic, and imagining the transformative possibilities for crip hacking, coding, and making as frictional access practices" (5). Many of us wrangle, hack, duct-tape, code, break, and rebuild access to, with, and through technology. It's important to recognize the crip skillsets involved in creating access, and I invite all feminist readers to participate in listening, learning, and sharing resources, skills, and tips on how to create access for the individual bodyminds that make up our movements. The more tools a feminist organization uses, the more people are welcome to participate, and the more our valuable knowledge and skills can be leveraged to break down unjust structures and build better futures.

KEYWORD: SEXY / FUNNY

Behind *Austin Power*'s silly and fuckable fembots lie two irrevocable cultural truths: sexism is funny, and sexism is sexy. Queer feminists Jodie Mitchell and Sofie Hagen's comedy trauma podcast *Secret Dinosaur Cult* (*SDC*) offers us a cripped feminist reframing of these patriarchal constructs, which we might write out like this: crip/queer feminists are funny, and crip/queer feminists are sexy. Exploring trauma, kink, dinosaurs, and gender identity, among other topics, this "podcast about dinosaurs and daddy issues" harnesses subversive feminist laughter as a tool of activism and community-building enabled through the digital media practice of podcasting.

Humor is identified as a tactic of resistance in *Mad Pride: A Celebration of Mad Culture* (2000), the editors insisting that "these writings mock conformity, resist 'normalisation' and refuse to be co-opted. They rejoice in madness from a standpoint of anger, humor and rebellion" (8). *SDC* employs humor as a method of resistance, rebellion, and self- and collective love. In the "Daddy Hole" segment of each episode, Hagen and Mitchell discuss how they filled the gap that their absent fathers left in their lives:

> "So Sofie, how did you fill your daddy hole this week?"
> "I bought a face steamer, and I steamed my face."
> "And did it work, did he come back?"
> "No."
> (2019b)

This segment is introduced by an improvised jingle that changes every episode as they riff off popular songs to sing about childhood trauma. "Never gonna give you up/ never gonna let you down" becomes "He's always gonna give you up/ he's always gonna let you down" (2019a), "I'm leaving on a jet plane/ don't know when I'll be back again" becomes "Dad's leaving in a big truck/ don't think I'll ever feel good enough" (2018b), while "Purple Rain" becomes a cry for "Child Support" (2019c).

My friends and I like to joke about how "fun" PMS/PMDD is, roll our eyes at the meditation videos our aunt sent us, make dark jokes about trauma and pain, and get sarcastic when yet another person asks, "Have you tried deep breathing?" Adopting a humorous approach to mental illness means trespassing on the social norms of "respectful" silence, of fearful misunderstandings and dangerous assumptions; of a psymedical culture in which doctors and medical professionals are the only experts allowed to shape our narratives. It means resisting a culture in which "pleasure" and "disability" are framed as incompatible. (How I would love to hear our subversive laughter reshaping the nervousness of a waiting room). Comedy has the potential to interrupt the institutional spaces of masculine psymedicine, to harness joy and pleasure and camaraderie as weapons in dangerous encounters with medical professionals and to bring us together as members of a shared community, identity, or experience. Exploring comedy means taking play seriously as a tactic of resistance. It gives us permission to take pleasure in our embodiment as disabled feminists rather than seeing our trauma turned into a spectacle for able-bodied viewers. *SDC* embodies Ann Cvetkovich's (2012) argument that "depression is not . . . wholly depressing" (2), but tied to righteous fury, love, and feminist connection.

"Girls with daddy issues" is a popular trope in mainstream media, and popular portrayals typically both minimize gendered trauma by laughing at traumatized women and celebrate sexual predation and rape culture. Deliberately invoking and then subverting this trope, *SDC* turns the laughter of the marginalized into a tool of resistance—in this podcast we are not laughing at trauma but laughing as a tactic of identification and community-building. We are feminists, we are disabled, we are trauma survivors. We have joined the dangerous "cult" of feminism. Crip feminist laughter offers a mode of speaking back to power, laughing not at the trauma but at the power structures that produce trauma. For example, the "Daddy Hole" musical revisions function as critiques of the patriarchy with its heteronormative narratives of love, romance, and family, which are also entangled with myths of capitalist and financial success. It's difficult for women and families to meet these prescriptive norms of success when men are routinely excused from doing childcare and domestic labor, when absent fathers refuse to pay child support, and when the state provides little to no support for single mothers. Indeed, poverty and the lack of support for Jodie's family in the UK is a recurring topic across the podcast (Sofie received more support in Denmark, but still discusses the lack of child support). These experiences resonate with North American society and the lack of social supports for single parents and lower-income families. Laughing becomes a mode of expressing collective anger at these unjust structures and in taking pleasure in that anger.

SDC is undeniably sexy, from discussions of kink to the hilarious (but also sexy) "bedtime story" that concludes each episode: dinosaur erotica. As a crip/queer listener and Twitter follower, I experience the "lust of recogni-tion" (Piepzna-Samarasinha 2018, 117) when Jodie worries about her/their lack of queer sexual experience (2018a) as a pansexual person, or when Jodie and Sophie discuss gender identity and express their discomfort with being identified under the category of "woman" (2018c). Sex is funny, even as it's pleasurable, and this laughing with other queer feminists about our wonder-ful, ridiculous, and fun sexual experiences renders hypermasculine sexuality flaccid and punctures the prescriptive madonna/whore complex in which women and fem(me)bots are supposed to be always sexually available while simultaneously reluctant. The serious, manly sex with hardheaded men who have perpetually hard dicks is jettisoned in favor of the playfulness of rose quartz dildos, roleplay gone awry, only being able to masturbate in a very specific position, and . . . dinosaurs.

There is an intimacy to listening to a podcast, to bearing witness to the conversation between two friends, and listening to the live cult members laughing and cheering and sometimes chatting with the podcasters. Maybe its hearing the timbre of their voices, the practice of sharing emotion with other "cult members," or the immediacy of unrehearsed moments, but *SDC* feels like a particularly vulnerable and intimate model of doing crip/queer digital feminism. Frequently listening to the podcast in the bath, which I use to moderate pain, I feel like I am part of the feminist cult, and my laughter and fury and passion and hurt connect me to other bodies. Chronic fem(me)bots can be funny and sexy; we take pleasure in many things—perhaps especially in dismantling the cisheteropatriarchy—and our pain and anger are entangled with love, connection, and intimacy between bodies and avatars and screens.

REFERENCES

Ahmed, Sara. 2010. *The Promise of Happiness*. Durham: Duke University Press.
———. 2017. *Living a Feminist Life*. Durham: Duke University Press.
Al Zidjaly, Najma. 2011. "Managing Social Exclusion through Technology: An Example of Art as Mediated Action." Disability Studies Quarterly 31(4). doi: http://dsq-sds.org/article/view/1716/1764.
Anable, Aubrey. 2018. *Playing with Feelings: Video Games and Affect*. Minneapolis: University of Minnesota Press.
Barbarin, Imani. Crutches and Spice. https://crutchesandspice.com/.
———. 2019a. "Bryan Cranston is Everything I Will Not Be Accepting from Abled Allies in 2019." Crutches and Spice. January 8, 2019.
———. 2019b. "6 Reasons Why Your Disabled Friend Has Stopped Responding to Your Invites." Crutches and Spice. 27 June 2019.
Cazdyn, Eric. 2012. *The Already Dead: The New Time of Politics, Culture, and Illness*. Durham: Duke University Press, 2012.
"chronic, adj." 2019. OED Online. September 2019. Oxford University Press.
Clare, Eli. 2017. *Brilliant Imperfection: Grappling with Cure*. Durham: Duke University Press.
Curtis, Ted et al. 2000. Mad Pride: A Celebration of Mad Culture. Chipmunka Publishing.
Cvetkovich, Ann. 2012. Depression: A Public Feeling. Oxford: Oxford University Press.

Ellcessor, Elizabeth. 2016. *Restricted Access: Media, Disability, and the Politics of Participation.* New York: NYU Press.

Flanagan, Mary. 2009. *Critical Play.* Cambridge: MIT Press.

Fritsch, Kelly, and Aimi Hamraie. 2019. "Crip Technoscience Manifesto." *Catalyst: Feminism, Theory, Technoscience* 5(1). doi: 10.28968/cftt.v5i1.29607.

Fritsch, Kelly, Clare O'Connor, and AK Thompson. 2016. *Keywords for Radicals: The Contested Vocabulary of Late Capitalism.* Chico: AK Press.

Garland-Thomson, Rosemarie. 1996. *Freakery: Cultural Spectacles of the Extraordinary Body.* New York: NYU Press.

———. 2009. *Staring: How We Look.* New York: Oxford University Press.

Haraway, Donna. 1991. "A Cyborg Manifesto: Science, Technology, and Socialist-Feminism in the Late Twentieth Century." In *Simians, Cyborgs and Women: The Reinvention of Nature.* Routledge.

Juhasz, Alexandra. 2017. "Affect Bleeds in Feminist Networks: An "Essay" in Six Parts." *Feminist Media Studies* 17(4): 660–87.

Kafer, Alison. 2013. *Feminist, Queer, Crip.* Bloomington: Indiana University Press.

Kellgren-Fozard, Jessica. 2017. "A Bad Day . . . with The Sims // Vlogmas Day 4." YouTube, 4 December 2017, https://www.youtube.com/watch?v=YU0qleTRp9k&t=460s.

———. 2019. "We Make Each Other in The Sims 4!!! // Jessie and Claud." YouTube, 2 April 2019, https://www.youtube.com/watch?v=VlltZBM9Tzc&t=538s.

———. 2017. "What Is a Spoonie?// The Spoon Theory." YouTube, 21 January 2017, https://www.youtube.com/watch?v=a2NGaG8mhjU&t=323s.

Kember, Sarah and Joanna Zylinska. 2012. *Life After New Media.* Cambridge: The MIT Press.

Mann, Benjamin W. 2018. "Survival, Disability Rights, and Solidarity: Advancing Cyberprotest Rhetoric through Disability March." *Disability Studies Quarterly 38(1)*. http://dsq-sds.org/article/view/5917/4886.

McRuer, Robert. 2006. *Crip Theory: Cultural Signs of Queerness and Disability.* New York: NYU Press.

Mitchell, David and Sharon Snyder. 2015. *The Biopolitics of Disability: Neoliberalism, Ablenationalism, and Peripheral Embodiment.* Ann Arbor: University of Michigan Press.

Mitchell, Jodie and Sofie Hagen. 2019a. "15. Compsognathus & School: It's Not Gay If It's a Dinosaur." Secret Dinosaur Cult. Podcast. January 7, 2019.

———. 2019b. "16. Dracorex Hogwartsia & Magic: My Vulva Looks Like 'The Scream.'" Secret Dinosaur Cult. Podcast. January 13, 2019.

———. 2019c. "22. Xiaotingia & Self Care: Extreme Lesbian Content published." Secret Dinosaur Cult. Podcast. March 4, 2019.

———. 2018a. "11. Patagotitan & Fat: Same Sadness, Less Pay." Secret Dinosaur Cult. Podcast. December 10, 2018.

———. 2018b. "12. Psittacosaurus & Virginity: Ice Dick! Ice Dick! Ice Dick!" Secret Dinosaur Cult. Podcast. December 17, 2018.

———. 2018c. "10. Sue the T-Rex & Gender Identity: The Feminazi App." Secret Dinosaur Cult. Podcast. December 3, 2018.

Munster, Anna. 2006. "Materializing New Media." Lebanon: Dartmouth College Press.

Noble, Safiya Umoja. 2018. *Algorithms of Oppression.* New York: NYU Press.

Piepzna-Samarasinha, Leah Lakshmi. 2018. *Care Work: Dreaming Disability Justice.* Vancouver: Arsenal Pulp Press.

Paasonen, Susanna, Ken Hillis, and Michael Petit, editors. 2015. *Networked Affect.* Cambridge: MIT Press.

Price, Margaret. 2011. *Mad at School: Rhetorics of Mental Disability and Academic Life.* Ann Arbor: The University of Michigan Press.

Rak, Julie, and Anna Poletti, eds. 2014. *Identity Technologies.* The University of Wisconsin Press.

Siebers, Tobin. 2008. *Disability Theory.* University of Michigan Press.

Transken, Si. 2005. "Creativity, Cultural Studies, and Potentially Fun Ways to Design and Produce Autobiographical Material from Subalterns' Locations." In *Auto/biography in Can-*

ada: Critical Direction, edited by Julie Rak, 145–72. Waterloo: Wilfrid Laurier University Press.

Trevision, Filippo. 2017. *Disability Rights Advocacy Online: Voice, Empowerment, and Global Connectivity.* Routledge.

Chapter Four

Virtual Dwelling

Feminist Orientations to Digital Communities

Brianna I. Wiens

REFLECTIONS ON VIRTUAL DWELLING

On October 15, 2017, I was sitting at the very back of a Toronto streetcar after squeezing through the crowded front doors, feeling someone's hand on the back of my upper thigh. Because this happens all too often on crowded transit cars, it was easy enough to walk to a different part of the vehicle. Rather than dwell on what happened in the physical space, I dwelled in virtual space. It was there, on Twitter, that I saw it for the first time: #MeToo. I scrolled past the first tweet from someone I went to high school with without feeling it sink in, but I saw it over and over again in my feed: from someone I went to undergrad with, from a childhood friend, from celebrities, and from a friend I had just made at a fitness studio. I paused, scrolled back up, and re-read each tweet. Some of them were a short 140 characters, some had parts one through seven. I switched over to Facebook. Long posts where some of my women and trans friends detailed their own experiences of sexual assault, and shorter posts that simply read, "me too." I tried not to think of my own experience, from only a moment ago, to another some seven years prior. I remember wanting to laugh at all these confessions. Not because I thought they were funny; not at all. But because I was so relieved. Perhaps you felt something similar. This chapter encourages you to pursue those individual tugs of affect, those bubbles of emotion that feel compelling or out of place, and to follow the example of the many feminists before us in sitting with feelings and bodily knowledges as acts of resistance and as sources of data.

This chapter is concerned with those #MeToo hashtags, the communities, and the affects they can create by developing a conceptual framework called *virtual dwelling*: focusing on what it is, what it does, and how it can move people towards action. In this chapter, I demonstrate how #MeToo created and continues to create a distinct kind of space for online dwelling that highlights both personal and collective affective lingerings as sites of political transformation. #MeToo and other feminist hashtags highlight how people can and have co-opted the white supremacist and misogynistic spaces of the internet via the use of hashtags for intersectional feminist and queer resistance against gender and sexual-based violence. Yet knowing how to co-opt these technological spaces, or when to do so, is not necessarily clear. Thus, what is required are acts of virtual dwelling, which ask that we linger in online spaces to sit with ideas, find out how tools work, how different tactics can be tools, and how they can be used in counter hegemonic ways to center marginalized voices and bring forth new ways of engaging in the world. Understanding virtual dwelling as a tactic helps us to better understand the significant impact of movements like #MeToo, #ShoutYourAbortion, #BlackLivesMatter, or #IdleNoMore. Activists dwell in these virtual spaces to better understand these movements, the technology they are circulated through, and the conditions that make the movements work. This dwelling is affective, personal, physical, and analytical. It's about paying attention to something for longer than someone may think we should be paying attention to it for, going against the flow, staying with something a little too long, and following it through, perhaps, to an "illogical" end. It entails going beyond what is considered standard, or "normal," because it is these norms that we wish to interrogate and be able to see through.

By situating myself and the individual subjectivities of social media posts and comments within broader sociopolitical and technocultural assemblages, I argue that the process of recognizing and documenting an orientation to these digital data is important for focusing attention on: (1) the relationship between or "intra-actions" (Barad 2003) of the researcher, research scene, participants, data, affects, and sociopolitical context; (2) the individual stories found through these data, not just the broader themes or trends of the aggregate; and (3) better conceptualizations of the interconnected domains of influence between individual spheres and their relationship to collective and then structural levels. As Patricia Hill Collins (1990; 2017) asserts, the individual level is just one of four areas (interpersonal, hegemonic, disciplinary, and structural) where domination and resistance occur, with each level contributing to existing power relations. Within this framework, I focus on individual or interpersonal acts to demonstrate how virtual dwelling offers a way to engage with and reveal complex relationships and responses to power within digital activism. Importantly, the social media posts that I highlight in this chapter come to me through my own interactions on social media. My dwelling begins subjectively, at the individual level, by following the metaphor of the witch and its very material impacts for women and women's resistance around. Although digital content can be analyzed via generalized trends and large randomized datasets, virtual dwelling refocuses attention to what might otherwise be lost or glossed over in big data. Practices of dwelling, although seemingly simple, matter for the ways in which they create opportunities to settle into smaller sets of data, recall past instances of related methods and practices, and to work with data on the research scene to follow them through to new ends.

VIRTUAL DWELLING AS FEMINIST PRAXIS

I and many others remember the experience of seeing #MeToo on social media for the first time so clearly not because we all have the exact same experience of assault, but because many have either experienced sexual violence of some kind or know someone who has, or because the stories we were seeing on social media forgot our communities and our people. From a hand on your ass on public transit to a catcall when you're walking home at night, to that time you try to forget (although it's more than likely that you try to forget each and every time)—they're the experiences we brush off because we don't think we'll be believed, because we're in too much shock, or because we've been told they're not serious enough, even though we know now, in theory, that every single time is serious enough, and our perpetrators will most likely never be held accountable.

The day after my first encounter of #MeToo I tried to go back and save all those tweets and Facebook posts from the previous day on that streetcar. I wanted to sit with those acknowledgements and self and public assertations of "me too" a little longer. Although I couldn't find all of the tweets, and I'm sure that Facebook's and Twitter's algorithms have buried them for good, what felt like a hundred more were in my social media feeds—and, indeed, 24-hours following the inciting tweet from celebrity-activist Alyssa Milano there were 109,451 #MeToo posts on Twitter (Main 2017 cited in Clark-Parsons 2019) and #MeToo was used or implied in over 12 million Facebook posts and comments (Park 2017 cited in Clark-Parsons 2019). I tried to read as many of them as I could, including the comments and who else had liked or retweeted the posts. Posts on my Twitter and Facebook feeds read:

@TwitterUser: #MeToo He didn't get that he'd done anything wrong, that he broke me. I lost a friend. Education is vital. It's been years & I'm still scared to tweet this. (October 18, 2017)[1]

@TwitterUser: #MeToo because when I was 18, starting my first year of college, a friend raped me while I was in and out of consciousness. I decided to not report it because I know it would have been an exhausting process for me to go through, and I couldn't handle that on top of the stress . . . (October 19, 2017)

Facebook User: As someone who is transgender—#MeToo—it's difficult to own your story and trust yourself when someone in a committed relationship commits these acts against you, but we must keep speaking. (October 17, 2017)

Facebook User: It might actually be harder to find a woman who hasn't been sexually harassed or assaulted. Now think about women with different ethnicities and races and disabled women. This needs to change. #Truth #MeToo. (October 18, 2017)

Of course, some of the comments were incredibly misogynistic, perpetuating rape culture:

@TwitterUser: There is only one man on Earth who is desperate enough to sexually assault (posted with a meme of Bill Clinton asking, "I didn't rape that ugly one did I?" and Chelsea Clinton responding, "that's mom"). (October 19, 2017)

Others showed a more subtle misogyny that still reinforced rape culture:

Facebook User: We have to think more critically. It's just sexist to say that all boys are rapists. Women are silenced by their own doing. From my personal experience, women should report the assaults. No one is preventing women from reporting assaults to the authorities. I know strong women, and these women have reported the assaults. Nothing will change if you don't report the assaults! (October 18, 2017)

Other posters commented back to these misogynist posts, alleviating the emotional and physical labor required to formulate a response and instead sharing the labor among many. Responding to the offensive meme of the Clintons, people on Twitter wrote:

@TwitterUser: Let's take a roll call of the men who have not inappropriately approached a girl or woman at some point in their life? (October 19, 2017)

@TwitterUser: I'm sorry. Women deserve better. I'm sad that women have to put up with this BS. May you be healed and grow stronger through this. (October 19, 2017)

@TwitterUser: Women and their attractiveness are not the cause of sexual harassment/assault. Men and society are the cause. Plain and simple, men need to respect, cherish, and honor women more and society needs to allow that to happen. Oh yeah, and me too. (October 20, 2017)

However, despite misogynistic comments and their responses, most of the posts worked to encourage those posting and joined with them in solidarity, offering words of support, love, friendship, and allyship. In response to the participant on Twitter speaking to the need for education to end sexual assault, people on Twitter shared the following:

> @TwitterUser: Powerful thread. Thank you for your courage. Peace to all. #NeverAgainIsNow. (October 18, 2017)

> @TwitterUser: I was moved by a beloved boss from my [job] post when I told him that a co-worker was punishing me because I refused. I felt alone. Thank you for doing this! (October 18, 2017)

> @TwitterUser: I think this [post] alone is a big enough statement. Now there are thousands speaking. I admire everyone on here who is speaking up, as well as those who are not. (October 18, 2017)

Before that day, I knew that social media's claim to fame was connection, but in the days following October 15, 2017, I felt it for the first time.

The internet can feel like it is overflowing with misogyny, racism, and homophobia. Because of this, there is a clear lack of space for marginalized voices to safely come together without the threats of racism, sexism, ableism, homophobia, and transphobia. But, reflecting on this sample of posts, the internet can also be a place of hope, a space to connect, confess, share, witness, educate, and learn with people across a variety of geographical, racial, gendered, ability, and, to a degree, class-based lines. #MeToo specifically speaks to me in ways that #BlackLivesMatter or #ShoutYourAbortion or #DisabledTwitter speaks to others, and I thus focus my attention and this entire project around the #MeToo movement because of these affective resonances. Each person has their own affective response to online spaces, and those individual responses are important for gesturing towards collective action.

Dwelling is like "staying with the trouble" (Haraway 2016), it's about assuming a responsibility and relationship to the present moment in order to be open to new ideas and knowledges within our networks of places, times, matters, and meanings—even if those ideas contradict other ideas we have previously held or ideas that seem antithetical to current power structures. Dwelling can also be about tracing the histories that inform the present moment and actions that seem novel but are really manifestations of earlier practices, as I illustrate later in this chapter with my own practice of virtually dwelling with the so-called "#MeToo witch hunt." Haraway writes that "our task is to make trouble, to stir up potent responses to devastating events, as well as to settle troubled waters and rebuild quiet places" (1), encouraging a reconceptualization of what it means to "make kin" so that we can recognize "the dynamic ongoing sym-chthonic forces and powers of which people are a part, within which ongoingness is at stake" (101). For Haraway, to "make kin" is to establish news lines of "response-ability" between beings to see the ways that they think and make together. Dwelling offers sanctioned time for learning to stay with the trouble of living, working, and playing in response-ability to the events unfolding around us. Through dwelling it becomes possible to see lines of communication between actants, connections and relationships of effect and affect between humans, ideas, technologies, and other things. Thus, in staying with the trouble—that is, in dwelling—Haraway calls for a praxis of "tentacular" thinking similar to that of assemblage or networking thinking that embraces presence and attention to the moment, for sticking it out in the "here and now" to trouble the waters of entrenched capitalist models that contribute to the destruction of collective organizing and change.

Here, we are pointed towards dwelling as a way of thinking, a way of changing our orientation to the current moment and place to consider in new ways the ideas laid out before us. Dwelling can encourage us to think with the moment, against the moment, or with the moment as it moves to new moments. Dwelling makes clear what Sarah Sharma (2020) calls "a feminism of the Broken Machine," which highlights and "uses the logic of the machines" to focus on "current power dynamics that are otherwise hard to pinpoint" (174). For Sharma, the Broken Machine, like Sara Ahmed's Killjoy and Donna Haraway's Cyborg, becomes worthy of attention once it begins to glitch, making itself known as it points out gendered power differentials and other hierarchical structures. As does dwelling, Sharma's Broken Machine creates space for new perspectives as they "flicker and burn out," becoming "powerful purveyors of mayhem and confusion" (174). These moments of glitch, especially as I found in following the figure of the witch in the #MeToo movement and the many ways it was used by both detractors and activists, initiate uncertainty and chaos and point towards layered and complex social and interpersonal relationships that become otherwise and change as their conditions do. The first questions of reflection we might consider when dwelling, then, are: how are the mayhem, confusion, and glitches indicated in the data changing the ways that people are relating to one another in affective and embodied ways on the research scene, and how are the data and the research scene creating a relationship with the researcher?

Simultaneously, dwelling can be understood as a way of being in flux, open to movement and change—the mayhem, confusion, and glitches—as we immerse ourselves into the scene and sit with the changing technological and natural landscape—whether that change happens on its own or through our doing. Different from critical analysis, where the goal is to evaluate a body of work (an artifact, a text, a film) and express an opinion on that work, dwelling demands a deeper, more proactive engagement with the subject of interest, its uses, the context, and the ways the subject has been taken up on embodied, affective, and intellectual levels, seeking an understanding of where we dwell. Dwelling is more radical in that it asks that we take up space and that we orient towards not just understanding, but also action. Through dwelling, we can see the tools at our disposal, how they have been used, and how we might use them differently in the future to provoke alternative programs and methods. As such, the second question we might ask when dwelling on the research scene is: how are data indicating what practices and tools are being taken up by participants to create or disrupt relationships?

Dwelling is, thus, also praxis: an action undertaken with the tools observed and acquired to sift through the scene, collect information, and then begin to understand that information. As an integral part of the artistic research method, dwelling also asks that we "pay attention to the specificities of the space that are overwritten by dominant perceptions and uses of it" (MacDonald 2018, 279). Dwelling is concerned with "access[ing] and convey[ing] [the] layered nature of space," and is "an embodied act that we do on a regular basis" (279). Such uses of dwelling include lingering with data to reconceptualize research as layered "scenes" (MacDonald and Wiens 2019) where research can be understood as "collections of material objects for researchers to study" while "also acknowledging researchers' bodies, voices, and gestures as essential forms of material data" (Wiens et al. 2020, 22). It is also a way to see how people have thought about and spent time with their own histories and experiences, and how those experiences have shaped other shared and individual stories.

Dwelling becomes a reflexive process, a tentacular theoretical intervention, highlighting different relational networks or assemblages that currently exist and that are actively coming into being in order to better understand experiences through affective and embodied time in a scene. This reflexive process helps to situate the researcher within the research scene, identifying (1) the personal relationship to the research in order to highlight (2) the importance of the individual behind each piece of data. Although dwelling may start off as a personal practice through lingering with different modes of thought, it allows for the creation of intimate connections. Because dwelling asks that we become familiar with a space in its current state, examining how previous interactions have created that space, it creates the conditions for reaching out to others through the space to find access to new people, data, cultures, organizations, and systems. In this way a variety of different kinds of relationships can be formed, helping in the formation of new communities, as seen in online hashtag communities. Dwelling also offers a way to take up and form relationships with space when that space has been denied within the institution—as a way to make yourself present in order to resituate, and to recast colonial, sexist, racist, and/or ableist histories.

In part, then, dwelling is also a method of coping. In sitting in spaces it is inevitable that we will dwell with past and present erasures, violences, and hurts of our own and/or others' stories and histories. Through preservation of representation in archives and in online spaces, which can be considered archives through the preservation of virtual data, and through agency in crafting current public discourses, dwelling can contribute to reckoning with individual and collective hurt, and, in taking time to recognize the pain and to hear stories, we may begin to be able to reconcile those past hurts and find ways to better cope within the present moment to envision different futures. The goal here is understanding and bringing attention to interrelated spaces; it is simultaneously a process of coping and working through the challenges that are sure to arise when there is a relationship between the researcher and the data. Thus, the third questions we can ask while virtually dwelling are: (a) how are social media participants reflecting on, building on, and/or draw-ing connections across sociopolitical, technocultural, and/or historical con-cepts, ideologies, and/or relationalities in the present moment? (b) how are both the researcher and social media participants coping with/on/through the research scene and the digital intimacies formed?

By theorizing feminist hashtag practices as spaces where virtual dwelling can be cultivated, I seek to contribute a more holistic understanding of our sociotechnical culture to underscore the importance of bringing together indi-vidual dwelling points to the collective in order to mobilize more inclusive, intersectional openings within our present moment to reimagine feminism's potential in the digital era. Each move is really a move to uncover another layer as I sit with these tools (hashtags, news media, journal articles) to think about their uses and their effects. In the next section, I outline what a reflec-tive process of virtual dwelling could look like using my own example of virtually dwelling with the claim that the #MeToo movement is a witch hunt. I follow the figure of the witch around to draw a theoretically and politically informed cartography of knowledges, histories, subjects, power relations, affects, and discourses that emerged through my relationship with the data. I illustrate this practice of dwelling by focusing on the figure of the witch, a highly political figure that has become prominent in discourses of #MeToo and feminist resistance more broadly.

WHOSE WORST NIGHTMARE?

In the weeks following the initial explosion of #MeToo posts, primarily on Twitter and Facebook between 2017 and 2018, I stayed tuned into the stories that were spilling out across social media platforms. On Twitter, a post in response to Alyssa Milano's October 15, 2017 #MeToo post caught my attention and held it, initiating the research bond of virtual dwelling between me and the digital research scene and impelling me into the #MeToo counter-public at an interpersonal level:

> @TwitterUser: It's a real tragedy how common it [sexual assault] is. For the monsters it's only a matter of minutes, for us, it's a lifetime of nightmares #MeToo. (October 23, 2017)

This lifetime of nightmares and the stuff that fills them—stuff like monsters, witches, darkness, shadows—speaks to a fear that cannot be so easily quelled. The fear evoked from these nightmares is not a fear that can be separated from everyday life; these are dreams that one cannot be so easily woken up from.

Curiously (or, perhaps not so curiously given the history and symbolism), in the year following #MeToo's viral surge on social media, posts on Facebook and Twitter took up the theme of the nightmare. And, for me, this is where the data glitched, causing confusion and mayhem in the ways that participants were able to relate to one another and to the social media scene: in some cases this became about the nightmare of the perpetrator of assault and not the waking nightmare of those who had survived sexual violence, harkening back to witch hunts and the terror that associated these quests for "justice." The ways that the #MeToo witch hunt became about re-traumatizing survivors of assault through claiming that they were falsely accusing men of assault caused a well-known history of witch persecution, the hunting and torturing of women assumed to have too much power and those who did not conform to gender standards, to flicker and be re-cast in favor of those in power. This historical intensity of feminist work speaks to the current activist moment, contributing to the development of the assemblage I began dwelling in here.

Conversations of a #MeToo witch hunt seemed to spike around the time of Brett Kavanaugh's nomination and eventual confirmation to the Supreme Court, and Dr. Christine Blasey Ford's testimony of sexual assault at the hands of Kavanaugh. These claims misappropriate the history of the witch, taking the gendered and racialized violence against women perceived to have power out of historical and cultural context (more on this to come):

> Facebook User: This is a nightmare. As a mother who loves her boys, it TERRIFIES ME that at ANY time ANY girls can make up ANY story about ANY boy that can be neither proved or disproved, and completely RUIN any boy's life. THAT. IS. SCARY. (September 17, 2018)

With same day responses including:

> Facebook User: This is spot on, it is terrifying. It's the #MeToo witch hunt

Facebook User: Hopefully by the time that your boys are old enough to deal with this sort of lunacy the hate for white heterosexual males will have come to an end. I know young men questioning their values as I write this. No more white guilt! #MeTooWitchHunt.

At the same time that this conversation was happening, others weighted in to reject the claims of a #MeToo witch hunt that supposedly attacked innocent boys and men, responding with:

Facebook User: Being falsely accused of rape is not as bad as actually being raped. Just see #MeToo or #WhyIDidntReport or #IBelieveChristineBlasey-Ford on Twitter. Not just #MeTooWitchHunt.

Facebook User: Nor is it as systemic, or as pervasive, nor is it to be conflated with, prioritized over, or is as bad as living in paranoid ideation of, or as fucking bad as actually being raped. #BelieveWomen #BelieveSurvivors.

Based on the lack of response to the two participants above pointing towards hashtags like #WhyIDidntReport and #BelieveSurvivors, it's likely that the conversation here did little to change the mind of the original poster or the commenters agreeing with the poster—at least not to the point of confession online. However, the importance of the conversation lies in the bridge that was built between echo chambers. It was an opportunity to follow a different set of hashtags that may not have come across these individuals' Facebook and Twitter feeds, and perhaps to extend beyond their own conversations. Further, it was this particular conversation that led to my own fascination with the #MeToo witch hunt, prompting me to dwell with these ideas and to interrogate the witch's power both historically and in the contemporary moment. This interaction also reaffirmed the discourse of the nightmare and drew my attention back to that Twitter post of nightmares and monsters from October 23, 2017—most likely because of my own experience of sexual violence.

Stories of harassment, and stories of disbelief at such harassment (of the witch hunts), continue to pour out across social media. The ways that stories move and affect people individually necessitates a greater understanding of the digital platforms and cultures that make individual testimony possible. In what follows, I demonstrate how dwelling with the idea of the witch opens up new questions about power dynamics. I start with a feminist response to the idea of "me too" as a witch hunt or nightmare, which was just one of many misogynistic reactions to #MeToo, because far too often the reaction is seen as secondary, as if the move to dwell and to heal is not as important as the intent to wound. Feminist responses to white supremacist capitalist patriarchal motions to harm are significant for the ways that they encourage individual and collective restorative solidarity. I also start with this response to misogyny rather than the instigating moment because: (a) there are too many instigating moments to count, and (b) in focusing on one particular response first it becomes clearer how each moment of misogyny necessitates new, context-dependent, and constantly evolving ways to think about responses, healing, and community. That is, how can intersectional feminists work with, transform, and utilize practices available within the leaky boundaries of social media spaces and rigid oppressions of the technologies themselves?

GETTING WITCHY WITH IT

On October 18, 2017, a mere three days after the viral spread of the #MeToo movement on Twitter—a movement that was founded by organizer and activist Tarana Burke for Black women and girls before being taken up by white celebrities—articles in *Chatelaine, Maclean's*, the *New York Times*, and the *Washington Post*, had already dispelled, critiqued, or analyzed the relationship between witch hunts and #MeToo, agreeing that it was an increasingly popular way to cite the movement. One article reminded readers, "It is Canada. It's the office you work in. The school you go to. The café you are sitting in right now. It's the streets you walk on every day. It is every industry. . . . Make no mistake: Sexual harassment is utterly ubiquitous and endemic to the culture we live in. This is not a witch hunt, it's a statement of pure, inescapable truth" (McLaren 2017, para. 3). Another article, in response to a well-known director's statement that the #MeToo movement, a witch hunt, was "sad for everyone," declared,

> When [Woody] Allen and other men warn of "a witch hunt atmosphere, a Salem atmosphere" what they mean is an atmosphere in which they're expected to comport themselves with the care, consideration and fear of consequences that the rest of us call basic professionalism and respect for shared humanity . . . Setting aside the gendered power differential inherent in real historical hunts . . . and the pathetic gall of men feeling hunted after millenniums of treating women like prey, I will let you guys have this one. Sure, if you insist it's a witch hunt. I'm a witch, and I'm hunting you. (West 2017, paras. 4, 6)

When "Allen and other men," including the 45th president of the United States, publicly condemn the naming of sexual abusers as a witch hunt, as harassment, their claims tap into a social consciousness and historical memories of false trials, of unjust persecution. Here it is significant to note that even before the reappearance of #MeToo on social media, Donald Trump had tweeted multiple times of being the subject of a witch hunt. These tweets date all the way back to his pre-inauguration days, where a "witch hunt" supposedly targeted Trump University, before ramping up once he took office to describe the investigation for Russian interference in the election. Repetition is key for persuasion, and by the time that Tarana Burke's #MeToo resurfaced on the Internet in October 2017, "witch hunts" had been already tweeted about by the American president over two dozen times and vocalized aloud in the news even more.[2] This tactic, a strategy called "firehosing," has been used to "quell dissent and control the political landscape" through, essentially, lying in order to inundate discourse with falsehoods to distract and mislead the general population (Paul and Matthews 2016 in Tran 2019, paras. 4–5). As of November 14, 2019, Trump had tweeted of a witch hunt over 300 times to denigrate political events ranging from talk show interviews to the Russia inquiry, to, yes, the #MeToo movement. Through doing so, 45 offered a familiar language to others through which to categorize news that they, too, dislike and believe to be untrue, or want to convince others to believe as false:

> @TwitterUser: This "movement" called #MeToo is clearly a #SympathySeeking movement! #MeTooWitchHunt. (September 18, 2018)

> @TwitterUser: Sure, every woman has a right, in my opinion, to be heard. But no one has a "right" to be believed #MeTooLiars #MeTooWitchHunt #Defamation. (September 19, 2018)

Because this rhetorical association to witch hunts was already in motion, and because Trump had endorsed and used the language of the witch hunt to discuss #MeToo, the groundwork was already laid for "Allen and other men," including the two who posted to Twitter above, to also implement that language as their own.

As more and more people publicly acknowledged "me too," misogynists continued to loudly claim that the "me too" movement was a type of witch hunt. But the witch hunt and subsequent uses of the concept are necessarily tied to a history of women who were arrested and killed for engaging in activities deemed unfit by patriarchal standards. In both instances, women become targets by challenging the status quo. Ignoring the actual history of witch hunts, their use of the concept became a commonplace metaphor for unjust accusations, seemingly having nothing to do with the real-life witch trials of days past.

And yet, through the repeated rhetorical linking of "harassment" and "witch hunt," "Allen and other men" appropriate the language of #MeToo to craft their own version of events where it is abusers who are the victims of #MeToo's unfolding events. In part, the use of this language works to remove gender and class as key factors in the historical pursuit and torture of witches. That is, in taking up this discourse of a witch hunt, "Allen and other men" erase the figure of the witch, the resister of heteropatriarchal norms, who has been held captive by the same sexist systems that uphold men like them. Calling on the witch as heretical and hysterical implies instead that the hunter, a figure of power, is the target of this unfair fight. This discursive reversal works by focusing only on the accusation, and by conflating both the accusation and the accuser as wrongful. Interestingly, as the forty-fifth president of the United States faced an impeachment inquiry (one that acquitted him from abuses of power), he continued his claims of a witch hunt in which he was the victim of a new accusation every week, while simultaneously positioning himself in the role of hunter pursuing anyone who opposes his racist and sexist political views (CBC News 2019). Clearly, invoking the figure of the witch is a political rhetorical tool. Forty-five's twisting of the story is yet again a twisting of history. But feminists have long been dwelling in the political and social moments they find themselves in, always using the tools available to them for their resistance.

WITCH, PLEASE

Dwelling with the history of the figure of the witch reveals a history steeped in resistance, protest, and revolt. Witches were people associated with femininity and nature, with repudiations of the masculine. Witches were women seen as having "too much" power (Rowlands 2013; Gasser 2017). Childbirth, menstruation, contraception, abortion, gynecology, healing, and herbology—work often considered to be in the realm of the feminine—were associated with witchcraft. Under patriarchy, witches were situated in opposition to men. Groups of women governing themselves matriarchally,[3] who organized separately from men's control, or who could not be disciplined by the patriarchy were called covens of witches. Black and Indigenous peoples who engaged in practices that emerged during enslavement and colonialism, like Santería, Voodoo, and Candomblé, have been labeled witches and violently persecuted (Joho and Sung 2020). Those who engaged in the work of the "feminine," who opposed patriarchal rule, or who defied what white supremacy dictated was acceptable have been criminalized, arrested, executed, or otherwise punished—to the degree that their histories have been strategically altered or, in some cases, completely erased (Gasser 2017). This has undermined the resistive, knowledgeable, powerful, and, in Indigenous and Black communities, self-determining nature of spirituality, "witchcraft" and magic, and the people who practiced those crafts.

Fundamentally, this was the goal of witch hunters as they hunted those with the least power, those who were already marginalized: women and other feminine presenting people and racialized people. For instance, the term "witchcraft" was used by European colonizers in an act of cultural genocide in order to demonize the traditions and spiritual practices of Indigenous peoples and Black people who had survived the Transatlantic Slave Trade (Joho and Sung 2020). Further, between 1638 and 1725, a period of time in New England when witch hunts and trials were a regular occurrence, an estimated 78 percent of those accused of witchcraft and executed were women and feminine presenting people, with men and enslaved people facing accusations and death because of their associations to women deemed guilty (Demos 2004; Karlsen 1998). Those accused of witchcraft were those who lived, even scarcely or through affiliation, outside the bounds of prescribed racialized and gendered social roles.

Consider midwives, who were accused of being witches to redirect authority to the Christian church and dismiss their expertise learned through oral histories (think: "old wives' tales") since women were not allowed into institutions of formal education. Think of this in contrast to the presumed "father" of gynecology, J. Marion Sims, who performed hundreds of nonconsenting surgeries on enslaved women for the sake of medical "innovation." Or, perhaps, Agnodice of Ancient Greece. Although known to be a practitioner of medicine, her very existence is debated. She is said to have disguised herself as a man, caring for women who were unfairly treated by male physicians during childbirth, becoming increasingly popular with her patients—so much so that, while still presenting as male, she was charged with adultery for engaging in affairs with her patients, for which she was later acquitted when her female patients came to her defense (Garza 1994). Recall also the iconic figure of Joan of Arc, who led French armies against the English and was ultimately burned at the stake at nineteen years old after a sentence of life imprisonment for dressing in what was considered men's clothing and because of her presumed connection to male-dominated authority of the church.[4] To this day, witches and those accused of being witches face violence. Still we see witch trials, resulting in violence and the murder of women and children (Amnesty International 2009; Migiro 2017). Witches were, and are, feminine-relegated figures who did not and do not conform to the kinds of patriarchal standards of their time.

And yet, from within these heteropatriarchal structures, activists have embraced the feminist power, like the power that the witch symbolizes, in order to speak back to such structures, challenging the white supremacist and heteropatriarchal standards of their time through feminist organizing. In the 1890s, African American women led by Ida B. Wells organized campaigns in the United States against rape and lynching, laying the groundwork for national organizations, like the National Coalition Against Domestic Violence, to emerge in later years (Greensite 2003). In 1968 W.I.T.C.H., Women's International Terrorist Conspiracy from Hell, also called Women Inspired to Tell their Collective History (and a number of different names, changing their name to suit the issue) stormed the streets of New York and later

Chicago to "hex the patriarchy," catcalling men who had made unwanted sexual moves on them, critiquing capitalism, and speaking out against marital rape (McGill 2016). Between 1969 and 1973 before *Roe v. Wade* made abortion legal across the United States, the Jane Collective, a feminist community of over one hundred women in Chicago, carried out an estimated 11,000 illegal abortions, learning through other women how to perform the procedures (Wilson 2015). In 1978, the first "Take Back the Night" march in San Francisco brought together over 5000 women from thirty states (Greensite 2003). And, more recently, in 2006 activist Tarana Burke founded the original "Me Too" movement, which focused on fostering solidarity among girls and women of color at her co-founded non-profit, Just Be Inc. (North 2018). Importantly, current feminist critiques of rape culture are part of this larger lineage of feminist political, medical, and social initiatives. The #MeToo movement advances these histories, acting as a networked social movement (Rentschler 2014; Clark 2016) that uses digital technologies to articulate lived experiences of sexual harassment and assault.

AND NOW, TECHNOLOGY'S DISAPPEARING ACT

However, despite the success of these technologies in circulating stories that help push for vital change, they still need to be interrogated, just as the systems that perpetuate misogyny and white supremacy need to be interrogated:

> @TwitterUser: When the majority of perpetrators of sexual violence walk free, calling #MeToo a witch hunt is tone deaf. We're in this situation because our justice systems have been failing victims since the beginning. (January 16, 2018)

Through my dwelling here, what has become clear is that the neoliberal racist sexist backlash to justice uses the same tools as feminist resistance (i.e., the witch) but different techniques, and that feminist organizing has always been about using the tools at hand, even as we queer those tools by using different techniques. Dwelling in spaces where digital activism is frequent is key for understanding the embodied and affective components to this claim. Through engaging in this particular process of dwelling, it may encourage others who approach the space to also dwell and, in doing so, to also begin their own processes of thinking differently and, subsequently, acting differently as they encounter new forms of relationality and different kinds of relationships. The longer or more concentratedly that we dwell within virtual spaces, the more information we're able to accrue and the better we are able to consider how our individual acts of dwelling serve, reflect, or intervene into the sociotechnical, political, and/or cultural scene which can prompt the process of moving from individual thought towards action, individually and collectively.

For me, in following the figure of the witch and its paratextual discourses in social media spaces and news sources, the experiences of dwelling that I lay out in this section demonstrate just how deeply infused technologies and platforms are with the political contexts from which they emerge, pointing to the ways that virtual dwelling can speak to not only interpersonal but hegemonic, disciplinary, and structural levels. In the days following the social media re-birth of #MeToo, writer Lindy West (2017), reflecting on the absurdity of #MeToo as a witch hunt, wrote the following: "I keep thinking about what #MeToo would look like if it wasn't a roll call of people who've experienced sexual predation, but a roll call of those who've experienced sexual predation and actually seen their perpetrator brought to justice, whether professionally, legally or even personally. The number would be minuscule. Facebook's algorithm would bury it" (para. 9). West's speculations of Facebook's algorithms here reflect the dark side of social media and the biases inherent in algorithms. Social media curates content based on what the platform assumes you want to see given what you have clicked on previously, all the while sorting out content considered unpopular. What is considered "unpopular" is up to the creators of such algorithms. Scholars and activists have been making similar observations for the past several years, arguing that we must be more aware of the ways in which data politics adversely affect Black, Indigenous, and racialized people and queer, trans, intersex, and gender non-conforming communities and women, particularly given the quickly shifting digital landscape where biometric, health, location, conversational, financial, and habitual data are easily stored, sold, and used (e.g., Noble 2018; Duarte 2017; Brown 2015; O'Riordan and Phillips 2007).

When Lindy West observed the link between social media's design politics and what social media participants experience, I saw the link between the sexist design of the platform and mainstream narratives about sexual assault, perpetuated by Facebook's algorithm. When a *New York Times* Gender Letter asked if algorithms could be sexist, I asked how the algorithmic favoring of mainstream narratives of the witch as predominantly white, full of corruption and malice *couldn't* be sexism and white supremacy. And so, while the witch was my anchor, the sexist and racist politics of technology became my wave. Through dwelling with technologies, we can start to see and articulate more clearly the power dynamics at play. If we brush off the glamor and look past the glow of enthusiastic discourses of technological ubiquity, we begin to see it: digital surveillance in border security; the performances of airport security; the biopolitics of pharmaceutical companies; predictive policing; sexist, racist and homophobic policies by technology and telecommunications companies; and the proliferation of digital health and administrative records. Technochauvinism, the belief that technology is always the answer

(Broussard 2018), and mediated misogyny (Banet-Weiser and Miltner 2016) are hard at work. False dichotomies between public and private, consent and privacy, and subjective and objective are re-worked to maintain age-old power dynamics in new-age form. The increased digitization of feminist and antifeminist movements has led to cyberbullying, censorship, and the silencing of marginalized groups including LGBTQ+; Black, Indigenous, and racialized communities; and women online.

Arguably, one of the reasons for this divisive digital landscape is the androcentric, racist conditions in which dominant media platforms are produced. As Judy Wajcman (2004, 2010) has argued, technical spaces have historically been created by men for men. The tech industry is overwhelmingly male: in 2015, men made up 90 percent of Twitter's engineering staff, and 85 percent of Facebook and 83 percent of Google's tech staff (Rushe 2014; Chemaly 2015). In 2017, the year that the MeToo hashtag would go viral, 63 percent of Facebook's staff, 56 percent of Google's staff, and 59 percent of Twitter's staff were white (Donnelly 2017). Evidently, there is a gendered and raced digital divide that polices the online world, where boundaries are based on the desires of those who design them. And yet, it seems to be that only once the most privileged of us speaks up against the sexism of technology that the media begins to really take notice.

Notably, statistics shared by these corporations don't bother to account for intersections of ability or sexual orientation. When (white) men create mediated spaces they inevitably create them for other men, predicating the exclusion of others. To put it bluntly, mediated misogyny and other forms of discrimination have existed as long as the internet, and its antecedents have existed for centuries longer. Algorithms weed out dissenting voices, particularly when those voices are marginalized. Take, for example, Apple's accusations of algorithmic sexism surrounding its credit card, the Apple Card. On November 15, 2019, Alisha Haridasani Gupta writing for the *New York Times* Gender Letter asked the question: are algorithms sexist? To which many of us even remotely interested in the intersections of gender studies and technology/digitality could respond: yes. Gupta outlines the "tweet-storm" brought on by a distinguished Danish software developer, David Heinemeier Hansson, when he called attention to the sexism underlying the Apple Card's programming after he had been given a credit line 20 times higher than his wife's—even though her credit score was better than his. Ironically, this is not a one-time affair, with Apple co-founder Steve Wozniak also confessing that his credit limit was 10 times that of his wife's. When New York State regulators opened an investigation, Apple, of course, pointed fingers at its banking partner, Goldman Sachs, and Goldman Sachs blamed the algorithm.

There is a history of recent research that outlines this kind of unassuming mediated discrimination. In 2007, Kate O'Riordan and David J. Phillips's collection of essays highlighted both the persistence of racism in online interactions and the constant homophobia that queer youth face in social interactions online, similar to the oppressions their material bodies face offline (Gosine 2007, 144). A few years later Simone Browne (2015) brought to light the ways in which surveillance is practiced on and resisted by Black bodies through surveillance technologies' long histories of policing those very bodies. Through an investigation of Google, Safiya Noble (2018) similarly reveals how algorithms are based on histories of racism and sexism, especially against women of color (type "Black girls" into Google, she suggests, and critically reflect on what you find). Thus, it should have come as no surprise that the Apple Card was also created with sexism and racism as its base. Two things are notable here: first, in blaming the algorithm as the culprit of sexism, as a crafty little formula it discursively constructs the algorithm as objective and apolitical. That is, it cannot be to blame for its subjective and material consequences in the world. Second, in ending the blame at the algorithm itself we are asked to forget, if not willfully ignore, the creators of such an algorithm, who have baked their own biases right into their product. Sexist and racist biases become, then, the foundation for technologies—the norm, the everyday lived experience. Returning to Alisha Haridasani Gupta's question in the *New York Times* asking if algorithms sexist, we should be able to even more soundly reply, "yes."

This rather bleak scene is one that feminist hashtag activism and #MeToo intervene into, with a hashtagging tactic and practice that can help us virtually dwell. Our current material-phenomenological conditions of oppression lead to particular material-phenomenological conditions of response, which currently includes a technological response. However, as we've seen above, digital technologies continue to create and re-create the gendered, raced, patriarchal aspects of its social world because of how the tech segregates based on data-gathering and targeting of bodies. This offers even more of a reason to dwell in spaces to then imagine together what best practices and orientations can be learned. And yet, the question remains, how do we challenge the discrimination and oppression right at the heart of a technology if even the co-founder of Apple can't seem to put his finger on it? What becomes clear is that each individual moment of virtual dwelling, even each layer as we go deeper, is not enough. We need to gather these individual dwelling spaces, home in on them, and then bring them together as a collective brood to indicate the importance of multiple perspectives so that us feminists can best determine how to re-tool and respond. While I have offered my version of dwelling, your version is different. What could happen if we brought our experiences of dwelling in the same spaces together?

By dwelling we can find ways to intervene into the history of male dominated control over technologies by better understanding the sociotechnical relations between assemblages and finding points to intervene along the way.

Knowledge in this context and history itself is white and male dominated. It is part of a larger epistemology, and as such we have to challenge the white supremacist patriarchal ways of knowing in order to challenge the technology. The affective material approach of virtually dwelling can help recognize the tactics used by racist capitalist cisheteropatriarchy, evidenced in my dwelling with the figure of the witch. Through lingering with the idea of #MeToo as a witch hunt, I was able to interrogate the politics of technology that enabled such misogyny to circulate. Since I dwelled there, I considered the history and activism of the witch, and am now able to better see what resistance to white supremacist and misogynistic forces can look like. Because the witch operates outside or in opposition to patriarchal space, this figure begs us to imagine what matriarchal space might be. If technology replicates patriarchal forms, how might we make or use technology to replicate matriarchal forms? Thus, although we began dwelling at an interpersonal level of power, this process enables us to think through other collective and structural domains. Virtual dwelling can help identify our different ideas and questions, finding different tactics and also more problems that we will need to mobilize solutions for in order to change the structures through which misogyny is perpetuated.

CONCLUSION

Hashtag activism is broad and the issues it contests are many. I've focused this chapter around #MeToo in the hopes that my conceptual orientation and analysis can also speak back to other hashtag movements, and in doing so continue the work of exploring the ways that such activisms create space. I take up #MeToo because, despite the "watershed" moment of its Twitter virality in 2017 (Ransom 2020; North 2018), I think there is more to be learned, especially given how easily white celebrities appropriated Tarana Burke's movement in those initial days, and the force with which Black social media participants righted that wrong. In 2021, almost four years after #MeToo's viral circulation on social media, we are grappling with a global health and economic pandemic due to the COVID-19 virus and a concurrent crisis of discrimination, domestic homicide, and racist and sexual violence as we abide by shelter-in-place regulations (Bain, Soore Dryden, and Walcott 2020; Qasim 2020; Patel 2020; Taub and Bradley 2020). Moreover, as those invested in a feminist politics of technology know, technologies and their data gathering, algorithmic filtering, surveillance tracking efforts are not neutral and have sociopolitical, technocultural, and corporeal consequences (Benjamin 2019; Browne 2015; Nakamura 2014; Noble 2018; O'Riordan and Phillips 2007).

I've offered here an example of the practice of my own virtual dwelling with #MeToo because by dwelling within feminist hashtag spaces we can better consider the material affective elements of feminist hashtag resistance, especially within the technical spaces of both constraint and possibility. Virtual dwelling is an effort to critically reflect with and through (or on our use of) technology, with the tools and practices at our disposal, in order to both imagine and propose change; virtual dwelling enables the learning of new onto-epistemologies for engaging with digital technologies, since those platforms and devices are racist, sexist, and ablest in design. Through spatially and temporally digitally dwelling in these online spaces, people can become open to reconsidering the kinds of logics they have become accustomed to in order to think about what new or different kinds of logics might also exist. It is possible, then, that we may find that these new or different logics are in fact the feminist, queer, crip, anti-racist, decolonial, eco logics that we've been told are "illogical" by current systems of domination, and that these logics are intertwined.

Virtually dwelling as a way to orient ourselves to digital spaces is important for the ways that it creates spaces to respond to and create new openings for resistance against the white supremacist heteropatriarchy. Feminist hashtagging is a particularly important site because it stands as an affective dwelling place, bringing together individuals into a collective gathering. Such performances of feminist hashtagging are located within temporal relations that reveal new ways of being, modes of organizing, and social relationships that did not exist before, even as they echo former movements and practices. In this way, dwelling with feminist hashtag campaigns offers insight into the embodied nature of our digitally constructed communication practices and the forms of memory, affect, performativity, and relationality they encompass—they not only interrogate what kinds of truths have been accepted, reified, and circulated in a culture that has long forgotten what truth might actually look like (e.g., "fake news"), but also creates and provide a springboard from which different sets of cultural truths can be revealed and disseminated.

It is no doubt that something about #MeToo "stuck"; the message resonated, it brought people together online, propelling action in the virtual and physical streets—it stuck so much that we saw violent responses to such moments. The characterizations of the #MeToo movement as a witch hunt are not new, and these invocations of witchiness matter. With the contemporary resurgence of "witch culture," what remains clear is that what is unknown or conceived of as undesirable under hegemonic conditions becomes feared and rejected. This fear or dislike of such activities is categorized as dangerous and unnatural, ultimately resulting in persecution. Invoking the witch hunt to describe the ways that #MeToo names abusers invokes a political history of resistance and revolt against the "imperialist white supremacist capitalist patriarchy" (bell hooks 2012). It is a protest that continues to conjure different ways of transgressing the androcentric norms of the times we live in. This is more than just a representation of "witchiness"; this calling up of the witch does something: it calls people to action (whether for or against protest), crafts stories about who and what constitutes resistance and what that resistance looks like, and situates people in a moment within a larger movement.

The widespread public reach of the Internet makes access to information about online movements possible and participating in and spreading the message is a tweet away. The emancipatory possibilities of a hashtag speak to its affective stickiness, as hashtags are circulated by feminist and queer users who seek to intervene into hegemonic on/offline public issues. It is crucial to identify where, when, and how something like #MeToo sticks. The kinds of histories and memories that are recalled, like the witch hunt, do something. Using virtual dwelling as an orientation for analyzing online social movements helps highlight both their possibilities via their affective "stickiness" and modes of critical praxis and the constraints posed by the embedded structural inequalities of the platform itself, including who is able to participate in this movement via access and who can navigate the largely white, male domain. Through dwelling in virtual spaces where #MeToo stories have been shared, we can better trace and orient ourselves to how hashtags, discourses, memes, photos, and texts gain and lose traction across different public, private, and global networks, media platforms, and offline communities. Virtually dwelling in these spaces over time helps to attune us to the spaces in order to learn, respond, and ultimately create new openings for resisting harmful norms because it asks that we slow down rather than speed up, which in itself is a way of slowing the capitalist machinery and subverting the goals of hegemonic white supremacist technologies. The practice of dwelling and imagining alternative logics is itself a redistribution of power: it suggests that the tools needed for resistance aren't so far out of our reach—they may already exist, out there to be reclaimed, co-opted, and wielded.

NOTES

1. Given the possibility of harm due to doxing, death, and rape threats against people, particularly women and gender non-conforming people, who speak out against rape culture and other forms of discrimination, I do not include the names of Facebook, Instagram, or Twitter participants who post about #MeToo unless those participants are celebrities or well-known activists who have previously spoken to the media about sexual harassment. I have also made subtle changes to the spelling and grammar of tweets and posts included in this chapter (unless they come from a celebrity or activist account) so that posts cannot easily be traced back to the participants.

2. See the Trump Twitter Archive to search through the forty-fifth president of the United States' tweets, including mentions of "witch hunt." http://www.trumptwitterarchive.com/.

3. Ironically, Google Docs is telling me that "matriarchally" is not a word, suggesting that I instead use "patriarchally."

4. See Feminists Do Media, @aesthetic.resistance, on Instagram for other detailed accounts of these witchy figures and other amplifications of marginalized voices.

REFERENCES

Amnesty International Press Release. 2009. "Gambia: Hundreds Accused of 'Witchcraft' and Poisoned in Government Campaign." *Amnesty International*. March 18, 2009.

Bain, Beverly, Omi Soore Dryden, and Rinaldo Walcott. 2020. "COVID-19 Discriminates Against Black Lives via Surveillance, Policing, and Lack of Data: U of T Experts." *University of Toronto News*. April 21, 2020.

Banet-Weiser, Sarah. 2018. *Empowered: Popular Feminism and Popular Misogyny*. Duke University Press.

Banet-Weiser, Sarah and Kate Miltner. 2016. "MasculinitySoFragile: Culture, Structure, and Networked Misogyny." *Feminist Media Studies* 16(1): 171–74.

Barad, Karen. 2003. "Posthumanist Performativity: Toward an Understanding of How Matter Comes to Matter." *Signs: Journal of Women in Culture and Society* 28(3): 801–31.

Benjamin, Ruha. 2019. *Race After Technology: Abolitionist Tools for the New Jim Code*. Polity Press.

Broussard, Meredith. 2018. *Artificial Unintelligence: How Computers Misunderstand the World*. MIT Press.

Browne, Simone. 2015. *Dark Matters: On the Surveillance of Blackness*. Duke University Press.

CBC News. 2019. "Trump complains about 'witch hunts' to oil and gas conference crowd.' *CBC News*. October 23, 2019.

Chemaly, Soraya. 2015. "Silicon Valley Sexism: Why It Matters that the Internet is Made by Men, for Men." *NewStatesman*. February 24, 2015.

Clark, Rosemary. 2016. "'Hope in a Hashtag': The Discursive Activism of #WhyIStayed." *Feminist Media Studies* 16(5): 788–804.

Demos, John Putnam. 2004. *Entertaining Satan: Witchcraft and the Culture of Early New England*. Oxford University Press.

Donnelly, Grace. 2017. "Google Diversity Report Shows Progress Hiring Women, Little Change for Minority Workers." *Fortune*. June 29, 2017.

Duarte, Marisa Elena Duarte. 2017. *Network Sovereignty: Building the Internet Across Indian Country*. University of Washington Press.

Gasser, Erika. 2017. *Vexed with Devils: Manhood and Witchcraft in Old and New England*. New York: New York University Press.

Garza, Hedda. 1994. *Women in Medicine*. New York: Franklin Watts.

Greensite, Gillian. 2003. "History of the Rape Crisis Movement." In *Support for Survivors: Training for Sexual Assault Counselors*. California Coalition Against Sexual Assault.

Gupta, Alisha Haridasani. 2019. "In Her Words: Are Algorithms Sexist?" *New York Times Gender Letter*. November 15, 2019.

Haraway, Donna J. 2016. *Staying With The Trouble: Making Kin in the Chthulucene*. Duke University Press.

hooks, bell. 2012. *Writing Beyond Race: Living Theory and Practice*. Routledge.

Joho, Jess and Morgan Sung. 2020. "How to be a witch without stealing other people's cultures." *Mashable*. October 31, 2020.

Karlsen, Carol F. 1998. *The Devil in the Shape of a Woman: Witchcraft in Colonial New England*. W.W. Norton & Company Ltd.

MacDonald, Shana. 2018. "The City (as) Place: Performative Re-mappings of Urban Space Through Artistic Research." In *Performance as Research: Knowledge, Methods, Impact*, edited by Annette Arlander, Bruce Barton, Melanie Dreyer-Lude, and Ben Spatz, 275–96. New York: Routledge.

MacDonald, Shana and Brianna I. Wiens. 2019. "Mobilizing the 'Multimangle': Why New Materialist Research Methods in Public Participatory Art Matter." *Leisure Sciences* 41(5): 366–84.

McGill, Mary. 2016. "Wicked W.I.T.C.H.: The 60s Feminist Protesters Who Hexed Patriarchy." *Vice Media*. October 28, 2016.

McLaren, Leah. 2017. "Me too: It's not just Hollywood, it's Canada." *MacLean's*. October 18, 2017.

———. 2017. "This Is Not A Witch Hunt—Sexual Harassment is Utterly Ubiquitous." *Chatelaine*. October 18, 2017, updated December 21, 2017.

Migiro, Katy. 2017. "Despite murderous attacks, Tanzania's 'witches' fight for land." *Reuters*. March 21, 2017.

Nakamura, Lisa. 2014. "Indigenous Circuits: Navajo Women and the Racialization of Early Electronic Manufacture." *American Quarterly* 66, no. 4: 919–41.

Noble, Safiya Umoja. 2018. *Algorithms of Oppression: How Search Engines Reinforce Racism*. New York University Press.

North, Anna. 2018. "The #MeToo Movement and its Evolution, explained." *Vox*. October 11, 2018.

O'Riordan, Kate and David J. Phillips, eds. 2007. *Queer Online: Media Technology & Sexuality*. Peter Lang Publishing Inc.

Patel, Raisa. 2020. "Minister says COVID-19 is empowering domestic violence abusers as raters rise in parts of Canada." *CBC*. April 27, 2020.

Pedwell, Carolyn and Anne Whitehead. 2012. "Affecting Feminism: Questions of Feeling in Feminist Theory." *Feminist Theory* 13(2): 115–29.

Qasim, Salaado. 2020. "How Racism Spread Around the World Alongside COVID-19." *World Economic Forum*. June 5, 2020.

Ransom, Jan. 2020. "Harvey Weinstein Is Found Guilty of Sex Crimes in #MeToo Watershed." *New York Times*. February 24, 2020.

Rentschler, Carrie A. 2014. "Rape Culture and the Feminist Politics of Social Media." *Girlhood Studies* 7(1): 65–82.

Rowlands, Alison. 2013. "Witchcraft and Gender in Early Modern Europe." *The Oxford Handbook of Witchcraft in Early Modern Europe and Colonial America*.

Rushe, Dominic. 2014. "Twitter's Diversity Report: White, Male and Just Like the Rest of Silicon Valley." *Guardian*. July 25, 2014.

Sharma, Sarah. 2020. "A Manifesto for the Broken Machine." *Camera Obscura* 104–35(2): 171–79.

Taub, Amanda and Jane Bradley. 2020. "As Domesic Abuse Rises, U.K. Failings Leave Victims in Peril." *New York Times*. July 2, 2020.

Tran, Lucky. 2019. "Firehosing: The Systemic Strategy that Anti-Vaxxers are Using to Spread Misinformation." *Guardian*. November 9, 2019.

Wajcman, Judy. 2004. *Technofeminism*. Polity Press.

———. 2010. "Feminist Theories of Technology." *Cambridge Journal of Economics 34*(1): 143–52.

West, Lindy. "Yes, This Is a Witch Hunt. I'm a Witch and I'm Hunting You." *New York Times*. October 17, 2017.

Wiens, Brianna. I, Stan Rucker, Jennifer Roberts-Smith, Milena Radzikowska, and Shana MacDonald. 2020. "Materializing Data: New Research Methods for Feminist Digital Humanities." *Digital Studies/Le Champ Numérique* 10(1): 1–22.

Wilson, Rachel. 2015. "How to Run a Back-Alley Abortion Service." *Vice Media*. November 12, 2015.

Chapter Five

Native and Indigenous Women's Cyber-Defense of Lands and Peoples

Marisa Elena Duarte

INTRODUCTION

The promise of network sovereignty for Native nations is that the increasing technical capacity of the nation—including Internet infrastructure, the peoples' digital acumen, and the tribe's knowledge economy—shapes the nation's capacity to exert their tribal sovereignty vis-a-vis the federal-trust relationship they bear with the United States (Duarte 2017a). [1] Contemporary information and communication technologies (ICTs) bear histories of colonization, sovereignty, and self-determination. Indigenous peoples' experiences of self-determination have thus resulted in place-based approaches to the reception, acceptance, interpretation, communication, transmission, and diffusion of information, inclusive of the data, devices, sociotechnical systems and assemblages through which information flows (Duarte 2017b; Littletree 2018; Vigil-Hayes et al. 2017; Showalter et al. 2019; Wemigwans 2018). Accordingly, Native and Indigenous peoples' approaches to information are adaptive, innovative, and often cohere to ancestral worldviews within a greater resistant imaginary (Belarde-Lewis 2013; Belarde-Lewis 2011; Rainie et al. 2017). The American will to settle and expand is the backdrop against which Native peoples springboard cybernetic place-making and digitally enhanced networks of belonging. This chapter contextualizes anecdotes, historical events, and prior studies indicating the digital tactics that Native and Indigenous women in particular forward against the backdrop of racism and sexism. Specifically, through consideration of journalistic and countersurveillance maneuvers around the 2016 #NODAPL movement, tribal network sovereignty, and data-driven tactics toward social change, this chapter reveals a Native and Indigenous women's cyber-defense of peoples and lands.

THE DECOLONIAL IMPETUS FOR
NATIVE WOMEN'S USES OF ICTS

Reproducing the tribal nation is serious business for Native and Indigenous women. It generates rifts between Indigenous and non-Indigenous women's rights movements. The term "feminist" has not historically been an easy moniker for Native and Indigenous women to accept. For the most part, Native and Indigenous women tech advocates do not as a group characterize their work as feminist, even though values of equity, representation, and women's participation in science, technology, engineering, and math play a role. Historically, non-Indigenous feminist movements have tended toward Euro-centricity, whiteness, and Settler norms toward liberal notions of progress. Contemporary self-described Native feminists tend to merge insights by women of color and feminists from the Global South with anti-sexist values embedded in their peoples' ancestral ways of knowing. Moments of solidarity with broader feminist movements occur through restoration of Indigenous land rights and environmental and racial justice. It is nevertheless clear that Native and Indigenous women activists and advocates employ ICTs to defend inherent tribal sovereignty; Indigenous women, motherhood, and sexuality; and healthy ecologies toward decolonial futures (Rule 2018).

Indigenous women are sexualized, assaulted, and denigrated at higher rates than almost any other social group. American practices and laws rely on a "metaphysics of Indian-hating" to harm Indigenous peoples and seize their rightful belongings. Various social groups stigmatize Indigenous women and children due to how they strengthen peoplehood (Anderson 2000; Drinnon 1997). Indeed, in twentieth-century American feminist Margaret Sanger's approach to birth control, social peace relies on the segregation and sterilization of "illiterates, paupers, unemployables, criminals, prostitutes, [and] dope-fiends" on "farms and open spaces," meaning, on stolen Indian lands, excluding groups "difficult to assimilate" (Sanger 1932). A generation later, the Indian Health Service began forcibly sterilizing Native American women. To this day, Native women, Latinas, and Black women are more likely to experience coercive sterilization or have their children taken from them while white women experience reproductive choice.

Lockean and Rousseauian ontologies tie American sovereignty—territorial expansion toward life, liberty, and the pursuit of happiness—with the property-owning rights of White male patriarchs. Their rights, as part of the civilizing social contract, were historically valued more than the lives of women, Black folks, and the "merciless Indian savages" inhabiting the American wilderness (Kendi 2016; Wunder 2000–2001). Another telling example of the structural divisiveness between white women and Native women is found in the early twentieth-century American women's suffrage movement. White women gained the right to vote in 1920, while Native women did not gain the right to vote in federal elections until 1924—and 1962 in some local state elections—through passage of an act politicians greatly debated for fear Native Americans would destabilize the white vote.

Presently, many Settlers believe genocide and assimilation have resolved the "Indian problem." This Settler ideology—that Indians are all dead or ought to be—is a form of ontopower: the power to produce the "particular form a life will take next" (Drinnon 1997; Slotkin 2000; Massumi 2014). Indeed, nineteenth-century American ethnographers built careers validating the delusion that Indians are barbarians—primitive, pre-modern, and prehistoric and deserving of political subjugation—even as their work depended on their legal and social bonds with Native women (Bruchac 2018). The Marshall Trilogy, a nineteenth-century series of Supreme Court trials, rendered inherently sovereign Native peoples "domestic dependent nations" under US congressional power. Thus tribal people are legally characterized as threats to national security while also stripped of their sovereign treaty rights. Settler ontopower thus justifies exploitation of Indigenous lands, waters, and people, including through technocratic means, as acts of US sovereignty. Interpreting the evolution of ICT infrastructure against eras of federal Indian policy since 1776—coexistence; treaties, removal and reservations; assimilation; reorganization; termination; and self-determination—reveals how Settler ontopower contributes to informatic exploitation. Exploitative actions include early US military surveillance of Indian movements via telegraph, Native family displacement amid widespread industrial-era uptake of telecommunications, manufacturing automation, and radio; algorithmic discrimination against Indigenous peoples via machine learning; and various forms of technological redlining.

Raised to the level of false consciousness, the Settler belief in the primitive and anti-technological nature of Indians shapes contemporary American practices such as extrajudicial killing of peoples; theft of lands, water, children, religious objects, bodies, and belongings; subjugation of Indigenous languages, histories and philosophies; and sexual assault of women. Indigenous women suffer an overlapping recursive pattern of colonization that is external, internal, social, and intertribal. Women of various tribes have endured, alongside men, elders, children, and all other beings, the physical violence, biological warfare, forced removal, dispossession, and loss of title that occurs when a dominant nation wars against a perceived barbaric tribe (Mihesuah 2003).

External colonization generates internal colonization, through which tribal leaders adopt aspects of US patriarchy (Anderson 2000; Mihesuah 2003). Whereas before, women and non-masculine gendered people within tribes sustained governance roles and relationships that worked within their Indigenous world-system, after US domination, all too often male leaders reconfigured self-governance systems to reflect the US Congress and executive branch, resulting in sexist male-dominated tribal councils and administrative branches (Keeler 2006). Sadly, the American Indian Movement (AIM) of the late 1960s and 1970s included a pattern of chauvinist spokesmen who womanized and demeaned women leadership (Mihesuah 2003). In the last decade or so, journalists and scholars have surfaced narratives revealing how sexism shaped activist attitudes and actions, including clues to crimes such as the feminicide of activist Anna Mae Aquash, who due to her romantic involvement with AIM leader Dennis Banks, was likely kidnapped and murdered by a group of activists including Native women.

Thus Indigenous women of many tribes lost title at least twice: once during the early American-Indian wars of dispossession, and once again through erosion of ancestral lifeways through the eras of American industrialization. A third form of colonization may be characterized as an aftereffect of the former two, in which women, including other women of color and Native women, subjugate Native and Indigenous women. Black and Latinx women have been culpable of denigrating and exoticizing Native and Indigenous women's physical bodies. A contemporary example includes rap star Nicki Minaj's use of a hypersexualized image of Disney's already hypersexualized caricature of Pocahontas to sell albums (Beck 2017). Another notable example includes the admonitions of Mexican and Hispanic matriarchs who caution their daughters to avoid the sun so as not to darken like "*una india*," while suppressing knowledge of their families' Indigenous lineage (Castro 2018; Gomez 2019). The appropriation of Native women's ways of knowing by Wiccans and New-Agers continues this pattern of exoticization. Native and Indigenous women also experience intertribal harassment, as the women and girls of one tribe habitually demean women and girls of another, and interpersonal harassment, as women and girls demean and sabotage each other within their own tribes and families (Bailey 2020; National Women's Association of Canada 2011). The lateral violence of this third form of colonization reveals itself in colorism, classism, sexism, hyper-sexualization, and pretendianism. Indeed, studying how women pretend to be Indigenous to benefit their careers or public profiles reveals structural and psychological dimensions of the racialized violence integral to its fraudulent effectiveness (Hopkins 2015; Keeler 2018).

To be clear, Native and Indigenous women in the US are not only raced/gendered subjects who deal with multicultural identity. They are also multinational subjects who, depending on their citizenship status, are more or less vulnerable to extrajudicial harm. Native Americans, as a kind of Indigenous peoples, are not ethnic minorities. They are an exceptional class of US citizens representing over 586 federally recognized Native nations and who, as tribal citizens, exercise sovereignty through tribal law, federal Indian law, and customary laws (Deloria and Lytle 1983). If tribally enrolled, Native American women are citizens of their tribal governments and of the US. Members of border tribes also have rights according to their tribes' legal and political relationships with, respectively, Mexico or Canada. Women who identify as Indigenous but are not tribally enrolled may also be alternatively a US citizen, an immigrant with status, or an undocumented migrant, and accordingly socially and politically vulnerable. The way Native and Indigenous women experience colonial border enforcement relates to their formation of a personal/political Indigenous consciousness. Combined with education, class, and community belonging, an Indigenous woman's political consciousness shapes her propensity for rights-based advocacy.

For Native and Indigenous women, their peoples' *inherent sovereignty* is tied to their capacity to reproduce their nations as matriarchs, knowledge keepers, leaders, healers, and educators. Tribes also function as kin- and place-based networks of belonging. For some tribes, key groups within these networks of belonging power the sociotechnical capacity—the network sovereignty—of their Native nation. However, tribal sovereignty within the US is not the sole endgame for Native peoples. Rather, it is one set of legal and political means through which to negotiate with the US on a nation-to-nation basis. Indeed, the US depends on its recognition as a sovereign by inherently sovereign tribes. This inherent sovereignty indicates an alternative pre-colonial and decolonial mode of self-governance through and beyond the legal boundaries of contemporary Native nations. Widespread ICTs enable the diffusion of information and knowledge beyond these boundaries.

Though not uniform, this decolonial mode shapes the emergence of relational collectives led by Native and Indigenous women. Indigenous women with educational training, disposable income, employability, available time off, and authority to speak with and on behalf of their people apply their digital facility toward radical creativity and collective action. They overtly and covertly sustain actions through tribes, intertribal organizations, grassroots social movements, academia, and state government to restore the dignity of Indigenous peoples broadly and Native and Indigenous women specifically. Native and Indigenous women who resist the colonial caste system also often find solidarity with anti-racist/anti-sexist movements, including groups that apply digital tactics toward social change. Thus, to assert their rights and ways of life, and regardless of their affiliation with feminism, small groups of Native and Indigenous women mobilize tribal sovereignty and decolonial knowledge via ICTs to pursue the means for justice. At present, digital mobilization occurs through at least one of three ways: leveraging the power of social media and independent journalism, promoting tribal network sovereignty, and applying data-driven tactics toward social change.

#NODAPL: DIGITAL MEDIA
MOBILIZATION BY WOMEN ACTIVISTS

The 2016–2017 protests of the illegal Dakota Access oil pipeline (DAPL) reveals how Indigenous peoples mobilized digital tactics in their struggle against Settler mass media and government surveillance. Native women as well as those moved by decolonial Indigenous matriarchal cosmovisions used digital media to boost the signal of the decolonial systems of knowledge needed to counteract Settler ontopower.

In the summer of 2016, residents and activists observed increasing militarization around the Standing Rock Sioux Tribe, whose lands are in what is now Morton County, North Dakota. The Morton County Sheriff collaborated with DAPL private security to barricade construction sites. Law enforcement turned away local drivers, and pulled over suspected rabble-rousers. Standing Rock Sioux women in the area circulated news in person and via social media among friends and relatives about the meaning of these unusual activities (Deschine-Parkhurst 2019). They contributed to the social media expressions of youth organizers and activists who were raising awareness of the environmental injustice.

Indigenous peoples and environmentalists all over the US heard their message. Standing Rock Sioux Tribe, the Army Corps of Engineers, and the US Justice Department contested the unlawful pipeline construction. Thousands of self-termed water protectors volunteered in growing protest camps. They worked in communal kitchens, ran security operations, set up supply chains, and cared for children among other chores. Women organized many of the everyday tasks. Women who had worked in tribal government leadership as well as tribal rights advocacy shaped #NODAPL tactics. Journalists reported on the work of LaDonna Bravebull Allard (Lakota), Phyllis Young (Lakota/Dakota), Faith Spotted Eagle (Dakota), Candi Brings Plenty (Lakota), and Madonna Thunder Hawk (Cheyenne River Sioux). Countless others contributed (Tallbear 2016; Indian Country Today 2017). Some of the women leaders had also been involved in direct action and organizing in the 1970s Wounded Knee and Alcatraz stand-offs, as well as the Women of All Red Nations (WARN). Others were active through the 2012 Idle No More movement, which also propelled through the digital acumen of First Nations women and allies. Young women were also involved in NoDAPL tactical planning through Red Warrior Camp, a group that strategized some of the more high-conflict stand-offs with law enforcement and private security. Scholars, journalists, activists, artists, writers, and others who stayed in the camps noted pervasive matrilineal practices shaping tactics, attitudes, and behaviors, including respect for elders; care and respect for future generations including the Lakota/Dakota/Nakota youth who initiated the movement in the spring of 2016; honor for the earth and cosmic care of Turtle Island; and holistic care for warriors of all genders (Privott 2019). Madonna Thunder Hawk partnered with Professor Elizabeth Castle to digitally archive these teachings and values through the *Water Protector's Community Oral History Project.*

From spring to fall of 2016 Indigenous peoples came together at the camps to weave a network of intertribal solidarity rooted in ancestral teachings. Morton County store owners refused to sell goods to the water protectors, even though the townspeople had voted to prevent the pipeline from being constructed through their community for fear of contamination. To stop DAPL drilling, water protectors marched in prayer around drilling pads protected by law enforcement and DAPL security armed with batons, pepper spray, tear gas, rubber bullets, rifles, bulletproof vests and gas masks, armored vehicles, drones, sound cannons, and water cannons. In response, Morton County law enforcement and DAPL security erected an observation station on top of Turtle Island, a high point in the landscape.

Turtle Island is the site of a Lakota/Dakota/Nakota cemetery, and bears spiritual and philosophical lessons about the origin of humanity. For Indigenous families with long histories in Indian Country, Turtle Island is a physical reference point in the Indigenous North American fabric of cosmovisions. That fall, when water protectors attempted to visit occupied Turtle Island in prayer, officers soaked them in pepper spray and tear gas. Their actions further radicalized Indigenous peoples whose ancestral teachings incorporate care for the earth as care for the people, past, present, and future.

By then, the FBI Terrorism Task Force was surveilling water protectors in the style of the United States' Counterintelligence Program (COINTELPRO), including tactics of sexual and emotional coercion. After Morton County Law Enforcement planted a firearm on and arrested water protector Red Fawn Fallis (Lakota), leaked information revealed that her new paramour was also an undercover FBI agent (Keeler 2018). The finding opened wounds in Indian Country harkening the founding of the thirteen colonies, when European men routinely targeted Indigenous women for protecting their peoples' ways of life, the sanctity of the earth, and for existing as women in command of their bodies, political will, and sexuality.

By December 4, 2016, over ten thousand outraged US veterans prepared to travel to Standing Rock, ready for front-line action against DAPL. To prepare, elders, including Faith Spotted Eagle, symbolically forgave veterans who represented historic US military actions against Indigenous peoples. Faced with an impending paramilitary standoff, DAPL temporarily stopped drilling, blaming the inclement weather. The Standing Rock Sioux Tribe leadership and camp leaders asked water protectors to return to their homes due to blizzard conditions and a persistent cough many feared was pneumonia, but which public health officials reported as overexposure to the rat poison Chloraphacinone, or Rozol. Residents and activists had observed planes drop chemicals over the campsites in the months prior. Over those months, hundreds of activists were arrested and charged with felonies though the crimes they committed were, for the most part, acts of civil disobedience worth no more than misdemeanors.

Activists and independent journalists developed tactics to disseminate news, documents, and critical communications about DAPL, waging an informatic war against Settler ontopower. According to military theorist Brian Massumi, ideally, "cognitive aspects of knowledge" are based on decisions that have already been made through accurate and actionable collection of data, and focus on the "availability, reliability . . . manipulability . . . [and] usability" of data. Ontopower, however, is the force of life that occurs in the pre-decision process, "before know-ability and action-ability have differentiated from one another" (Massumi 2014, 71). A social group that harnesses ontopower defines perception ahead of time, and instigates fear of the unknown through establishing infrastructures of preemptive action and deterrence such that any challenge to their distorted sense of reality is cognitively processed as a threat to the entire social order. Centuries of US policies positioning American Indians and Indigenous women in particular as threats to civil society turn institutions that are supposed to uphold legal and social justice for all into systems that erode Native American legal, civil, and human rights. This was observable in the scant distorted media coverage and unwarranted government surveillance of the water protectors and in the law enforcement and private security assaults and maltreatment of elders, women, and youth at the camps.

From the spring of 2016 through the beginning of the bitterly cold winter, finding news and information about the injustices at Standing Rock depended on one's proximity to relatives at the sites as well as independent and citizen journalists in Indian Country. Since the spring, those of us in tribal communities observed our young people taking more and longer trips to the camps. Indigenous friends and relatives described the brutality of the Morton County officers as well as hopeful lessons learned within the camps. In September, Morton County law enforcement arrested *Democracy Now* journalist Amy Goodman for reporting DAPL's use of attack dogs. The story went viral by virtue of Goodman's name, and the charges were dropped. When DAPL security and Morton County officers began using the LRAD, or sound cannon, in September of 2016, the absence of mass media coverage was notable.

One day in the late summer of 2016, officers and DAPL security entrapped a camp of activists and elders. Young people jumped on top of buses, live-streaming footage of officers forcing elders in prayer out of tipis at gunpoint. People hid under vehicles to escape being beaten with batons. Snipers shot rubber bullets at the people recording atop the buses. During the livestreams, viewers commented, "where is the media?" Journalist Tristan Ahtone (Kiowa), wrote "It's as if the press is engaged in a group conspiracy to expedite the destruction of America's indigenous people by ignoring their dynamic lives, voices, struggles, and contributions" (Ahtone 2016).

Facebook intermittently blocked certain activists from posting updates. In an area with limited Internet access, citizen journalists climbed "Facebook Hill" to upload videos. Groups like Unicorn Riot, Digital Smoke Signals, Last Real Indians, and Longhouse Media, which had consisted of media folks more accustomed to wireless networking, blog posts, and coverage of Native arts and entertainment, found themselves working like war-zone reporters, wearing bandanas to prevent inhalation of tear gas while narrating crises through the lenses of cameras affixed to smart-phones. They faced arrest like Goodman, but without dedicated legal representation or recognition of journalistic rights. After months of investigative reporting, when journalist Jenni Monet (Laguna) was arrested in February 2017, she applied her connections in the media to raise awareness of the wrongful targeting of journalists by corporate private security, and indicated the specific maltreatment of Indigenous water protectors (Tolan 2017). Myron Dewey (Newe-Numah/Paiute-Shoshone), proprietor of Digital Smoke Signals, an Indigenous-owned and -operated wireless networking and media company, developed onsite expertise as a commercial drone pilot. He used long-range high-definition cameras to record evidence of brutality against activists amid illegal construction. DAPL security took to shooting down Dewey's drones. Dewey's activist "eye-in-the-sky" videos were eventually used in court to defend water protectors (Dewey 2017).

In October 2016 after weeks of intensifying violence major news organizations finally sent reporters who mostly broadcast in the no-travel zone hundreds of feet behind DAPL armored vehicles. Their vantage point was completely unlike that of reporters inside the camps, who documented overt acts of violence by law enforcement and corporate security. The mass media framed the DAPL conflicts as a tragedy of Indian loss, the need for stronger American courts, or worse, a story about how Indians deserve to be assaulted because it is illegal to threaten US energy infrastructure. Colonial ontopower thrives on the mental habits of social groups who are habitually arrogant, close-minded, and intellectually lazy, and who profit by refuting or ignoring the suffering of oppressed peoples around them (Medina 2013). Ontopower grants institutional leaders the legitimacy to respond to the suffering of oppressed peoples when the suffering is economically or politically profitable. Thus the bloodshed and suffering of the water protectors—including the hopeful stories of women's resistance and environmental justice—lay hidden from public view for many months.

In 2009, military theorist Antoine Bousqet (2009) characterized a shift from linear models of informatic warfare to non-linear chaoplexity, in which machine learning yields patterns in a swarm of actors and information. Predictive analytics requires an assemblage of reliable on-the-ground human intelligence, digital data points, artificial intelligence, scientific interpretation, and rights to surveille in the interest of national security (Chertoff 2017). Any scientist can attest: garbage in, garbage out.

But under the logic of colonial ontopower just the possibility of an opposing social group with potentially more information than Settler law enforcement represents an actionable threat. In the fall, hundreds of thousands of activists used Facebook to deceptively "check in" to the camp locations in an effort to flood the Morton County Sheriff with false information. The sheriff's department used their Facebook page to post alleged evidence of water protector law-breaking, including photos of crudely made stone hatchets termed "weapons," and rationalizations of police brutality. Indigenous Environmental Network activist Dallas Goldtooth (Dakota/Diné) used his social media accounts to thank people for their support while assuring them that their "check-in" wasn't likely going to change the course of action for Morton County law enforcement.

However, true to chaoplexic warfare, Goldtooth was unable to infer how intelligence analysts at different levels of government would interpret this surge in social media attention. Department of Homeland Security intelligence analysts pursued protocols to surveil the social media of Standing Rock activists. The FBI Terrorism Task Force contacted associated activists in the Washington area. Each week independent investigative journalists submitted Freedom of Information Act requests through Muckrock, revealing patterns around upper midwest law enforcement purchases and uses of Stingray cell tower interceptors and surveillance gear. Ontopower thrives on paranoia. Who was watching who? Which party would strike the first blow, from what direction would it come, and how badly would it hurt?

The majority of privileged Americans do not suffer daily hyper-surveillance, and research suggests they do not experience the physical sensations as deeply as those who do: Black Americans, undocumented people, Muslims, formerly incarcerated people, and Native Americans (Benjamin 2019). Most American citizens do not know that COINTELPRO targeted Black American and American Indian activists in the 1960s and 1970s, resulting in the murder of Black Panther leaders and the 1973 stand-off at Wounded Knee. For American Indians, hyper-surveillance occurs as federal agencies collect data at varying points in a single month, such as through visits with Child Protective Services, public health agencies, parole boards, and social services. Read against maps of the natural resources embedded in Native lands, it's easy to see why the mainstream media frames narratives of Indians as impoverished and ill-prepared denizens of an oil-hungry advancing technoscientific nation-state. There are few knowledge institutions from which Native Americans can challenge this trope. During the construction of DAPL, social media and independent journalism represented a key means through which women water protectors could counteract Settler mentalities through distribution of a socially just vision grounded in the "interconnectedness of life and the sanctity of that holism . . . the ancient and sacred authority of women as life-givers/mothers to protect the water . . . [and] the historic collective survival of Indigenous women and their children in the face of colonial/patriarchal violence" (Privott 2019, 76). The strength of their accounts continues as Indigenous women who were active in the NoDAPL movement leverage digital means to raise awareness of the environmental destructiveness of pipelines and the associated logics of colonial patriarchy.

TRIBAL NETWORK SOVEREIGNTY: NATIVE WOMEN PROPEL INFRASTRUCTURAL DEPLOYMENT

For the above-mentioned reasons, Zapatista leader Subcomandante Marcos (1997) encouraged Indigenous activists to build communication networks from below. Indigenous means of media production and information distribution are key to Indigenous solidarity. In 2009, at a National Congress of American Indians (NCAI) Tribal Leader/Scholar Forum in River Bend, South Dakota, during a plenary session an elder matriarch stood and asserted inherent sovereign rights over the air waves above Indian land, adding that the tribes needed to work through and beyond the Federal Communications Commission (FCC) to assert the tribal-centric construction of Internet infrastructure toward self-determination. She was expressing symbolic support for moves already underway through a network of Native technology policy advocates.

Among them, influential Native women, including Valerie Fast Horse (Couer d'Alene), Traci Morris (Chickasaw), Loris Taylor (Hopi/Acoma), and Kimball Sekaquaptewa (Hopi), had been working through the NCAI, a Native advisory group to the FCC, Native Public Media (NPM) and their tribes to advocate for a national regulatory tribal priority to airwaves over Indian lands. They had researched, written policy, and raised awareness of the challenges to and ways of overcoming the digital divide in Indian Country. Those with technical skill constructed their own fiber-optic networks. In the 2000s, Valerie Fast Horse aided the Couer d'Alene tribe in constructing Internet infrastructure to disseminate Indigenous news, media and education through Rezkast, a tribal video-sharing channel. Through the 2010s Fast Horse implemented a fiber-to-the-home plan that would serve as a model to many tribes seeking to construct a similar infrastructure. Kimball Sekaquaptewa coordinated an intertribal network backbone project for several Pueblos in New Mexico, and developed the technology outreach plan for Santa Fe Indian School. During those years, Morris partnered with Taylor to organize a national conference on tribal telecommunications, which proved highly influential in the burgeoning sub-industry. Morris now directs the American Indian Policy Institute, and advocates for high-speed Internet capacity in Indian Country, including policies affecting rights-of-way and spectrum allocation. Through NPM, Taylor raises awareness of the link between sovereign Native American rights to free expression through radio, TV, and news with network sovereignty: regulatory command of towers, fiber, and service providers. Through meetings with tribal leaders, technologists and industry professionals, for decades these individuals have worked to bridge the digital divide in Indian Country in spite of male-dominated federal policy arenas and the "bro-grammer" "dude-core" tech and telecom industry.

Tribes experience overt racism/sexism when deploying Internet infrastructure and service. For years, residents of the Nez Perce tribe in Idaho were receiving dangerously inadequate responses from the sheriff's department. The region has a history of violent white supremacist organizations who focus on Indian-hating. It is common knowledge among tribal peoples that many members of local law enforcement are white supremacists. Yet in the 2010s when the Nez Perce tribe's Internet Service Provider became operational, folks in the area—including the sheriff's department—realized that they would be willing to overlook a whites-only ethos if it meant that they would get better Internet and cellular coverage through the tribe. The regional telecom monopolies were not improving ICT infrastructure in rural Idaho, citing low demand and income. Rural poor white families were experiencing the same technological redlining that rural poor Nez Perce families were experiencing. The tribe, however, had the sovereign authority to acquire the licensing and capital for a large-scale Internet build-out. Danae Wilson (Nez Perce) now leads the tribe's technology services department, and at a 2015 Tribal Telecom and Technology summit, shared this humorous anecdote: for a moment in history, every time a racist officer turned on his or her mobile phone, the first image that flashed on the screen was the Nez Perce tribal logo (Wilson 2015). Under the logic of Settler-colonial ontopower, it is satisfyingly just for tribes to run internet service providers (ISPs). It is also right that Wilson holds an influential policy-making position through the Idaho Broadband Task Force and the Affiliated Tribes of Northwest Indians. Her good humor, commitment to the tribe, and dedication to technological capacity-building is rooted in philosophies of what it means to be Indigenous.

Tribal network sovereignty has a decolonial root. It divests the coercive power of foreign Internet and telecom companies on Indian land, and asserts tribal command of a technically sophisticated infrastructure and regulatory authority. It integrates community tech labor, state and national policy leadership, and supranational governance of Internet protocol. In the US and Canada, it is distinct from community-based broadband as it is premised on federal government negotiations shaped by two centuries of colonial treaties and policies. Tribal Internet infrastructure deployment reveals the materiality of network sovereignty alongside tensions around security, access, privacy, and creative expression under colonial domination. The cost of tribally owned, operated or regulated ISPs or, alternatively, network backbones, is high, requiring millions of dollars in capital investments. Tribal governments regularly appoint ICT champions to advance grants and loans, and to advocate for tribal priorities through the Federal Communications Commission (FCC) and the National Telecommunications and Information Administration. Competition for spectrum licenses is stiff; often tribes cannot afford them, and cannot compete with private industry broadband providers. As with any natural resource in Indian Country, the prevailing Settler-colonial assumption is that airwaves are free for sale to the highest bidder, even though those airwaves are above sovereign Indian lands. Native women tech advocates challenge this assumption on legal, policy, and moral grounds. For example, tribal telecom policy advocate Darrah Blackwater (Diné) pointedly titled a recent editorial "Time for the "well-meaning man" to return spectrum rights to tribes" (Blackwater 2019).

DATA-DRIVEN TACTICS FOR CHANGE: GENDER RIGHTS AND SELF-DETERMINATION

Beyond digital literacy, a digitally enabled "subversive lucidity" frees Indigenous thought leaders to push ideas through assemblages of devices, social groups, and institutions toward social change (Medina 2013). As technologically advanced nations have adapted to an ethos of big data, Indigenous peoples have begun asserting Indigenous data sovereignty: the repatriation of large data sets for tribal or Indigenous community decision-making (Kukutai and Taylor 2016). With this movement is a call by Indigenous scientists and researchers to develop skill as data analysts—or as movement leaders suggest, "data warriors"—including the adoption of data-fied ways of knowing—lab-like observations, categorization, taxonomies, quantification, spreadsheets, and machine-learning—toward distinctly Indigenous ends (Walters and Anderson 2013; Walters et al. 2021). Many leaders in the call for data-driven tactics for change are highly educated Native and Indigenous women, including Maggie Walters (Palawa), Stephanie O'Carroll (Ahtna Native Village of Kluti-Kaah), and Desi Lonebear-Rodriguez (Northern Cheyenne). Evidence shows that social movement activists and policy-mak-

ers are also conscientious of how information flows and large data sets acti-
vate social transformation.

At a grassroots level, Indigenous peoples' digital tactics are shaped by the
resistances essential to their endurance within their geopolitical locale (Duarte 2017). In one context Indigenous environmental rights groups might
widely photograph their plight, as in the case of the 2012 Idle No More flash
mobs, while in another, Indigenous environmental rights activists might
avoid digital dissemination, as in the case of the Rio Yaqui Justicia and
Namakasia movements. Tactics reveal a conscientiousness of the power of
datafication toward social change.

When in early 2012 Canadian Prime Minister Harper threatened First
Nation's treaty rights to pave the way for oil and natural resource extraction,
Jessica Gordon (Saulteaux-Cree), Sylvia McAdam (Cree), Nina Wilson (Na-
kota/Plains Cree), and Sheelah McLean combined their social media, activist, educator, and web programming skill to craft the messaging of the Idle
No More movement (Caven 2013). First Nations and US-Canada Indigenous
rights activists hosted teach-ins and circulated memes, news, and policy updates tagged #idlenomore through their social media networks. Idle No More
supporters hosted prayer rallies and flash mobs in which Indigenous individuals spontaneously sang hand drum songs and recorded each other at locations signifying the neoliberal erosion of Native homelands: border cross-
ings, public parks, malls, interstate highways, major city intersections, and
town squares. The datafication of the hand drum song, harkened through a
flash mob—a distinctive orchestration of human action dependent on hyper-
mediated social networking—indicated widespread diffusion of smart
phones and text messaging among networks of Indigenous peoples. It also
indicated a consciousness of the power of information diffusion, and a willingness to digitize semi-private Indigenous ways of knowing into a public
political stance.

Around the same time, in Rio Yaqui, the sacred ancestral homelands of
the Pascua Yaqui Tribe and Yaqui people of Sonora, corporations and
governments illegally constructed a dam pumping hectares of water from the
river in direct violation of Yaqui water rights. Though the Supreme Court
found the dam illegal, the theft continued. In acknowledgement of the illicit
use of the water and to coerce impoverished Yaqui families to sign permits
for dam construction, the governor of Sonora pledged to pay the tuition of all
Yaqui students enrolled in state colleges, but then reneged after students
enrolled, leaving families deeper in debt. Environmental groups coordinated
a blockade of the single north-south interstate through the region. Semi-truck
drivers carrying produce and goods from southern Mexico to US ports of
entry turned their engines off at Vicam Switch, in the heart of the Yaqui
autonomous zone.

News of the blockades was hardly noticed in the English-speaking Indigenous world, much less by US mass media. Language and culture barriers between English-speaking Indigenous networks and Spanish-speaking Indigenous networks made transnational solidarity unlikely. Thus while First Nations and Native American activists engaged in bold and strategic digital dissemination of content about Idle No More, engaging a politics of recognition and shame, Rio Yaqui activists kept digital communications to a relative minimum. This was in part due to limited Internet connectivity, and due to the governor of Sonora and corporate partners paying goons to kidnap, intimidate, and assault activists. In 2012, a state propaganda campaign was in full swing. From 2010–2016, human rights observers collected information for use in what has become a five-fold human rights legal strategy against the state of Mexico. In 2016, the state hired private security forces to kidnap Anabela Carlon Flores (Yaqui), a human rights attorney and outspoken environmental rights advocate, and kept her husband for ten days in an attempt to terrorize the #rioyaqui and #namakasia movement (Telesur 2016). With transnational drug-trafficking organizations threatening Mexican journalists, and state-sponsored private security threatening activists, the cost of social media mobilization was high for activists. To this day, most now prefer face-to-face communications. The circumstances reveal an Indigenous conscientiousness about another aspect of information diffusion: surveillance, misinformation and disinformation, and the erosion of tribal privacy and security.

Thus, matters of Indigenous Internet connectivity, access, and use are not simply about acquiring capital and know-how to construct an ISP, deploy devices, or teach digital literacy. Rather, they are about framing Internet connectivity, access, and use as a matter of human rights and self-determination (Casaperalta 2020). They are about providing Indigenous peoples with the means to disseminate critical information as a matter of choice. This in particular is when collectives of Native and Indigenous women intertwine Indigenous feminist praxis and digital tactics. In that sense, tribal network sovereignty is one strategy in a repertoire of digital strategies to empower Indigenous peoples to determine their quality-of-life, one that upholds Native and Indigenous women, children, elders, and queer and two-spirit relatives. Tactical information flow is another strategy.

A 2016 social network analysis of over eleven thousand tweets associated with Native rights advocates' user accounts revealed that advocates were more likely to continually tweet about systemic injustice of American Indians; systemic injustice of First Nations peoples, including missing and murdered Indigenous women; police brutality; colonialism; and appropriation and representation. Furthermore, Native rights advocates applied the hashtag #indigenous like a glue to carry tweets from one sub-community to another (Vigil-Hayes et al. 2017). A follow-up study tracing the content of tweets associated with 108 Native American women candidates running in the 2018 mid-term US elections revealed a discursive standoff between feminist hashtags supporting democratic and liberal Native American women candidates and pro-MAGA hashtags. Issue groups supporting the candidates used Twitter to demand government attention to high rates of missing and murdered Indigenous women, gender equity and representation in public office, and environmental justice (Vigil-Hayes et al. 2019). That year, voters elected a record number of Native American women as governors and congresswomen. Notably, Peggy Flanagan (DLF-MN, White Earth Band of Ojibwe), Sharice Davis (D-KS, Ho-Chunk), and Deb Haaland (D-NM, Laguna Pueblo) among several other Democratic Party Native American women candidates had active social media campaigns organized around their platforms, all of which included progressive stances on gender equity, antiracism, and Indigenous representation. Poignantly, within her first few months in office, Haaland introduced Savanna's Act to increase data- and information-sharing toward investigating cases of missing and murdered Indigenous women. In the summer of 2020, after the first round of COVID-19 outbreaks and community lock-downs in the US revealed the extent of the digital divide in Indian Country, Haaland also partnered with Elizabeth Warren (D-MA) to introduce the DIGITAL Reservations Act. If passed, the act would subsidize large-scale Internet infrastructure build-outs across Indian Country, including training, capacity-building, and economic development. In sum, these studies reveal savvy use of social media and information diffusion by Native rights advocates and Native American women policy-makers. Thus, though Indigenous peoples are often structurally constrained to participate in the "spielraum"—the interstices between competing hegemonic political powers—through data-driven tactics, Native and Indigenous women with political and digital acumen and data-analytic skill set the agenda for national conversations around issues affecting Indigenous peoples (Tully 2009).

CONCLUSION

This chapter offers an overview of the efforts of Native and Indigenous women who apply their digital prowess to assert the rights of Native and Indigenous peoples, waters and lands. They persist in spite of working through a male-dominated tech industry and policy circles, an anti-Indian sexist surveillant social order, and the everyday challenges to body, family, tribe, and mind that characterize life as an Indigenous woman. Their work does not go unnoticed. In 2020 members of FemTechNet among others co-authored the Data ManifestNo with the feminist politics of Audra Simpson (Mohawk) in mind (Cifor et al. 2019). In their stance against wrongful use of surveillance and violent datafication they cite Simpson's (2014) theory of ethnographic refusal. Long practiced by Indigenous women, ethnographic refusal is the use of social and emotional intelligence and political lucidity to prevent exploitative information-gathering through tactics of silence, withholding, selective informing, persuasion, intelligence-gathering, and coded communication. This politics of refusal is about sustaining the privacy and security of tribal governments, social movements, and Indigenous peoples who must strengthen internal ways of knowing under coloniality while strategically sharing information with Settlers and allies toward greater long-term goals. Thus, in a data-driven networked Settler society, Indigenous feminist ethnographic refusal is also about saying yes to a politically savvy Indigenous command of digital and analog information-sharing.

Certainly this kind of intelligence yields challenges. Indigenous writers, researchers, artists, and activists interrogate tensions between ancestral ways of knowing and hyper-mediated relationships. They question how trust can be established among kin when younger generations now treat hundreds of social media friends and followers as their online family. In her retrospective reflection on the Idle No More movement, Leanne Betasamosake Simpson (Anishinaabe) (2017) challenges the value of a media coordinated around instant gratification and surveillance in a movement rooted in ancestral relationship-building, the integrity of talking story, honoring the earth, and the confidentiality of trustworthy Indigenous activist families. These are profound ontological and moral questions for Native and Indigenous women as they impart a peoples' way of life. These concerns relate to the nature of gendered collective labor among self-governing peoples, and in many cases, matrilineal peoples. They relate to the dangers of surveillance, especially as social media enables cyberstalking, grooming, and behaviors predicating violence against Indigenous girls and women (Bailey and Shayan 2016). The deleterious power of digital information-sharing is apparent.

This chapter has provided examples of how Native and Indigenous women harness informatic power toward what Vine Deloria Jr. (Dakota) (1978) referred to as the American Indian right to know. As the US breaks every promise they make to American Indians, the need for information professionals to uphold an American Indian's right to know the causes of their oppression increases. Indigenous peoples connect with each other beyond the interface of their smartphones, capture evidence of their oppression with drones outfitted with high definition cameras and GPS devices, circulate documentaries through social networking sites, and practice a kind of decolonial citizen journalism through an assemblage of digital tactics, devices, and know-how. The most expert among these advise the construction of large-scale infrastructures. The most agile travel to sites of protest, carrying digital toolkits. As Indigenous peoples of North America, we are all epistemic border-crossers, mobilizing across jurisdictions and sharing information beneath the surveillant gaze of the US, Canada, and Mexico. While tribal governments strengthen their national ICT infrastructures, critical Indigenous Internet users distribute memes, hashtags, and news designed to subvert Settler ontopower, practicing grassroots self-determination. For Native and Indigenous women who contribute to the cyber-defense of their peoples, lands, and waters, their know-how may be rooted in a generational adaptation to digital techniques, but their resiliency lives in their political will: women who assert what it means to be Indigenous in the long cycles of belonging that ground Indian Country.

NOTE

1. This work is in part funded by an NSF grant and was based on talks previously delivered at NAISA 2017 as well as at the University of Pennsylvania Center for Advanced Research in Global Communication October 2019 symposium on Data and Dominion.

REFERENCES

Ahtone, Tristan. 2016. "How media did and did not report on Standing Rock." *Al Jazeera*, December 14, 2016.

Anderson, Kim. 2020. *A Recognition of Being: Reconstructing Native Womanhood.* Toronto: Second Story Press.

Bailey, Jane, Shayan, Sarah. 2016. "Missing and Murdered Indigenous Women Crisis: Technological Dimensions." *Canadian Journal of Women and the Law* 28(2): 321–41.

Bailey, Kerry A. 2020. "Indigenous students: resilient and empowered in the midst of racism and lateral violence." *Ethnic and Racial Studies* 43(6): 1032–51.

Beck, Abaki. 2017. "Rendered Invisible: Pocahontas is not a sex symbol." *Bitch Media*, November 20, 2017.

Belarde-Lewis, Miranda. "Sharing the Private in Public: Indigenous Cultural Property and Online Media." *Proceedings of the 2011 iConference*: 16–24.

———. 2013. "From Six Directions: Documenting and Protecting Zuni Knowledge in Multiple Environments." PhD dissertation, University of Washington.

Benjamin, Ruha (ed). 2019. *Captivating Technology: Race, Carceral Technoscience, and Liberatory Imagination in Everyday Life.* Durham: Duke University Press.

Blackwater, Darrah. 2019. "Time for the 'well-meaning man' to return spectrum rights to Native American tribes. *Indian Country Today.* November 29, 2019.

Bousqet, Antoine. 2009. *The Scientific Way of Warfare: Order and Chaos on the Battlefields of Modernity.* New York: Columbia University Press.

Bruchac, Margaret. 2018. *Savage Kin: Indigenous Informants and American Anthropologists.* Tucson: University of Arizona Press.

Canek-Pena Vargas, Marco and Greg Ruggiero (eds). 2007. *The Speed of Dreams: Selected Writings 2001–2007.* San Francisco: City Lights Books.

Casaperalta, Edyael. 2020. "Tribes have chance to control spectrum, build broadband networks." *The Daily Yonder.* April 3, 2020.

Castro, Giselle. "Why Understanding Colorism Within the Latino Community is So Important." *HipLatina.* July 31, 2018.

Caven, Febna. "Being Idle No More: The Women Behind the Movement." *Cultural Survival.* March 2013.

Cervantes, Andrea Gomez. 2019. "'Looking Mexican': Indigenous and non-Indigenous Latina/o Immigrants and the Racialization of Illegality in the Midwest." *Social Problems.* doi: 10.1093/socpro/spz048.

Chertoff, Michael. 2017. *"Unlocking the Privacy-Security Debate."* Paper presented at the Arizona State University Global Security Initiative, Phoenix, Arizona. March 17, 2017.

Cifor, Marika, Patricia Garcia, TL Cowan, Jasmine Rault, Tonia Sutherland, Anita Say Chan, Jennifer Rode, Anna Lauren Hoffman, Niloufar Salehi, and Lisa Nakamura. 2019. *Feminist Data Manifest-No.* www.manifestno.com.

Deloria, Vine, Jr. 1978. *"The Right to Know."* Paper presented at the U.S. Department of the Interior, Office of Library and Information Services, Washington, D.C.

Deloria, Vine, Jr. and Clifford Lytle. 1983. *American Indians, American Justice.* Austin: University of Texas Press.

Deschine-Parkhurst, Nicholet. 2019. "From #MniWiconi to #StandwithStandingRock: The #NoDAPL Movement and Disruption of Physical and Virtual Spaces." Paper presented at *Indigenous Peoples Rise Up: The Global Ascendancy of Social Media Activism*, Flagstaff, Arizona. October 17, 2019.

Dewey, Myron. 2017. "Awake: A Dream from Standing Rock," Lecture and film screening at the *Human Rights Film Forum*. Tempe, Arizona. November 13, 2017.

Drinnon, Richard. 1997. *Facing West: The Metaphysics of Indian-Hating and Empire-Building.* Norman: University of Oklahoma Press.

Duarte, Marisa. 2017a. *Network Sovereignty: Building the Internet Across Indian Country.* Seattle: University of Washington Press.

———. 2017b. "Connected Activism: Indigenous Uses of Social Media for Shaping Political Change." *Australasian Journal of Information Systems* 21: 1–21.

Hopkins, Ruth. 2015. "My Native Identity Isn't Your Plaything: Stop with the Mascots and 'Pocahotties.'" *Guardian.* June 19, 2015.

ICT Staff. 2017. "Women Water Protectors of NoDAPL." *Indian Country Today.* March 8, 2017.

Keeler, Jacqueline. 2006. "Fire Thunder Impeachment and the Rights of Women." *TiospayeNow.* July 8, 2006.

———. 2018. "'I was born free'—Red Fawn and State-sponsored Sexual Assault of Native Women at Standing Rock." *TiospayeNow.* July 13, 2018.

———. 2018. "Elizabeth Warren and White Attachment to Native Identity." *TioSpayeNow.* October 2, 2018.

Kendi, Ibram X. 2016. *Stamped from the Beginning: The Definitive History of Racist Ideas in America.* New York: Bold Type Books.

Kukutai, Tahu and John Taylor. 2016. *Indigenous Data Sovereignty: Toward an Agenda.* Canberra: ANU Press.

"Lawyer for Yaqui Tribe Fighting Mexico's DAPL Kidnapped." *TelesurHD.* December 15, 2016.

Littletree, Sandy. 2018. "'Let Me Tell You About Indian Libraries': Self-Determination, Leadership, and Vision: The Basis of Tribal Library Development in the United States." PhD dissertation, University of Washington.

Massumi, Brian. 2014. *Ontopower: War, Powers, and the State of Perception.* Durham: Duke University Press.

Medina, Jose. 2012. *The Epistemology of Resistance: Gender and Racial Oppression, Epistemic Injustice, and Resistant Imaginations.* Cambridge: Oxford University Press.

Mihesuah, Devon. 2003. *Indigenous American Women: Decolonization, Empowerment, Activism.* Lincoln: University of Nebraska Press.

Native Women's Association of Canada. 2011. *Aboriginal Lateral Violence.* Ottawa: NWAC.

Privott, Meredith. 2019. "An Ethos of Responsibility and Indigenous Women Water Protectors in the #NoDAPL Movement." *American Indian Quarterly* 43(1): 74–100.

Rainie, Stephanie Carroll, Jennifer Lee Schultz, Eileen Briggs, Patricia Riggs, and Nancy Lynn Palmanteer-Holder. 2017. "Data as Strategic Resource: Self-determination and the Data Challenge for United States Indigenous Nations." *International Indigenous Policy Journal* 8(2). doi:10.18584/iipj.2017.8.2.1.

Rule, Elizabeth. 2018. "Seals, Selfies, and the Settler State: Indigenous Motherhood and Gendered Violence in Canada." *American Quarterly* 70(4): 741–54.

Sanger, Margaret. 1932. "My Way to Peace." MSM Margaret Sanger Papers, Library of Congress 130: 198.

Showalter, Esther, Nicole Moghaddas, Morgan Vigil-Hayes, Ellen Zegura, and Elizabeth Belding. 2019. "Indigenous Internet: Nuances of Native American Internet Use." *Proceedings of the Tenth International Conference on Information and Communication Technologies and Development*: 1–4.

Simpson, Audra. 2014. *Mohawk Interruptus: Political Life Across the Borders of Settler States.* Duke University Press.

Simpson, Leanne Betasamosake. 2017. "Constellations of Coresistance." *As We Have Always Done: Indigenous Freedom Through Radical Resistance.* Minneapolis: University of Minnesota Press.

Slotkin, Richard. 2000. *Regeneration Through Violence: The Mythology of the American Frontier, 1600–1860.* Norman: University of Oklahoma Press.

Tallbear, Kim. 2016. "Badass (Indigenous) Women Caretake Relations: #NoDAPL, #IdleNoMore, #BlackLivesMatter." *Society for Cultural Anthropology*, December 22, 2016.

Tolan, Sandy. 2017. "Journalist faces charges after arrest while covering Dakota Access Pipeline protest." *Los Angeles Times,* February 5, 2017.

Tully, James. 2009. *Public Philosophy in a New Key, Volume 2: Imperialism and Civic Freedom.* Cambridge: Cambridge University Press.

Vigil-Hayes, Morgan, Nicholet Deschine-Parkhurst, and Marisa Duarte. 2019. "Complex, Contemporary, and Unconventional: Characterizing the Tweets of the# NativeVote Movement and Native American Candidates through the 2018 US Midterm Elections." *Proceedings of the 2019 ACM Conference on Human-Computer Interaction:* 1–27.

Vigil-Hayes, Morgan, Marisa Duarte, Nicholet Deschine-Parkhust, and Elizabeth Belding. 2017. "#indigenous: Tracking the Connective Actions of Native American Advocates on Twitter." *Proceedings of the 2017 ACM Conference on Computer Supported Cooperative Work and Social Computing*: 1387–99.

Walters, Maggie and Chris Andersen. 2013. *Indigenous Statistics: A Quantitative Research Methodology.* Routledge.

Walters, Maggie, Tahu Kukutai, Stephanie Russo Carroll, and Desi Rodriguez-Lonebear. 2021. *Indigenous Data Sovereignty and Policy.* Routledge.

Water Protector's Community Oral History Project. https://waterprotectorscommunity.org.

Wemigwans, Jennifer. 2018. *A Digital Bundle: Protecting and Promoting Indigenous Knowledge Online.* University of Regina Press.

Wilson, Danae. 2015. Panel presentation. *Tribal Telecom and Technology Conference.* May 7, 2015. Albuquerque, New Mexico.

Wunder, John R. 2000–2001. "Merciless Indian Savages and the Declaration of Independence: Native Americans Translate the Ecunnaunuxulgee Document." *American Indian Law Review* 25: 65–92.

Chapter Six

"Being Seen for Who I Am"

*Counterpublic Trans Intelligibility
and Queer Worldmaking on YouTube*

Ace J. Eckstein

Within mainstream culture, trans men can only be understood in relation to cisgender men. In this cultural moment of increased transgender visibility, there is a push to make trans people intelligible within normative understandings of gender. Even affirming discourses such as "trans men are real men" imply that the unmarked cisgender man is the standard for comparison. This normative framework centers cisgender experiences of manhood. By contrast, the transmasculine YouTube community centers trans experiences of manhood—a necessary first step for trans men to become recognizable not despite of, but through their transness.

The Internet has been a crucial site of transgender community in large part due to the lack of large groups of transgender people in close geographic proximity to one another. Facebook groups, Tumblr blogs, and resource websites, among many other Internet platforms have all contributed to transgender people connecting in ways that transcend geographic divides. However, as Laura Horak (2014) notes, "Though trans people also use platforms like Vimeo and Tumblr, YouTube has almost single-handedly transformed the trans mediascape" (572). By creating, viewing, and subscribing to trans YouTube channels, commenting on trans videos, and connecting with trans YouTube users through features such as private messages, trans people use YouTube to access resources and build community.

YouTube also provides an alternative to mainstream media, and the distinctions between trans self-presentation on YouTube and trans representation in mainstream media are worth noting. The self-publishing nature of

YouTube allows transgender people to tell their own stories without the need for a cisgender mediator or tailoring to a cisgender audience as would be required to engage with mainstream media. Many genres of YouTube videos focus on the everyday over the sensational that often is featured in mainstream media. Trans use of YouTube mediates trans experiences, lives, and communities. This mediation matters in that videos can circulate widely, constituting counterpublics—as Michael Warner (2002) suggests—"through mere attention" (419). As such, trans YouTube counterpublics include all those who watch the videos or interact with them passively as well as those who participate more actively through creating videos, commenting on videos, or subscribing to channels.

Trans men in particular have taken to YouTube, creating YouTube channels to document their physical transition. These transition channels generally feature the channel's creator by himself speaking into a webcam or video camera on a tripod. The first video tends to coincide with the beginning of the channel creator's physical transition, often in anticipation of his first injection of testosterone. The trans men in the videos tend to speak to an implied audience of other transgender men. Across most channels, trans men provide regular updates about the physical changes due to testosterone, often using the camera to try to show changes. The frequency of video updates as well as content beyond updates on the physical transition varies between channels. Many trans men include reflections on their transition process and stories related to other aspects of their lives in addition to documenting physical changes.

This essay is based on the close reading of five trans men's—James, Kyle, Melvin, Raúl, and Michael—transition channels. All five channels' creators identify as trans men and are pursuing medical transition in some form. This is important to highlight the specificity of particular trans experiences. Trans women, for example, may use YouTube in qualitatively different ways than trans men. These five channels also represent diverse experiences within the transmasculine YouTube community, specifically in terms of race, masculinity, and popularity of the channel. I analyzed each channel in their entirety, totaling 419 videos. Although not exhaustive, the goal of this archive is to establish a range of intersectional transmasculine performances that both overlap and are distinctive within this genre.

Based on this analysis, I argue that trans men on YouTube collectively constitute transmasculinity as an intelligible subjectivity; this constitution is the work of queer worldmaking (Berlant and Warner 1998; West et al. 2013) and can only be accomplished in the context of transmasculine YouTube as a counterpublic. As opposed to a recognition based on individual schemas, intelligibility is based in socially recognized norms (Butler 1990), leading to the necessity of counterpublics to enable alternative intelligibilities. Counterpublics typically do not (re)produce the same norms as dominant publics. In this case, the transmasculine YouTube community does not (re)produce the same norms of gender and, more specifically, masculinity; the centering of trans men is key to the establishment of alternative norms. The trans men imagine and enact a queerer world in the counterpublic space of YouTube through the production of transmasculine intelligibility.

To date, counterpublics theory has largely been concerned with discursive practices that constitute and emerge from counterpublics (e.g., Fraser 1990; Warner 2002). While norms and normativity are implicated in these theories, they are rarely addressed head on and are often norms of dominant publics that function to exclude counterpublics. This chapter contributes to counterpublics theory by explicitly linking the discursive practices of trans men on YouTube to the norms they (re)produce *within* the counterpublic. These counterpublic norms are generative, imagining and enacting a world from the perspective of an alternative set of norms. In this case, the transmasculine YouTube counterpublic imagines and enacts a world in which trans men are intelligible not despite of but through their transness; this intelligibility can only be constituted from a set of gender norms that centers transmasculine identities and experiences. A queer worldmaking perspective brings into focus questions of norms, normativity, and standards of queerness. Importantly, trans men on YouTube are not exemplars of queer purity, rejecting all semblance of normativity. Rather, the way trans men on YouTube queer norms of gender and masculinity is exemplary of queer worldmaking.

COUNTERPUBLICS, WORLDMAKING, AND INTELLIGIBILITY

Worldmaking, counterpublics, and intelligibility are the what, where, and how of transmasculine subjectivity on YouTube, enacting new possibilities for trans men both within and beyond YouTube specifically. Trans men perform worldmaking insofar as they actualize these possibilities. YouTube's worldmaking potential is bound with its counterpublic enclave capacities where trans men can—to an extent—create their own norms of gender. These uniquely transmasculine norms of gender enable the intelligibility of trans men, which is a key way that trans men on YouTube actualize queerer worlds.

Worldmaking is, in part, interested in how discourse shapes the possibilities of the social world. Warner (2002) contends that all publics engage in worldmaking in that "all discourse or performance addressed to a public must characterize the world in which it attempts to circulate, and it must attempt to realize that world through address" (422). Berlant and Warner (1998) call for attention to queer culture as a "worldmaking project," asserting the need "to recognize that queer culture constitutes itself in many ways other than through the official publics of opinion culture and the state, or through the privatized forms normally associated with sexuality" (558). To this end, counterpublic spaces—like the transmasculine YouTube counterpublic—are key sites of generating queer culture. Trans men on YouTube characterize and attempt to realize a world that centers trans men as wholly intelligible subjects.

The Internet in particular provides an alternative platform for increased potential for participation and queer worldmaking. Robert Glenn Howard (2008) theorizes participatory media enabled by the Internet as "a vernacular web of communication performance that hybridizes the institutional and non-institutional" (491). These participatory media "hybridize multiple agencies in the texts that they produce. Rejecting reified notions of pure or authentic vernacular, participation in this web can be seen to open up new venues for transformative public discourse" (491). Given the possibilities of the vernacular web, the Internet is a site particularly ripe for rhetorical inquires of queer worldmaking. West et al. (2013) define queer worldmaking as "practices and relationships that contest the logics of compulsory heteronormativities" (56). The Internet is uniquely situated to contribute to such practices in that it transcends the temporal constraints of face-to-face interaction, and, especially in the case of video sharing platforms like YouTube, can foster queer worldmaking.

Trans people in particular have taken to the Internet in pursuit of queer worldmaking projects. K. J. Rawson (2004) characterizes three worlds constituted through trans engagement with the Internet: a world where trans lives count, a world where everyone makes history, and a world of shared experience. Of particular interest here is Rawson's last world wherein he examines trans people documenting their lives on YouTube as one exemplar. Trans people using platforms such as YouTube, Rawson (2014) asserts, "create a world where trans people can share their experiences, recognize their shared experiences, and contribute to the development of community knowledges" (56). In a world where trans people are ostracized and have their experiences delegitimized, such acts of community and recognition are radical and political acts. Trans use of YouTube is exemplary of how participatory media of the Internet can constitute counterpublic spaces that do the work of queer worldmaking.

On YouTube, trans men's communities reflect a counterpublic, publics that "[come] into being through an address to indefinite strangers" (Warner 2002, 424). By the nature of YouTube videos' circulation, trans men address indefinite strangers; anyone with an Internet connection *can* access transition channels. However, "counterpublic discourse also addresses those strangers as being not just anybody. They are social marked by their participation in this kind of discourse; ordinary people are presumed not to want to be mistaken for the kind of person that would participate in this kind of talk, or be present in this kind of scene" (Warner 2002, 424). The strangers that are addressed by trans men on YouTube are largely presumed to be trans men, shown through discourse practices such as trans men addressing their audience as "brothers" or not defining transmasculine-specific vocabulary and experiences that other trans men would presumably know and others would likely need more explanation to understand, marking their alterity and counterpublicity. But counterpublics do more than mark their publicity as subordinate; they enable certain discursive actions within their boundaries and with broader publics. Trans men on YouTube invent and circulate discourses of transmasculinity and manhood that run counter to normative discourses, forming the basis of the reformulation of gender intelligibility and subjectivity. The transmasculine subject that emerges through YouTube may not be an explicit or direct challenge to normative conceptions of gender, but the intelligibility of transmasculinity in the counterpublic enclave of You-Tube enables a new imaginary of how trans men can be recognized.

In constructing this new imaginary, the transmasculine YouTube counterpublic is steeped in potentiality. The transmasculine YouTube counterpublic allows trans men to make rhetorical moves that work collectively to constitute the intelligibility of trans men. This subjectivity largely lives in the realm of possibility, as it is not yet fully realizable within dominant publics, but this does not diminish its significance. Rather, the intelligibility of transmasculinity that emerges on YouTube allows trans men to imagine their own subjectivity otherwise, a necessary first step to toward broader recognizability. According to Butler (1990), "'[i]ntelligible' genders are those which in some sense institute and maintain relations of coherence and continuity among sex, gender, sexual practice, and desire" (23). Trans men disrupt this normative coherence between sex and gender. From this perspective, the idea of an intelligible transmasculine subject—that is a subject whose transness is recognizable as a part of his gendered subjectivity—is an impossibility. However, this definition relies exclusively on the social norms of mainstream culture. The transmasculine YouTube counterpublic constitutes alternative gender norms not reliant on sex-gender coherence.

In highlighting the relationship between norms and recognition, Butler (2004) asserts that "[t]he norm governs intelligibility, allows for certain kinds of practices and action to become recognizable as such, imposing a grid of legibility on the social and defining the parameters of what will and will not appear within the domain of the social" (42). The implications of this "grid of legibility" are paramount because it is fundamentally linked to recognizable personhood. As Butler (2004) writes,

> The very criterion by which we judge a person to be a gendered being, a criterion that posits coherent gender as a presupposition of humanness, is not only one which, justly or unjustly governs the recognizability of the human, but one that informs the ways we do or do not recognize ourselves at the level

of feeling, desire, and the body, at moments before the mirror, in the moments before the window, in the times that one turns to psychologists, to psychiatrists, to medical and legal professionals to negotiate what may feel like the unrecognizability of one's gender and, hence, the unrecognizability of one's personhood. (58)

While gender and personhood are not one in the same, their recognizability is fundamentally intertwined. In a world where a trans person's gender is unrecognized or can only be recognized through medical and legal stigma, the recognition on YouTube makes it a space that validates trans men's personhood. "To put it bluntly," Horak (2014) states, "these videos save trans lives" (581). Such validation is essential; it makes life more livable.

Butler (1991) famously asserts that "gender is a kind of imitation for which there is no original . . . it is a kind of imitation that produces the very notion of the original as an effect and consequence of the imitation itself" (22). Gender norms constitute gendered subjects in such a way that cisgender men are produced as the naturalized original that trans men imitate. Seemingly caught between illegibility and imitation, trans men's subjectivity highlights the need to investigate the margin of the intelligible. Butler (1990) leaves open the possibility of alternative and subversive grids of legibility: the "persistence and proliferation" of gender identities that are illegible within dominant norms "provide critical opportunities to expose the limits and regulatory aims of that domain of intelligibility, and, hence to open up within the very terms of the matrix of intelligibility rival and subversive matrices of gender disorder" (24). The realization of this possibility always exceeds the individual; the necessary proliferation of illegible gender identities cannot happen in isolation. The transmasculine YouTube counterpublic is thus an exemplar of transforming gender intelligibility.

YouTube provides the space for trans men to connect, collectively constituting an alternative set of gender norms that determine intelligibility. Central to the theory of gender performativity is that one is constantly doing gender and that "one does not 'do' one's gender alone (Butler 2004). One is always 'doing' with or for another, even if the other is only imaginary" (1). The indefinite strangers of the transmasculine YouTube counterpublic are the imagined other: "The act of self-reporting and the act of self-observation takes place in relation to a certain audience, with a certain audience as the imagined recipient, before a certain audience for whom a verbal and visual picture of selfhood is being produced" (Butler 2004, 67). In this case, the audience is imagined as other trans men, which stands in stark contrast to the generalized imagined other of dominant publics whose presumed naturalized cisness marks trans men as unintelligible subjects. Understanding each trans man's construction of self on YouTube in relation to this imagined audience of trans men contextualizes it within work of intelligibility. That is to say, the subjectivity concretized in each individual's videos is inextricably tied to the imagined audience of trans men.

CARVING SPACE FOR TRANSMASCULINE SUBJECTIVITY

Each of the five trans men whose YouTube channels I explore identify and relate to their transness and their masculinity differently, and many of their understandings of self in relation to these two aspects of their identity change over the course of their videos. However, what is consistent across the five channels is reflexivity about their gendered identities and subjectivities. Most frequently this occurs in the form of providing an account of an everyday life experience that caused gender to be particularly or differently salient. These acts of reporting through YouTube do something more than repeat or reflect the gendering that transpired. Rather, if we are to take seriously Butler's assertion that we are always doing gender for an other, then the YouTube videos' imagined counterpublic of trans men engenders a different doing of gender than the presumably cisgender actors of the story that is shared.

These reflections of identity largely fall into three categories: *rhetorics of openness, rhetorics of stealth*, and *rhetorics of passing*. Each of these rhetorics consists of discourses that situate a channel creator's transness as a part of their larger identity. In examining these three categories, I argue that each of the rhetorics—even when seemingly contradictory—collectively constitute an intelligible transmasculine subject. As Butler (2003) writes, "When we do act and speak, we not only disclose ourselves but act on the schemes of intelligibility that govern who will be a speaking being, subjecting them to rupture or revision, consolidating their norms, or contesting their hegemony" (132). Trans men's disclosure to an imagined audience of other trans men both contests the hegemony of gender intelligibility based on sex-gender coherence and recognizes, and thus confers, the specifically trans subjectivity necessary to and constituted by these reflections.

Rhetorics of Openness

Rhetorics of openness are anecdotes and reflections that convey disclosure of trans identity, that is, to be open about one's transness. For example, James discusses how when he began college, he wanted to be stealth, but made him more dysphoric because he felt like he had to fit the box of a cisgender male. He realized while being stealth that he wasn't presenting as himself because he was so concerned with not being seen as trans. He has since become very open about his trans identity. He reflects, "Now I feel like I am actually being seen for who I am." For James, intelligibility strictly within the framework of cisgender masculinity limited his sense of authenticity. This was also reflected in a video in which James discussed why he chose to leave his job as a counselor for an all-boys farming camp after a week of staff training. The camp administration was requiring that he be stealth to the campers. During the week of training, James started to feel a lot of anxiety about being surrounded by all cisgender men. He realized that a big part of his identity as a man is that he had to "become a man." To "become a man" highlights transition, and for James to say that becoming a man continues to be a large part of his identity as a man suggests that that transition is not merely a means to an end. Rather, James marks—for him—a difference between cisgender manhood and trans manhood. Notably, this difference is not one of lack; trans manhood is not less desirable for James than cisgender manhood. This distinction cannot be understood within the sex-gender coherence framework of intelligibility, which constitutes cisgender manhood as natural and primary.

Kyle, like James, rejects the idea that intelligibility within a cisgender framework can authentically capture his subjectivity. In a video entitled, "FTM Vlog: The Guy That Doesn't Quite Fit the Mold," Kyle, in anticipation of beginning testosterone hormone replacement therapy, reflects, "Obviously once I start transitioning, to other people, to society at large, I will be read as male and I'm ok with that, but people that know me, people that will come to know me will know that I am transgender, not 100% male, not 100% female." The "mold" that Kyle references in the title of his video is constituted by gender norms; it represents normatively intelligible gender. For Kyle, when people read him as male within that framework, he feels inauthentic. While expressing being "ok" with these moments of passing, he seemingly feels it is a compromise or concession because a more authentic reading of his subjectivity is inaccessible to society at large.

Each of these reflections deals with a tension between normative intelligibility and authenticity. There is largely no reference point for this tension outside of trans experience. Speaking into this tension to an imagined audience of other trans men calls into being a subject position that does not exclusively rely on binary cisgender norms of maleness, manhood, and masculinity; that is to say, a transmasculine subjectivity rehearsed and performed within subaltern counterpublic space.

Rhetorics of Stealth

Rhetorics of stealth are anecdotes and reflections of a choice not to disclose trans identity when perceived as a cisgender man. In a video entitled "I'm invisible!," Michael elatedly describes going to a bar and flirting with a group of lesbians who all ignored him. He realized through this interaction that they were ignoring him because they were reading him as a—presumably cis—man. The title is particularly telling in this instance. Contrary to the dominant attitude towards the importance of visibility and the stifling effects of invisibility, Michael seems to be suggesting that invisibility is, in fact, a positive. Invisibility means a kind of blending in that it allows Michael to experience a legibility that affirms his identity.

In a separate video, Michael remarks, "I live my life as a male. I am seen, regarded, and treated as a male, not a trans guy." Throughout his videos, Michael seems to desire and enjoy legibility within normative frameworks of gender intelligibility. That the transmasculine YouTube community constitutes an intelligible transmasculine subject outside the normative framework of gender intelligibility does not mean that trans men cannot or should not desire intelligibility within the normative framework. Rather, the transmasculine YouTube counterpublic constitutes an alternative subjectivity that opens up possibilities for how trans men can become legible beyond the normative framework. The assumptions of a framework of intelligibility grounded in sex-gender coherence will always preclude trans men on some level. Michael gestures towards this when he goes on to say, "I'm at the point in my transition that I'm a man and the transition left is about me deciding what kind of man I am going to be . . . How am I going to be the man I want to be without detrimental actions i.e. outing myself when outing myself could be detrimental to whatever situation I am in . . . Lord knows we can lose our jobs over nothing." Even within the relative freedom that Michael finds in being stealth, there are limits to his normative intelligibility.

In discussing his thought process around going stealth, Melvin reflects that he was thinking of doing so because he identifies more as male than as transgender. Melvin comments that he used to take issue with people who went stealth because he felt like they were turning their back on the trans community, but he has since realized that it is not about that. Melvin says that he would probably keep his YouTube channel when he goes stealth because he values the community it fosters. His reflection raises interesting tensions around stealth and visibility. YouTube becomes an interesting limbo space for those who are stealth in that maintaining a YouTube channel or even keeping it up without continuing to post retains a community of visibility. Understandings of the transmasculine YouTube counterpublic as an enclave are particularly prevalent in moments like these in that YouTube appears to be a distinctly separate space from "the public," one where one's performance of identity can be markedly different. Because a framework of transmasculine intelligibility does not rely on the same rigid norms that a normative framework of gender intelligibility does, there is potentially space for being intelligible as a stealth trans man. This allows for the kinds of reflections that Michael and Melvin both have in their videos without seeming to compromise an understanding of themselves as stealth.

Rhetorics of Passing

Many trans men on YouTube provide suggestions to other trans guys on how to pass, that is, to be perceived as (cis) men by others. James and Michael both did this in their channels. James gave several specific tips about how to bind one's chest, how to speak in a lower vocal register, and other ways to alter one's physical appearance, but he emphasized a main theme of confidence as the most important. He encourages trans guys to leave the house feeling like "I am male" regardless of others' reactions. He also suggests trans guys go for a walk alone in order to escape people reading them as female and as a way to "be at home in your masculinity." Passing is framed as something that is for the outside world, recognizing that there are limits to how much individuals may be able to pass. Although James does not make this explicit, providing this advice on YouTube and in this framework suggests that YouTube provides a sort of refuge in which one's gender is intelligible in ways that may not be possible in the cisgender world. Michael's tips on passing are very different than James' video. Michael focuses on various "gear" that trans men use to alter their appearance such as chest compression binders Michael commented "the other piece of gear that no trans guy can be without is some sort of stand-to-pee device . . . when you walk into the men's room it is nice to go up to the urinal and do your business like the man that you are." This piece of advice seems to suggest that manhood is tied to urinal usage and that a stand-to-pee device is thus necessary for trans men to be intelligible as men.

Advice around passing should be understood in terms of counterpublicity. As Warner (2002) writes, "[a] counterpublic maintains at some level, conscious or not, an awareness of its subordinate status" (423–24). Michael's advice perhaps suggests a more conscious awareness of subordinate status. YouTube may be a refuge for many trans men, but YouTube does not and cannot exist entirely separated from trans men's everyday lives. Nancy Fraser (1990) writes of a dual character of counterpublics: "On the one hand, they function as spaces of withdrawal and regroupment; on the other hand, they also function as bases and training grounds for agitational activities directed toward wider publics" (68). Advice around passing, including Michael's advice, recognizes that trans men live in a world in which transmasculinity is not intelligible and supplies trans men with techniques for navigating dominant publics. YouTube is the rehearsal space. Further, a generous read of Michael's advice would see it as a reclamation of manhood, a stand-to-pee device allows at least some trans men to access legibility in the typically highly normatively gendered context of a public restroom. Giving this advice in the context of the transmasculine YouTube counterpublic, on some level, marks the transness of the act.

Discourses of passing can also complement discourses of openness. Kyle and James both decenter the importance of passing, although it is important to note that their attitudes toward passing—as with several of the other trans men included in this analysis—fluctuate over the course of their channel. After two and a half years on testosterone, Kyle reflects that he only passes 30–40 percent of the time. He says that he has never passed 100 percent of the time. It used to upset him, but he has gotten to a point where it has stopped bothering him because he is happy with how he looks and sounds. James, on the other hand, reflects that it is very easy for him to pass since he has been on testosterone and had top surgery.[1] For James, passing can be helpful, but it can also erase his trans experiences. He concludes that he does not want to be seen as a cisgender man. Both Kyle and James disrupt the idea that passing is the *telos* of transition. Instead, YouTube becomes a space where trans experience and subjectivity can be understood more capaciously. The way in which rhetorics of passing and discourses of openness can over-lap points to a uniquely transmasculine intelligibility in that these two dis-courses are mutually exclusive in dominant publics.

NORMS OF TRANSMASCULINITY

To claim that YouTube allows a space for transmasculinity to become an intelligible subjectivity is necessarily to claim that there are norms of trans-masculinity circulating through YouTube. Just as norms grounded in sex-gender coherence produce a grid of legibility that precludes many, including trans men, from being fully intelligible in dominant publics, norms of trans-masculinity on YouTube also produce a grid of legibility that is not fully inclusive. To begin, there is an expectation that trans men on YouTube are seeking medical transition through testosterone and top surgery. Medical transition plays a central role in the content of the videos as well as in structuring the timeline of the channels. For example, many videos are titled or begin with a statement of how long the channel creator has been on testosterone and/or how long it has been since they had top surgery. There are some trans men on YouTube who do not transition medically or who "detransition," meaning that they stop medical transition once they have started. However, this is a clear minority. Moreover, trans men on YouTube not pursuing medical transition are largely having entirely different and pre-dominantly separate conversations. This pervasive expectation of medical transition means that an intelligible transmasculine body is constituted through and constrained by medical transition.

A second pervasive norm of trans men on YouTube is that of whiteness. A norm of whiteness is constituted algorithmically, among many other ways. In order to find trans men of color's YouTube channels, you must mark your search terms. This was reflected in my own methods. I found the three channels of white trans men—James, Michael, and Kyle—first. I began with James' channel because it was the most frequently recommended channel on online discussion boards pertaining to trans men. From there, I found Michael and Kyle's channels through the "recommended videos" section of YouTube. The recommended videos section as well as the first several pages of results of a racially unmarked search for "ftm" yielded exclusively channels of white trans men. In order to find Melvin's channel, I had to use a racially marked search for "Black ftm." Likewise, to find Raúl's channel, I used the search term "Latino ftm." Horak (2014) notes, "tens of thousands of trans people of color do post videos and many actively comment on each others' videos. These networks are invisible, though, to anyone who does not specifically seek them out" (576). This upholds broader social normalization of unmarked whiteness that constitutes naturalized racial hierarchies.

Both Raúl and Melvin mark and push back against the norm of whiteness among trans men on YouTube. In his first video, Raúl reflected on the reason why he wanted to create a YouTube channel to document his transition saying, "I think the reason I am doing this is to provide some sort of support to other trans guys who are from the Hispanic community because, ya know, it's hard. I'm not going to say it's harder but it might be a little bit harder when you come from a different culture and we don't really hear a lot about it." Raúl's channel then becomes a way to speak into that silence.

Melvin offers reflections on both the racialized and gendered aspects of his changing embodiment and intelligibility. He notes that he felt either invisible or hyper visible, especially in interactions with white men or white women. As he was read more consistently as male, he was frustrated over how he was being treated being read as a Black man. Melvin recounts several experiences of this maltreatment including white men walking up behind him and intimidating him and hearing a man lock the doors of his car as Melvin walked past. Melvin's reflections clearly point to the ways in which changing intelligibility within normative frameworks as a passing cisgender man is racialized. White trans men can experience passing male intelligibility as largely alleviating the stares and such of hyper visibility, but that is not the case for trans men of color, and Black trans men in particular. Like in other discourses of openness, stealth, and passing discussed above, addressing these reflections to the imagined audience of the transmasculine YouTube counterpublic functions to bring them into the realm of transmasculine intelligibility. In this case, reflections that mark the racialized specificity of transmasculine experience as part of the discursive milieu constituting an intelligible transmasculine subjectivity also functions to resist the norm of whiteness in transmasculinity on YouTube.

A third norm of transmasculinity is lack of femininity. Although there is a wide range of masculinity represented by trans men on YouTube, there is a level of reflexive awareness among more effeminate trans men that their masculinity is not valued in the same way. Of the channels used for this analysis, this is most clearly seen in several of Kyle's reflections. Before beginning testosterone, Kyle shares struggles with whether or not he was "really trans" and if he should medically transition. He reflects, "I realized when I came out as trans, whether I realized it or not, whether I wanted to or not, I tried to cram myself into a stereotype. I tried to make myself very masculine, and that's not me." More specifically, Kyle shares that he had put on make-up for the first time in a long time and enjoyed it. Afterwards, he struggled with how he could be trans if he enjoyed putting on make-up. This norm runs parallel to dominant norms of masculinity as mutually exclusive from femininity. However, in a later video, Kyle addresses the specificity of this norm being enforced within transmasculine communities. He reflects that he has heard from cisgender people that "I yearn to be a man, but I don't conduct myself as a man should." Kyle expects this kind of comment from the cisgender community because they do not have to think about gender on a regular basis. However, he was flabbergasted that he would receive such ridicule from within the transmasculine community. He went on to say,

> I don't care if I parade down the fucking street in a goddamn mini skirt and knee-high boots fully made up, I'm still a trans guy. And just because you say I'm not or that I make you look bad, I'm still a trans guy. And just because you say I'm not or that I make you look bad, fuck you because it's not up to you how anyone presents their gender or whether anyone is transgender . . . All you're doing is perpetuating this horribly destructive patriarchal mindset that our society shoves down our throats since conception.

Kyle directly pushes back against the lateral gender policing within the transmasculine community. His assertion "I'm still a trans guy" carves out space for effeminate trans men to be intelligible as trans men.

Marking and pushing back against norms of transmasculinity on YouTube points to the need for even more capacious standards of transmasculine intelligibility. Moreover, these moments of pushback function performatively to begin to carve out that more capacious space of intelligibility. Butler (2004) suggests, "to the extent that gender norms are *reproduced*, they are invoked and cited as bodily practices that also have the capacity to alter norms in the course of their citation" (52). Through their bodily presence and discursive marking of their experiences, trans men like Raúl, Melvin, and Kyle harness the resistive power to alter norms within the transmasculine YouTube counterpublic.

LIVABLE LIVES

Even with this counterpublic set of gender norms, trans men on YouTube do not reject all semblance of dominant normativity. Michael's assertion that trans men need stand-to-pee devices in order to pee like men stands out as an example with strong normative elements. Transmasculine intelligibility constituted within the transmasculine YouTube counterpublic does not and cannot entirely escape normativity's grasp. To this end, trans men on YouTube and the emergent intelligible transmasculine subject are productively understood within the framework of an impure transgender politics (West 2014). West suggests, "we undersell the potential, even if it is only momentary and in small ways, for practices to queer existing normativities, which may settle back into previously received relations of meaning but not without some disturbance and accommodation" (167). An impure transgender politics returns us to the generative power of (imperfectly) queering norms.

The power of the queered norms in the transmasculine YouTube counterpublic is tied to the connection between intelligibility and livability, wherein, "[one] may feel that without recognizability[one] cannot live. But [one] may also feel that the terms by which [one is] recognized make life unlivable" (Butler 2004, 4). Trans men in dominant publics largely exist at the nexus of craving recognition and rejecting the terms of recognition that are deeply seeded in sex-gender coherence. By queering gender norms, trans men reconfigure the terms of recognition. In so doing, the transmasculine YouTube counterpublic becomes a space that is uniquely livable for trans men. The recognition of the ways YouTube makes life more livable for trans men suggests that the work of counterpublics should not only be understood in terms of effects and influence on dominant publics, but also in terms of what is accomplished within the counterpublic.

Queer worldmaking is invested in the constitution of queer culture, partly constituted through queer subjectivities. This investment does not preclude the possibility for queer culture to trouble and complicate dominant publics. In fact, queer worldmaking necessarily does so. However, the starting point is the realm of queer counterpublics and the relations among queer people and practices. As such, naming transmasculinity as an intelligible subjectivity within the specific context of the transmasculine YouTube counterpublic matters as it signals the depth and nature of the recognition of trans men by trans men in that space, as well as the importance of YouTube for trans men. The recognition and legitimacy conferred through intelligibility of transmasculinity moves beyond what is possible within dominant publics. The counterpublic space of YouTube is the actualization of possibility for trans men, and, to quote Butler (2004), "possibility is not a luxury; it is as crucial as bread" (29).

NOTE

1. "Top surgery" is the vernacular within trans masculine communities referring to a double mastectomy surgery that masculinizes the chest.

REFERENCES

Berlant, Lauren and Michael Warner. 1998. "Sex in Public." *Critical Inquiry 24:* 547–66.

Butler, Judith. 1990. *Gender Trouble.* New York: Routledge.

———. 1991. "Imitation and gender Insubordination." In *Inside/out: Lesbian Theories, Gay Theories*, edited by Diana Fuss, 13–31. New York: Routledge.

———. 1993. *Bodies that Matter: On the Discursive Limits of Sex.* New York: Routledge.

———. 2003. *Giving an Account of Oneself.* New York: Fordham University Press.

———. 2004. *Undoing Gender.* New York: Routledge.

Fraser, Nancy. 1990. "Rethinking the Public Sphere: A Contribution to the Critique of Actually Existing Democracy." *Social Text* 25/26: 56–80.

Horak, Laura. 2014. "Trans on YouTube: Intimacy, Visibility, Temporality." *Transgender Studies Quarterly* 1: 572–85.

Howard, Robert Glenn. 2008. "The Vernacular Web of Participatory Media." *Critical Studies in Media Communication* 25: 490–513.

Rawson, K.J. 2014. "Transgender Worldmaking in Cyberspace: Historical Activism on the Internet." *QED: A Journal of GLBTQ Worldmaking* 1(2): 38–60.

Warner, Michael. 2002. "Publics and Counterpublics." *Quarterly Journal of Speech* 88: 413–25.

West, Isaac. 2014. *Transforming Citizenships: Transgender Articulations of the Law.* New York: New York University Press.

West, Isaac, Michaela Frischherz, Allison Panther, and Richard Brophy. 2013. "Queer Worldmaking in the 'It Gets Better' Campaign." *QED: A Journal of GLBTQ Worldmaking* 1(1): 49–85.

Chapter Seven

Online (Indian/South Asian) Digital Protest Publics Negotiating #POC, #BIPOC, and #anticaste

Radhika Gajjala, Sarah Ford,
Vijeta Kumar, and Sujatha Subramanian

INTRODUCTION

This chapter is based on the various ongoing research projects of the authors and is structured around underlying identity labels that surface within hashtag activist publics with a focus on the labels that emerge around women and the intersectional categories that differentiate "women's experience" as shaped by race, caste, geography, class, sexuality, gendering, and more. We focus our thinking through the label "women of color" as it is applied to immigrant women living in the global north locations and how Indian histories of colorism and casteism complicate the use of the identity label "women of color" in hashtag publics, especially on Twitter. In order to think through these nuances of identity we bring together seemingly disparate contexts of struggles around racism built into digital crafting communities on the one hand and casteism in Indian society on the other, as they make evident the ruptures, complicities, and tensions in gendered South Asian digital publics.

We seek to highlight interactions between Indian women from India and South Asian women immigrants living outside of the region of South Asia. These latter groups are part of larger multiply geographically, historically/temporally tiered South Asian diasporas made of immigrant communities from that region. Transnational South Asians living in U.K., Europe, Australia, New Zealand, and particularly in the U.S. have come to self-identify as "brown" and as "people of color" (POC). Yet this category in itself elides histories of casteism within such immigrant communities while also potentially flattening and erasing histories of racism rooted in African American slavery and genocide of Indigenous communities.

We ask in this research: when and how—through Indian digital diaspora—might the coalition between the formations of "BIPOC" and "Anticaste" make political sense? These coalitions exist, but they need to be discussed through specific intersections and further nuanced as not all upper caste recent immigrant issues reveal caste struggles within South Asian diasporas. Thus, how do we read these different ruptures (fissures as well as coalitions) by taking caste into account? And where, in all these configurations, do different geographically situated indigenous communities fit in with their distinctive histories?

While access to the Internet and social media has been lauded as an advantage for people of all walks of life, high measures of inequality still exist online. This inequality includes not just the digital divide of who has access to what technology but a wide host of issues, including how people use the Internet. In looking at the South Asian digital diaspora, this chapter examines how race and caste feature in the way women of South Asian origin—both from the geographical region and with varied histories of migration from the region—operate in these social spaces with a focus on two paths of Internet usage: "digital domesticity" and "digital streets" (Gajjala 2019). We do this to differentiate more private, traditionally gendered and house-hold oriented digital publics from more public corporate, philanthropic, and/or social justice-oriented influencer and activist digital publics. We particularly focus on the nuances of South Asian digital publics around the politics of "brown" and "people of color" as mobilized by South Asians in diaspora—by discussing this in relation to hashtags such as #BIPOC (Black, Indigenous, people of color) and #anticaste.

Digital domesticity relies on a contemporary neoliberal, individualized, entrepreneurial ethos where such reproductive labor is sourced into the gig-economy through online sociality. Digital streets, on the other hand, are similarly situated in this neoliberal socio-political ethos, but they provide a context for the emergence of explicitly political subjectivities online. Online activism exists in both of these spaces and creates and negotiates transnational and global visibility for activists. South Asian feminist academics exploring these spaces must take into account the various historical and local contexts of the activist movements.

Positionality and Context

Women, people of color in the global north, and people of the oppressed castes in South Asia (among others) have historically been "deleted" as narrators of history and producers of knowledge because of the ways that dominant knowledge structures privilege the naming of authorship, despite feminists interventions. The coauthors' politics of location are therefore important and the four coauthors worked in dialogue to theorize with each other in the production of this chapter. Much of what is theorized in this article is based on each of our academic and personal life experiences through the layered identities that we inhabit. For instance, when we speak of queer, trans, Black, Indigenous, people of color (POC), women (white, cis, Indian, US citizen and so on), Dalit, and Bahujan-led hashtag activist movements in digital publics, we each enter and examine these through our own experiences of them as participants/audiences from our multiple social and geographical locations and from the standpoint of feminist researchers and/or activists. While we each have been (inter-generationally and across geography, race and caste locations) immersed in digital worlds and internet spaces for work and leisure, some of us claim the label of activist or researcher while others may claim both. Thus, we will be mentioning each of our social locations as relevant and noting any differences in standpoints in relation to particular themes even as the article is written in a collective "we."

Debates around who can and cannot theorize and represent Dalit, Bahujan, trans, queer, and other marginalized populations in India have been very visible in social media and social networking sites (Soundarrajan 2018; Kumar and Sharmishta 2019; Ayyar and Gajjala 2019) as more and more activist individuals and groups from these social locations emerge. These important debates rehash many of the points from feminist researchers of color since the late 1980s, where questions about "speaking for Others" (Alcoff 1992), being "under the western eye" (Mohanty 1986), and regarding subalternity and voice (Spivak 1999) were discussed in relation to mostly white feminist scholarship.

These call-outs from Dalit, Bahujan, Adivasi, Queer, and Trans folks among others are essential for demonstrating how Indian and transnational postcolonial feminists might aspire to be anti-caste. The debates are now globally visible through transnational digital publics with regard to the upper caste upper class who are called out as "Savarna" feminists,[1] a category one of our authors would come under. While our writing speaks to the histories of caste and other oppressions and incidences of these struggles as documented in existing literature, we ourselves are not claiming the position of researchers of caste or of queer and/or trans communities who can speak about or to the lived experiences of individuals in these communities.

We do, however, believe that researchers from upper caste locations within and outside South Asian academia as well as those situated in the western academy must carry the responsibility for by-passing these issues and failing to acknowledge how integral they are to the structural and everyday oppressions within present day South Asian communities, whether in India or elsewhere. The references to caste should not remain consigned to history where casteism is assumed to be only a premodern fact. Casteism is experienced every day and not even class mobility (as left leaning progressive upper caste intellectuals and activists had hoped) erases the everydayness of how it operates on the bodies and minds of those marked implicitly or explicitly as from untouchable caste and other minoritized and out-caste origins. As Thirumal and Christy (2018) have pointed out, even contemporary higher education institutions in India are steeped in an "Affective Economy of Caste." Whether through formal or more informal measures, such as procedures that delay Dalit-Bahujan scholarships, Dalit-Bahujan scholars are devalued in favor of the more "normalized presence of savarna aesthetic and taste" (Thirumal and Christy 2018).

Thus, we as scholars can look at specific hashtag publics through our own experiences of them, experiences which are not necessarily reflective of those involved in these spaces. We do this with a degree of suspicion directed at ourselves regarding each of our own location-induced oversight. We also draw on dialogues with a range of Indian activists and researchers, to whom we are highly indebted for taking the time to help us continue processes of unlearning (Gajjala 2019).

METHODOLOGY

We examine the dynamic scenario of hashtag activist publics along vectors of race, caste, gender, sexuality, and geography by foregrounding them in the tensions that emerge through encounters in multiple offline contexts of activism around issues impacting Indian women and women of Indian descent across Indian diasporas. The sites we focus on are particular hashtag publics formed around activist movements from India and those formed around Black, Indigenous, people of color fiber crafters (such as those who clearly identify as Black, Indigenous, and [South]Asian knitters or spinners or crocheters on Instagram or on Ravelry for instance) in global north regions. The gendered nature of fiber crafting means these digital spaces of knitters, crochets, and weavers (amongst other forms) means these spaces are often dominated by women. We draw on these two specific scenarios because the core problematics and theoretical issues in this chapter became evident through each of our immersion in either or both spaces as a highly invested participant. Both areas of interest connect in the overall focus on gender, labor, and affect in relation to digital mediation. There is much overlap in theoretical, social, and political concerns that emerge from each of the coauthors' participation and research within these sites.

This chapter is not an attempt to speak for or describe the subjective experiences of the activists themselves, nor is it an attempt to speak for or displace the voices of the groups of activists and the people they represent.

Neither is it our intention to separate the hashtags as objects from the speaking agent even as in later discussion we question individual claims of ownership of hashtags. The hashtags are integrally connected to the human communities and their causes. In the contemporary social media ethos they speak for themselves and their causes. Instead we look at the hashtags they use, the discourse these hashtags produce, and the digital publics that emerge from them.

The first author initiated the scraping of data using netlytics.org tools and brand24 tools in order to follow up on her digital ethnographic immersion writing about fiber crafters and people of color (Saxena 2019) before inviting the other three coauthors, all of whom claim different identity positions Collectively, the coauthors draw on activist experiences where some of them (particularly Vijeta and Sujatha) have used social media tools to create awareness and influence both local and transnational opinion around gender, caste, queer, and trans communities as well as on United States-based feminist research around race, trans, and queer communities. Others have conducted participant research in social media activist space. These investigations led us to look at issues that connect across contexts of activisms as the categories of activism and identity both connect and clash across contexts. When examining the terrain of social media we find there is both context collapse as well as encounters that are able to bring together seemingly unconnected contexts. In this chapter we draw on examples from two seemingly unconnected contexts—one from the gender activism of Indian digital publics on Twitter and one from the activism of Black, Indigenous, and people of color (BIPOCs) on Instagram. The different gendered experiences are closed connected demonstrations of how women are treated online and how they must navigate that treatment in order to make their voices heard.

GENDERED (INDIAN/INDIAN DIASPORIC) HASHTAG PUBLICS

In what follows we focus on specific hashtag events that are part of larger social justice movements that appeared in 2018 and 2019 hashtag publics (Rambukkana 2015). On the one hand the hashtag publics (such as #LoSHA, #stoptransbill2018, #metooindia, and others [Kumar and Sharmista 2019]) we discuss here have been brought into being by online protest movements from the geographical region of India, whether seemingly spontaneous or with careful planning and strategy on the part of activist and nonprofit groups. These movements may come from India or they are located in concerns based in Indian institutional contexts. Each movement serves to highlight intersections, tensions, and nuances of gender, caste, and queer identities that have always existed within Indian national social geographies but remain relatively invisible in transnational/global feminist theory and feminist activist discourse. On the other hand while hashtag publics such as #bipocknitters, #bipocspinners, #diversknitty, and many others mostly focus on global-north based issues around racism, diversity, and inclusion in fiber communities, they include participants based in the South Asian diaspora with varied histories of migration, and thus various social and political contexts around the world. These also raise questions regarding how "women of color" and "brown" have been mobilized in global north contexts at the same time as the post 1990s (neoliberalized and pro-Hindu fundamentalist) Indian diasporic politics have contributed to a renewed erasure of caste discrimination (Zwick et al. 2018; Rao 2018) that continues even in such immigrant and diasporic communities.

This chapter is both descriptive and contemplative as we think through emergent practices and the performative action in these digital spaces as they reveal how the production of affect draws on exhausting emotional labor on the part of the activists and influencers. The development of (tacit almost) instincts for affective interaction—where the triggering or the sourcing of affect towards specific goals requires both external performance of affect and of triggering particular affects to garner responses—are central to these strategies and tactics. Affective interaction and performative techniques/tactics for social media engagement thus are important in this space. This production of affect also relies on machine to machine communication enabled through visible and invisible software/app affordances, features, and architectures.

PRODUCING AFFECTIVE INTENSITIES: PERFORMING THE ACTIVIST VOICE

What we refer to here as "affective intensities" is based in what Karatzogianni and Kuntsman (2012) describe as the "affective fabrics of digital cultures" that are "the lived and deeply felt everyday sociality of connections, ruptures, emotions, words, politics and sensory energies" (3). These intense and ephemeral affective responses are material in how they shape and impact our daily life. They serve to mobilize people into reflexivity and action around certain causes, and digital communication networks have been crucial for the development of such a political voice. These "affective intensities are co-constituted as part of digital-material everyday spatialities" (Sumartojo et al. 2016, 38), but we note specifically how they are triggered through what Kuntsman (2012) refers to as the "digital terrains of politics" (2). Recent examples include rape in India, sexual harassment in the Indian academy, the various protests against trans discrimination bills in India, the beef ban (that implicitly targets Muslims, Christians, and Dalits), and the forced shut down and isolation of Kashmir, all of which serve to criminalize and oppress marginalized groups.

Yet this production of affect in digital publics is not new or unique to social justice-oriented internet spaces. The digital performance of demanding attention, instigating a response, and strategically calling upon others to consider the issue is in itself amoral. The affects triggering emotions of anger, anguish, and even self-righteousness embedded in these voices of outrage and contestation are not always or necessarily morally based in an egalitarian politics but may be based in the desire for attention or the social clout of calling awareness to an issue that may or may not relate to the voice demanding attention. This affect is, in Papacharissi's (2015) words, "the energy that drives, neutralizes, or entraps networked publics" (7). Most speaking tactics adopted by activists work to produce affective intensities and contribute to forming publics through speech acts that are "networked together, through affectively charged discourses about events that command our attention in everyday life" (Papacharissi 2015, 7). This is done through the adoption of discursive and performative strategies—some of which are standardized once they are seen to be effective.

Keeping in mind these affective dynamics of online engagement, we set up the context for each set of hashtag publics we refer to as "scene one" and "scene two" to emphasize the performative nature of these online publics even as they are embedded in our everyday life and use of digital gadgets. The focal point in both of the scenes is the interplay of the activisms by global-north based people of color in digital publics and the various India-based identity categories. This interplay of identities has implications for how we think of the layering of identities within transnationalized Indian digital publics.

In looking at particular activist hashtags on Twitter and on Instagram, it became clear that the transnationalism, coalitions, debates, and nuances produced in these spaces can invigorate and question the ways in which transnational and post-colonial feminist scholarship has engaged with issues around global south gender, queer, and feminist activism. These spaces may also implicitly elide the question of caste and other layered histories in discussion racial formations of women of color in global north contexts. In short, the issue of caste in transnational South Asian diasporas has remained invisible despite attention to other hierarchies and issues of identity.

Another issue that clearly emerged was around the complexity of "brown" as a racial category as well as around the model minority politics of "POC" (leading to the formation of "BIPOC," or "Black/Indigenous People of Color). Brown as a discursive and political descriptor of "race"—even as it is not legally recognized as "race" in any nation-state worldwide—does not just refer to South Asians (Prashad 2000; Silva 2016). Although brown has been a label adopted for South Asians, people of Latin descent, and/or people from the Middle East living in the U.S., the term was re-charged and has gained particular political valence leading to political aligning of people of color around brown skinned bodies in a post 9/11 era of brown visibility as potentially "terrorist." The lumping together of "brown" serves to erase the different lived existences of those grouped under the label and, because of the association with "brown" and "terrorist," has served to further dehumanize and malign entire groups of people.

Diverseknitty and Bipoc: Who Owns a Hashtag?

In this chapter we have selected particular hashtag events/encounters to flesh out the interplay between antiracism and anticasteism in digital publics. This discussion, while it is not always purely about the Indian or South Asian digital publics, is meant as a theoretical intervention to raise questions for further investigation in relation to South Asian digital diasporas. We urge fellow transnational/postcolonial feminists and international feminists located in various geographical regions to rethink what it means for us to be truly intersectional as we form coalitions as women of color in the global north while also coming to voice (academic or activist) through complex histories of privilege and loss. We do this for scene one below through a focus on activism around minoritized identity groups living in India. For scene two we draw on the voices of mostly South Asian descent—with varied historical itineraries away from "home" and then into the digital. The intention here is not to discount any of the activists in any of these settings but to reveal the larger tangle that we are made aware of when all these individual voices emerge in publics.

Scene One

Indian digital publics since the 1990s have tended to be dominated by South Asians/Indians in diaspora, with the politics and sociality visible in such spaces reflecting this digital population even while there were several Indians from India in digital spaces. Today, the questions that 1990s scholars and policy makers posed about access to technology in South Asia/India are not directly relevant anymore—no longer is the divide determined by a clear distinction between the connected and unconnected. Sociotechnical access to digital worlds, whether through leisure activities or because of work demands, has led to Indians from India being online alongside diasporic South Asians. In past work Radhika Gajjala (2012) has argued that this entry into the digital immersion *placed* these young workers in "digital diasporas" (95). Digital diaspora is thus not just how immigrants and diasporic populations use the Internet and associated digital tools to connect back to "home" regions while not being physical in the home region in order to form "imagined community" (Anderson 1983). The place they attach their affects to is a place that is both "real" in a physically geographical sense and imagined as in temporally frozen in a nostalgic past.

"Placement" in digital diasporas is enabled through digital existence and, while it is indeed a "locus around modes of interaction and imaginative practices" (Dasgupta 2017, 29), it also results in acts of placemaking. Implicit in articulation of Indian digital diasporas—which are multiple, varied and diverse—is a notion of place-making within digital infrastructures where the technical interface and socio-cultural contexts as well as layered literacies of access construct an architecture for the placement of affective-materiality through our everyday weavings in and out of the online and offline.

In our encounters online and as a result of this reorganization through human-machine interaction and digital infrastructures, digital place becomes inhabited. This notion of place-making is predicated on the everydayness of how we are immersed in and interact within online settings while we simultaneously also inhabit physical space. The conceptualization of place-making in this way relies on Doreen Massey's (1991) work on a "sense" of place. The notion of place-making in digital space and networks however is definitely *not* framed by the often rehearsed and tired binary of virtual as opposed to "the real." We use 2012 as a marker to signal the shift in Indian digital publics to what we've called "digital street" activisms (Gajjala 2015; 2019). This is not to imply there was no feminist, queer, or anti-caste activism in internet space directly from India prior to 2012 but to note the flurry of activity, speed, and the affective energy that can categorize social media and hashtag activity post 2012. The work these post-2012 hashtag publics are doing is less likely to be noted by mainstream Indian media or academic discussion in global north universities and relies upon Twitter activists to reach global audiences.

For instance, while Dalit, Bahujan, and Adivasi women's activisms in India have a long history and the Indian feminism movement itself has begun to acknowledge the need to take into account Dalit feminist interventions (see Rege 2006; Menon 2015; Guru 1995; Kannabiran and Kannabiran 1997), this awareness has not spread to transnational feminist publics or into diasporic Indian spaces with equal force. Much of the feminist curriculum in gender studies programs in the west/global north tokenizes upper caste brown feminist work and claim intersectionality of syllabi. Some transnational feminists have signaled the Dalit, Bahujan, and Adivasi feminist voices in specific instances (Dutta 2018; Loomba 2009; Banerjee 2014) and pointed to how the Indian women's movement's nationalist origins are problematic in regard to those marginalized along caste, region, and sexuality and transgender identities. Yet there is a difference between talking about caste as a third-world issue and talking about caste as a transnational issue, which may erase layered histories of migration, caste, class, and religion.

It is the labor of activists, allies, and influencers in digital feminist, queer, and anticaste campaigns that occur through hashtag publics that serves to internationalize and trans-nationalize the problematics of how postcolonial feminism has articulated the "women's question" (Chatterjee 2010) in an implicitly "upper"[2] caste Hindu Indian context while erasing the issue of the "Dalit woman's question" (Arya and Rathore 2019) and eliding intersections of trans, queer, and caste issues. While some Indian feminists have worked in collaboration with or in tension with Dalit feminists, women of color feminists of South Asian descent writing in global north academic settings have been less likely to reflect on the layered contradictions around their mostly "upper" caste locations and how they work through their status as women of color is shaped by their status as coming from upper caste privilege historically. For instance, "brown" is more easily homogenized as a racial category[3] defining—among many others including Hispanic populations—the experiences of immigrant and diasporic women of South Asian descent. Caste is less visible or acknowledged as a category of difference to be noted while addressing concerns around social justice and oppression of South Asian women transnationally. While such critique has been ongoing within Indian gender and caste activist communities for some time, what we are pointing to is that these issues surface as "new" in digital publics and open up further possibilities for critique and collaboration transnationally than in previous times. The consciousness of such tension is being produced by the hashtag publics enabled through the labor of a select few activists from marginal social categories in the Twittersphere. Current generations hailing from Indian and South Asian origins—whether living in India or diasporic—are bound to encounter these contestations when hanging out in digital space. It is the labor—physical, affective, technical, and social—of social media using activists and influencers (whether Black, Indigenous, Dalit, Bahujan, queer, trans, disabled, and many others) that serves to create global awareness of histories of oppression and contemporary nuances of intersectional identity politics that have been implicitly and explicitly suppressed in a majority of work around postcolonial feminisms, transnational feminisms, and global feminisms.

Digital Streets

The conceptualization of "digital streets" in this work has its origins in email exchanges with Radhika Gajjala in early 2015. Being continually networked through the use of digital gadgets in young people's everyday life plays a role in the spread of feminist, queer, trans, and anticaste consciousness, among others. What was happening was that several young women were beginning to develop and perform their political subjectivities through engagement with each other and the outside world through these digital gadgets, such as cell phones (Srila 2014). In earlier research one of the coauthors had come across the production of global youth cultures in digital environments that reach various local physical geographies, but she had assumed, based on those ethnographies, that much of what resulted from "hanging out" in digital networks was leisure related or related to professional growth. Around 2012—perhaps as a consequence of the activism around the Delhi Rape protests—there was a visible increase in outspoken feminist language from young women in India in digital publics. This move to locate within region and nation activism into digital publics serves to transnationalize critiques of dominant Indian nationalist narratives by placing them on a global stage.

Scene Two

The mostly global-north based DIY fiber crafting scene online has a fairly long history, possibly going back to the 1980s and 1990s; however, the history we briefly narrate here starts by contextualizing these online communities from approximately 2005 when a knitting renaissance exemplified by "not your grandma's knitting" (Fields 2014) as well as "stitch and bitch groups" (Stoller 2003) were becoming visible through online podcasts and blogs. As we describe this scene as a digital public, it is important to keep in mind that the online publics are formed by various geographically located fiber crafting groups and communities in several regions across North America, Europe, Norway, Iceland, UK, Australia, and New Zealand with an occasional participant from mostly metropolitan global south regions. These online publics and communities are integrally linked to the offline fiber meets, local yarn shops (LYS), and various crafting guilds as well as to domesticity traditionally considered to be women's space of work and leisure. We turn to events in this fiber crafting spaces to talk about how the label of "brown" works to bring together many generations and several kinds of immigrant and diasporic bodies into a category for forming alliances.

In these locations where whiteness (Harris 1993) dominates the everyday habitus, a majority of the people of color experience some form of signaling from those who consider themselves to be the core citizenry that they (the people of color) are "outsiders." This sort of marginalization is rooted in histories of nation formation in these various regions. We argue they are all part of the global north ethos of digital domesticity that is integrally intertwined with neoliberal entrepreneurship based on the promise offered by platforms/startups such as etsy.com; that individuals expressing themselves through the creative arts and crafts can "make a living, making the things they love," while the online platform serves "to connect makers with buyers from around the world" (Adam 2011, 1).

Scholars have noted how Instagram has aided bloggers in furthering their entrepreneurial ventures (Abidin 2014) and how the "cultural power of the mamasphere" (Wilson and Chivers 2015, 232) has shaped labors of domesticity and child-reading are shared, put on display, and narrated through digital packets of information. Domesticity in such a scenario is not that of the idealized and glamorized American housewife as such domesticity also comes with the pressures and responsibility of also becoming "mamaprenuers" who manage economic risk and precarity through home-based entrepreneurship. Further, as Jessie Daniels' (2012) work shows, the feminist underpinnings of women bloggers is at odds with the capitalistic pressure to make their digital presence marketable in such a way that often requires a reinforcement of traditional roles as mothers and wives. The "mamasphere" then models a certain kind of white aspiring upward mobility. Even when people of color participate it is likely to appear white in the digital sphere and through textual/discursive space.

Similar to the "mamasphere," the feminized spaces of fiber crafters on and offline do include some men, queer people, and trans persons, as well as people of color (particularly those coming from cultures that see fiber crafting as an expression of cultural identity), and people with disabilities. It was not until 2016 and a public intensity of activity on Instagram became visible that the presence of marginalized persons in digital fiber crafting spaces become evident as these people began to write about their disabilities, illnesses, sexuality and gender. The predominant discourse, pre-2016, seemed mostly "white" in spaces such Etsy, the popular fiber crafting site ravelry.com, and the larger blogosphere, even when people of color were in these portals.

It was around 2015 and later that the formations of people of color coalitions occurring through hashtags became clearly and distinctly visible as POC fiber crafters began to connect across ranges of geography and various situated persons. The surge of awareness in experience through hashtags such as #BIPOCknitters, #BIPOCmakers, #BIPOCspinners, #Blackpeopleknit came about when, in 2019, a fiber crafter (of Indian origin but located in Australia)[4] visibly expressed her outrage over a blog post by a well-known and high profile fiber-crafting entrepreneur geographically located in the U.S. Apart from and along with the hashtags accentuating #BIPOC members of the DIY global north oriented crafting scene, there was one hashtag that kept popping up on several of those who spoke of various kinds of diversity in the fiber community. This was #diverseknitty, started by a self-identified gay male who goes under the public Instagram handle of "Sockmitician."

The issue of diversity and the debates around the exclusion of BIPOC and/or protests against racism in the knitting community became visible among the fiber community's digital publics around January of 2019 through a *Vox* article by a self-identified "mixed race Indian." This article details the events in which blogger Karen Templer, a white woman fiber/knitting entrepreneur known for her products and designs, described her Indian friend's parents' invitation to join them on a trip to India as equivalent to being offered "a seat on a flight to Mars." The author of the *Vox* article "The Knitting Community is Reckoning with Racism" notes this post was "for every conversation a white person has ever had with me about their "fascination" with my dad's home country" (Saxena 2019). However, the affective intensities protesting this characterization and labelling it racist took on momentum after a comment on the blog post called out Karen Templer's characterization as "feed[ing] into a colonial/imperialist mindset toward India and other non-Western countries." While Templer has since apologized via a blog post analyzing why and how her previous blog post was inappropriate, the event triggered outbursts and conversations around racism in the fiber community. People of color in global north geographical locations started to narrate their experiences of being treated differently or outright badly in their local yarn stores. The hashtags #BIPOCknitters and #diverseknitty were used to build community and raise consciousness around the issue of people of color and how they were treated by mostly white entrepreneurs. Instagram particularly was alive with these discussions. Even though #diverseknitty was attributed as being started by a white gay knitter, several of the Black, Indigenous, and people of color fiber community members started to use the hashtag along with variations of the #BIPOC. Offline, in various fiber market events, the awareness of racial discrimination was being made public and several entrepreneurs who had online and offline presences started to put up diversity policies and make statements of inclusion in their social media posts as well.

There was also defensiveness on the part of several white fiber crafters who felt that their peaceful knitting space was now being politicized. There was policing and criticism that all these people who were "bringing politics" into the knitting community were taking away the focus from the actual craft and the peace within the fiber community. Several white fiber crafters claimed that they felt bullied. At least one such crafter/entrepreneur devoted a whole episode of her podcast/YouTube video to talking about how Instagram had become a hostile space and how there was so much bullying going on in Instagram. It was clear from such posts, podcasts, and even some articles on known alt-right websites that the alt-right community did have a presence in the fiber community. Later, in the summer of 2019, in response to various kinds of racist and hate groups–related activity on Ravelry.com (which has sometimes been described as a Facebook for knitters and other fiber crafters), the website came up with a policy to ban support of designs such as the "MAGA" (Make America Great Again) hats and discussion around them, characterizing them as fostering hate. They explained that they "cannot provide a space that is inclusive of all and also allow support for open white supremacy," noting in a full explanation that users who supported the President's administration could still use Ravelry, they just "can't talk about it here" (Ravelry 2019).[5] This was followed by proclamations of #Istandwithravelry, later #Isupportravelry, due to the ableist language of "I stand." However, much to the shock of several allies and BIPOC in the fiber community, the person who began the #diverseknitty movement posted his disapproval of Ravelry's policy in poem form on an Instagram post and a blog post. This in turn caused great outrage amongst the fiber crafters of color and their allies.

While this scenario itself deserves much more elaboration and discussion it is not in the scope of the current chapter to go into further detail here.

CONCLUSION

In continuing research on the two projects—diversity in fiber crafting communities and digital domesticity and the project looking at activism through hashtag publics from India—we also see the interlacing nuances, contradictions, and tensions around the ideas of "POC" and "anticaste"—and the #BIPOC ideas which show these terms change in different locations and lives. These ruptures reveal how postcolonial/transnational feminist work that maps implicitly through a Nehruvian and Gandhi Indian nationalist framework of dealing with the (cis) "woman's question" and eliding the "caste question" has been limited by caste, class, and religion.

The discussion that we opened up here in this chapter is meant to highlight the need for nuance and great care in how we as academics describe the onsite coalitions of activists working towards a common goal while staying alert to various historical and local contexts. This analysis reveals a shift in how activists strategically reach for the global/transnational "theater" (Spivak 1996 invoking Spivak's article) from not only early times of internet activisms that harks back to the time of the Zapatista movements and their western internet spokespeople such as Harry Cleaver but also from the ways in which local issues came to transnational notice purely through mainstream media or through niche academic writing. An understanding of differently— based in material and historical trajectories—socio-historical situatedness of hashtag projects is necessary. BIPOC struggles are just as distinct and contextual as anticaste feminist struggles.

And thus, to return to one of our first question: is it possible to be in both struggles—#POC and #Anticaste—or in just one? Similarly, as Dalit and Bahujan feminists have noted upper caste/upper class "WOC" and "POC" also tend to elide caste issues or even contribute to explicitly casteist projects. It might be that #BIPOC and #anticaste coalitions connect more clearly as they represent populations in both the South Asian context and in the

Euro-American contexts who are historically and institutionally marginalized.

NOTES

1. Savarna refers to, "in Sanskrit, literally 'those with Varna.' Thus, the term refers to members of the Caste System, and especially those in the higher-ranking Varnas." B. R. Ambedkar, "The Annihilation of Caste," in Dr. Babasaheb Ambedkar: Writings and Speeches, vol. 1(Bombay: Education Department, Government of Maharashtra, 1979), 25–96.

2. Because the term *upper caste* reinforces a hierarchy that assumes that some castes are better by birth than others, we have used scare quotes to denote "upper" as a referent but also as a problematic referent.

3. Savarna refers to, "in Sanskrit, literally 'those with Varna.' Thus, the term refers to members of the Caste System, and especially those in the higher-ranking Varnas."

4. Although this critique was initiated by some others at the actual blog site it was amplified significantly by the outrage expressed by several others and this particular fiber crafter of Indian descent was one of several of these.

5. https://www.ravelry.com/content/no-trump.

REFERENCES

Abidin, Crystal. 2014. "#In\$tagLam: Instagram as a Repository of Taste, a Burgeoning Marketplace, a War of Eyeballs" in *Mobile Media Making in an Age of Smartphones*, edited by Marsha Berry and Max Schleser. New York: Palgrave Pivot. doi:https://doi.org/10.1057/9781137469816_11.

Adam, Timothy. 2011 *How to Make Money Using Etsy: A Guide to the Online Marketplace for Crafts and Handmade Products*. Wiley. Kindle Edition

Alcoff, Linda. 1991. "The Problem of Speaking for Others." *Cultural Critique* 20 (Winter): 5–32.

Ambedkar, B. R. 1979. "Riddles in Hinduism." In *Dr. Babasaheb Ambedkar: Writings and Speeches, Vol. 4*. Bombay: Education Department, Government of Maharashtra.

Ambedkar, B.R 1979. "The Annihilation of Caste." In *Dr. Babasaheb Ambedkar: Writings and Speeches Volume 1*, 25–96. Bombay: Education Department, Government of Maharashtra.

Anderson, Benedict. 1983. *Imagined Communities*. London: Verso.

Arya, Sunaina and Rathore, Aakash Singh. 2019. *Dalit Feminist Theory: A Reader*. London: Taylor & Francis.

Ayyar, Varsha & Khandare, Lalit. 2013. "Mapping Color and Caste Discrimination in Indian Society." In*The Melanin Millennium: Skin Color as 21st Century International Discourse*, edited by Ronald E. Hall, 71–95. New York: Springer.

Banerjee, Amrita. 2014. "Race and a Transnational Reproductive Caste System: Indian Transnational Surrogacy." *Hypatia* 29(1). doi:https://doi.org/10.1111/hypa.12056.

Balaji, Murali. 2014. "Not Caste In Color: Dispelling Myths in Our Classrooms." *Huffington Post,* January 23, 2014.

Chatterjee, Partha. 2010. "The Nationalist Resolution of the Women's Question" in Empire and Nation: Selected Essays, 116–35. New York: Columbia University Press.

Clough, Patricia. 2012. "War by Other Means: What Difference Do (es) the Graphic (s) Make?" In *Digital Cultures and the Politics of Emotion: Feelings, Affect and Technological Change*, edited by Athina Karatzogianni and Adi Kuntsman, 21–32. Basingstoke: Palgrave MacMillan.

Gajjala, Radhika and Yeon Ju Oh *Cyberfeminism 2.0*, edited by. New York: Peter Lang Publishing.

Dasgupta, Rohit K. 2017. *Digital Queer Cultures in Indian: Politics, Intimacies and Belonging*. London: Routledge India.

Dey, Adrija. 2019. "Sites of Exception: Gender Violence, Digital Activism, and Nirbhaya's Zone of Anomie in India." *Violence Against Women*. August. doi:10.1177/ 1077801219862633.

Dutta, Debolina. 2018. "Of Festivals, Rights and Public Life: Sex Workers' Activism in India as Affirmative Sabotage." *Australian Feminist Law Journal* 44(2): 221–43.

Fields, Corey D. 2014. "Not Your Grandma's Knitting: The Role of Identity Processes in the Transformation of Cultural Practices." *Social Psychology Quarterly* 77(2): 150–65.

Gajjala, Radhika, 2012. *Cyberculture and the Subaltern: Weavings of the Virtual and Real.* Lanham: Lexington Books.

Gajjala, Radhika. 2015. "When your Seams Get Undone, Do You Learn to Sew or to Kill Monsters?" *The Communication Review* 18(1): 23–26.

Gajjala, Radhika and Ayyar, Varsha. 2019. *Digital Diasporas: Labor, Affect in Gendered Indian Digital Publics.* London: Rowman and Littlefield International.

Guha, Pallavi. 2017. "Mind the Gap: Connecting News and Information to Build an Agenda Against Rape and Sexual Assault in India." PhD diss., University of Maryland.

Gupta, Dipankar. 2004. "Caste is not Race: But, Let's Go to the UN Forum Anyway" in *Caste, Race and Determination*, edited by Suhkadeo Thorat and Umakant. Jaipur, India: Rawat Publications.

Guru, Gopal. 1995. "Dalit Women Talk Differently." *Economic and Political Weekly,* October 14, 1995.

Harris, Cheryl. 1993. "Whiteness as Property." *Harvard Law Review* 106 (8): 1707–91.

Kannibaran, Vasantha and Kannibaran, Kalpana. 1997. "Looking at Ourselves: The Women's Movement in Hyderabad," in *Feminist Genealogies, Colonial Legacies, Democratic Future* edited by M. Jacqui Alexander and Chandra Talpade Mohanty. New York and London: Routledge.

Karatzogianni, Athina and Kuntsman, Adi. 2012. *Digital Cultures and the Politics of Emotion: Feelings, Affect and Technological Change.* Basingstoke: Palgrave MacMillan.

Kumar, Vijeta and Sharmishta. 2019. "If 'Untouchability' at Sabarimala Makes You Angry, Then Welcome to the World of Dalit Women.."*News 18,* January 7, 2019.

Loomba, Ania. 2009. "Race and the Possibilities of Comparative Critique." *New Literary History* 40(3): 501–22. The Johns Hopkins University Press.

Massey, Doreen. 1994. "A Global Sense of Place." in *Space, Place and Gender*, 146–56. Minneapolis: University of Minnesota Press.

Menon, Nivedita. 2015. "Is Feminism About Women: A Critical View on Intersectionality from India." *Economic and Political Weekly* 50(17): 37–44.

Misrahi-Barak, Judith, K. Satyanarayana, Nicole Thiara. 2019. *Dalit Text: Aesthetics and Politics Re-imagined.* United Kingdom: Taylor & Francis.

Mohanty, Chadra Talpade. 1984. "Under Western Eyes: Feminist Scholarship and Colonial Discourse." *boundary 2* 12(3): 333–58.

Noble, Safiya. 2018. *Algorithms of Oppression: How Search Engines Reinforce Racism.* New York: New York University Press.

Papacharissi, Ziza. 2015. *Affective Publics: Sentiment, Technology, and Politics.* New York, NY: Oxford University Press.

Parameswaran, Radhika & Cardoza, Kavitha. 2009. "Melanin on the Margins: Advertising and the Cultural Politics of Fair/Light/White Beauty in India." *Journalism & Communication Monographs* 11(3): 213–274.

Pawar, Urmila and Moon, Meenakshi. 2014. *We Also Made History: Women in the Ambedkarie Movement.* New Delhi: Zubaan.

Platt, Louise C. 2019. "Crafting Place: Women's Everyday Creativity in Placemaking Processes." *European Journal of Cultural Studies* 22(3): 362–77.

Prashad, Vijay. 2000. *The Karma of Brown Folk.* Minnesota: University of Minnesota Press.

Punathambekar, Aswin & Mohan, Sriram, 2019. *Global Digital Cultures: Perspectives from South Asia.* Ann Arbor: University of Michigan Press.

Rambukkana, Nathan. 2015. *Hashtag Publics.* New York: Peter Lang.

Rao, Pallavi. 2018. "The Five-Point Indian: Caste, Masculinity, and English Language in the Paratexts of Chetan Bhagat." *Journal of Communication Inquiry* 42(1): 91–113.

Rege, Sharmila. 2006. *Writing Caste, Writing Gender.* New Delhi: Zubaan Books.

Roy, Srila. 2014. "New Activist Subjects: The Changing Feminist Field of Kolkata, India." *Feminist Studies* 40(3). 628–56.

Saxena, Jaya. 2019. "The Knitting Community is Reckoning with Racism." *Vox.* February 25, 2019.

Silva, Kumarini. 2016. *Brown Threat: Identification in the Security State.* Minneapolis: University of Minnesota Press.

Soundararajan, Thenmozhi. 2018. "Twitter's Caste Problem." *The New York Times.* December 3, 2018.

Spivak, Gayatri Chakrovorty. 1996. "'Woman' as theatre: United Nations Conference on Women, Beijing 1995." *Radical Philosophy* 75 (Jan/Feb 1996).

Spivak, Gayatri Chakravorty. 1999. *A Critique of Postcolonial Reason: Toward a History of the Vanishing Present.* Massachusetts: Harvard University Press.

Subramanian, Sujatha. 2015. "From the Streets to the Web: Looking at Feminist Activism on Social Media." *Review of Women's Studies, Economic & Political Weekly* 50(17): 71–78.

Subramanian, Sujatha. Interview with Radhika Gajjala. "Conversation with Sujatha Subramanian." *Cyberdiva's.* Podcast audio, July 13, 2019.https://anchor.fm/radhika-gajjala/episodes/ Conversation-with-Ms--Sujatha-Subramanian-e4k4vk.

Sumatojo, Shanti, Pink, Sarah, Lupton, Deborah, and Heyes, Christine LaBond. 2016. "The Affective Intensities of Datafied Space." *Emotion, Space and Society* 21: 33–40.

Stoller, Debbie. 2003. *Stitch 'n Bitch: The Knitter's Handbook.* New York: Workman Publishing Company.

Templer, Karen. 2019. "2019: My Year of Color."https://fringeassociation.com/2019/01/07/ 2019-my-year-of-color/.

Thirumal, P. and Christy Carmel. 2018. "Why Indian Universities are Places Where Savarnas Get Affection and Dalit-Bahujans Experience Distance." *Economic and Political Weekly* 53(5)https://www.epw.in/engage/article/why-indian-universities-are-places-where-savarnas-get-affection-and-dalit-bahujans.

Tuck, Eve and Waynge K. Yang. 2012. "Decolonization is Not a Metaphor." *Decolonization: Indigeneity, Education & Society* 1(1): 1–40.

Wilson, Julie and Emily Yochim Chivers. 2015. "Pining Happiness: Affect, Social Media and the Work of Mothers" in *Cupcakes, Pinterest and Ladyporn: Feminized Popular Culture in the Early Twenty-First Century*, edited by Elana Levine. Champaign: University of Illinois Press.

Zwick-Maitreyi, M., T. Soundararajan, N. Dar, R.F. Bheel, and P. Balakrishnan. 2018. *Caste in the United States: A Survey of Caste among South Asian Americans.* Equality Labs, USA.

Chapter Eight

Affect Amplifiers

*Feminist Activists and Digital
Cartographies of Feminicide*

Helena Suárez Val

We remember.

INTRODUCTION

As part of a long genealogy of street protests, performances, hashtag campaigns, advocacy, and multifarious online and offline actions, feminist activists across Latin America have been denouncing feminicide—gender-related violent deaths of women—by recording cases and creating digital cartographies of this form of violence. Such works have emerged in Ecuador, Mexico, and Spain, amongst others. My own project, *Feminicidio Uruguay,* which I started in 2015, maps cases in Uruguay to date. In this chapter, I present an examination of digital cartographies of feminicide as *feminist affect amplifiers*: digital artefacts through which data about cases—modulated through feminist knowledges, emotions, and affects—are recirculated in/to the world to create social change.

Feminist activists are well aware of the "enormous political potential" of public displays of grief "bound up with [. . .] outrage in the face of injustice" (Butler 2009, 39). In this sense, digital records and cartographies of feminicide are a form of research-creation where data about violence becomes public displays of feminist activists' emotional and affective—and political—responses to feminicide. Thus, the chapter locates the practices of recording, mapping, and visualising feminicide as a form of digital feminist activism that can be understood as part of a feminist affective politics, work-

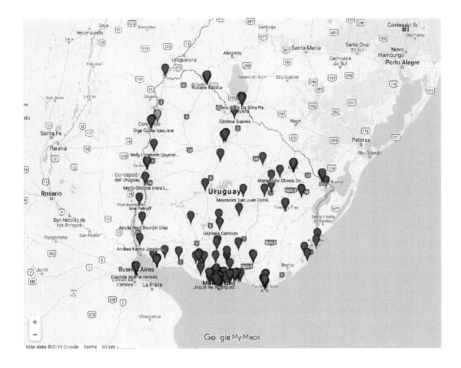

Figure 8.1.   Feminicidio Uruguay, an ongoing record of cases of feminicide created in Google Maps. Screenshot by author.

ing towards, or, to invoke Anderson (2006, 2009), hoping to end violence against women.

Situating the Research

Feminicide names the gender-related violent deaths of women, the tip of the iceberg in a continuum of violence that is "terrorizing women" in the Americas (Fregoso and Bejarano 2010). While there are several works from various disciplines addressing definitional and legal aspects of feminicide, including a focus on measurement and related art and activism, scholarship is only recently exploring the data and cartographic practices developing around the issue. Some exceptions include a collection entitled "Establishing a Femicide Watch in Every Country" (Hemblade et al. 2017); Engle Merry and Bibler Coutin's (2014) anthropological analysis of measurement technologies applied to violence against women; Chenou and Cepeda-Másmela's (2019) examination of data activism on gender violence in Argentina; Lan, Prado, and Vera's (2019) work on mapping spaces of fear; Marchese's (2019) critique of mapping feminicide in Mexico; and Goldsman's (2018) brief review of feminicide data visualizations. Meanwhile, scholarship on emotions and activism has studied the role of emotions in social movements and the emotional and political impacts of activism, but the affective practices of feminist activists recording and mapping cases of feminicide have not been studied specifically. Thus, this work contributes to scholarship on emotion, affect, and social movements and data activism by offering an approach to critically analyze the practices of feminist activists creating digital cartographies of feminicide.

Historically, feminist and women's movements in Latin America have persistently campaigned to end violence against women at local and regional levels. However, since Lagarde (2010, xv) translated Radford and Russel's (1992) "femicide" into Spanish as *feminicidio* (feminicide)—adding the particle "ni" to stress the gender aspect and frame gender-related murders of women as a human rights' violation—the term has "[brought] a unifying conceptual lens and discursive coherence to multiple national contexts in which women are being murdered due to their gender" (Bueno-Hansen 2010, 292). Manifesting as murdering[1] or otherwise causing the violent death of women and girls[2] or female gendered others, feminicide names the lethal end of a continuum of violence that women experience throughout their lives. As an evolving category, appearing as *femicide, feminicide,* or *feminicidal violence* depending on local or national realities, the term has empowered feminist activists to collaborate and support each other's activism, and campaign at a transnational level.

Lagarde (2008, 217) claims social silence, inattention, the idea that there are more urgent problems, and shame and anger that compel not to transform reality, but to minimize the issue dismissing "the dead" as not that many, all contribute to feminicide. She describes how mobilization against feminicide in Ciudad Juárez in the 1990s started because of the alarm about crimes against girls and women, with horror and consternation becoming denunciation and a demand for justice, and anti-violence groups emerging out of indignation (Lagarde y de los Ríos 2008, 209, 211). In Uruguay, it was concrete experiences of pain, inequality, and injustice that moved women into action against domestic violence in the 1990s (Clavero White 2012), and against feminicide most recently. These descriptions show how the innovative discourse that feminist activists take to the public sphere is not limited to academic or legal definitions: feminist activists protesting feminicide operate within a charged affective atmosphere that implicates emotions and affects, producing a range of political effects in the world.

Affective atmospheres are "singular affective qualities that express a certain world," creating "a space of intensity" through which represented objects, for example, feminicide, "will be apprehended and will take on a certain meaning" (Anderson 2009, 79). While bodies are "caught up" in them (Ahmed [2004] 2014, 222), atmospheres "are not necessarily sensible phenomena" (Anderson 2009, 78), meaning that not all bodies feel an atmosphere, "feel the same way about an atmosphere, or even feel an atmosphere in the same way," (Ahmed [2004] 2014, 218). Bodies must sensitize themselves to an atmosphere, must make sense of it. Attunement to an atmosphere "requires emotional labor" (Ahmed [2004] 2014, 224), a process not limited to accepting or rejecting prescribed "feeling rules" (Hochschild [1983] 2012), since atmospheres can be "circumvented and circulated," and they can be modulated: "[b]y creating and arranging light, sounds, symbols, texts and much more, atmospheres are 'enhanced,' 'transformed,' intensified,' [or] 'shaped'" (Anderson 2009, 80, following Böhme 2006). This notion invites us to trace the affective flows surrounding feminicide across the publics and counterpublics that make up the public sphere.

Segato (2010, 75) has proposed that feminicide be understood as a form of *expressive violence* that is enunciated along vertical and horizontal axes of communication. In cases of feminicide, the perpetrator's vertical, fatal address is to his victim, but "[t]hose who give meaning to the scene are other men"; horizontally, the perpetrator is addressing his peers, feminicide acting as a pledge to guarantee masculinity (Segato 2010, 76–77). However, in an opposite movement, feminicide, as an "empowered term" (Bueno-Hansen 2010), expresses a rallying call to feminist activists in Latin America. In a different sense, *to express* means to "represent (a number, relation, or property) by a figure, symbol, or formula" ("Express" 2019), while in Spanish an *expressive* oral, written, musical, or artistic manifestation is one that vividly conveys the feelings of the person who manifested themselves through it ("expresivo, va" 2019). Ahmed ([2004] 2014, 6) has analyzed how the *press* in the word *impression* facilitates "associate[ing] the experience of having an emotion with the very affect of one surface upon another, an affect that leaves its mark or trace." Similarly, if an atmosphere presses upon us (see Anderson 2009, 77), we can attempt to press back by *expressing* our thoughts or feelings. Joining these senses, we can conceive digital cartographies as attempts to press on the atmosphere surrounding feminicide, to modulate "affective and ethical dispositions through a selective and differential framing of violence" (Butler 2009, 1). Thus, figures and symbols on a map become digital marks or traces, expressing feminicide as gender-related violence, while also vibrantly expressing feminist activists' emotions and affects of grief, outrage, hope against violence against women.

Thinking affect and emotion through hope, Anderson (2006) suggests that "[b]eing political *affectively* must [. . .] involve building a protest against the affectivities of suffering into a set of techniques that also hope to cultivate 'good encounters' and anticipate 'something better'" (749). Joining this idea with Ahmed's ([2004] 2014) observation that "feelings do not reside in subjects or objects, but are produced as effects of circulation" (5), I propose that by learning and building a set of techniques for creating and circulating affective digital cartographies, feminist activists are being political affectively. By "shar[ing a] (multi-medial) vocabulary of 'emotions' or 'feelings'" (Schröer 2017, 154) about feminicide, activists come together as a feminist community of affect. Wetherell and Beer (2014) have noted the political relevance of "[a]ffective-discursive practices such as 'doing righteous indignation' or 'doing being the victim.'" Similarly, feminist activists "do protesting feminicide," an affective-discursive practice that includes creating and arranging symbols and texts about feminicide into maps and webpages. The resulting digital cartographies constitute affective data visualizations (be it markers on a map, entries on a database, or a webpage)—public displays of affect and emotion that tell the lives of murdered women and frame individual cases as part of a systemic pattern of violence against women, hoping to intensify and transform the affective atmosphere surrounding feminicide.

Methodology

During the *Primer Encuentro de Feminismos del Uruguay,*[3] held in Montevideo in November 2014, feminists at the gathering learned of another case of feminicide. After weeks of salacious speculation, the media reported that missing 15-year-old Yamila Rodríguez had been found, murdered by her sister's partner, who had been sexually abusing her. As the *Encuentro* ended, participants assembled in Montevideo's main square, to make a loud demonstration against feminicide, protesting: the media's misogynist portrayal of women as complicit in their own murders, social indifference, and government inaction. By January 2015, *Feministas en Alerta y en las Calles*[4] had been set up to coordinate a street protest following each case of feminicide. This group, of which I was part, organized logistics, made banners, distributed information on *Facebook*, and recorded details of each case on a collaborative *Google Spreadsheet*. As part of a range of actions for 25 November, *International Day to Eliminate Violence against Women*, I transferred thirty-three cases from the spreadsheet onto *Google Maps*, creating the first version of *Feminicidio Uruguay*.

While conducting this research, I kept updating the map with cases (and still do), each time understanding a bit more, though not completely, what I am doing and how . . . Immersed as I am, I started from autoethnography to critically examine my cartographic practices. I also sought other examples of activist digital cartographies or records of feminicide, to analyze the ways in which they were made, and their text descriptions and iconography. Finally, to share and compare my experiences, learnings, and insights, I set up semi-structured interviews[5] through *Skype* video with four feminist activists[6] doing similar works: Haydée Gallego,[7] who created *Quiénes eran,*[8] a memorial webpage listing cases of feminicide in Uruguay 2001–2014 (Centro Interdisciplinario "Caminos" n.d.); Ivonne Ramírez,[9] maker of *Ellas Tienen Nom-*

bre, an ongoing mapping of feminicide in Ciudad Juárez, Mexico (Ramírez n.d.);[10] Gabriela Ruales,[11] member of *Geografía Crítica,* a collective who made *Violencia Feminicida en Ecuador*, a map and infographic recording cases between 2014–2016[12] (Colectivo de Geografía Crítica del Ecuador 2016); and Alicia,[13] a volunteer with a Spanish-based NGO who maintains a feminicide database. All interviews took place between March and June 2017, but our activist conversations have continued since.[14]

FEMINICIDE AND AFFECTIVE ATMOSPHERES

An aesthetic object's atmosphere both "belong[s] to the perceiving subject" and "'emanate[s]' from the ensemble of elements that make up [the aesthetic object]" (Anderson 2009, 79). If we consider data visualizations as aesthetic objects, through making digital cartographies of cases of feminicide— ensembles of discourses, technology, design elements, and politics express- ing the lethal violence of gender relations—feminist activists attempt to ex- press what has remained invisible in the world, hoping to unsettle the prevail- ing atmosphere.

Figure 8.2.   Ni Una Menos street demo in Montevideo, Uruguay, June 3, 2015. Photo: Agustín Sorgin https://flic.kr/p/tQLDv8 (Creative Commons Licence).

Prevailing Atmospheres

Seven in ten Uruguayan women have experienced some form of gender-related violence, and nearly half who have ever been in a relationship reported being subjected to violence by their partner or ex-partner (CNCLVD and SIPIAV, 2013). Moreover, Uruguay ranks high in the rates of feminicide in Latin America and the Caribbean (see UN ECLAC n.d.), and according to official figures (2012–2016) just over two thirds of all women murdered in the country were killed in domestic or intimate-partner homicides (Gambetta Sacías and Coraza Ferrari 2017, 30). Yet, until feminist activists started loudly expressing the issue, most people found these facts surprisingly unexpected. Regarding violence against women, Uruguay seemed to enjoy a public sphere atmosphere of silence, denial, and ignorant bliss, increasingly subverted by feminist activists' affective interventions.

Feminicide has united Latin American feminist activists into continent-wide protests against this and all forms of violence against women. In Uruguay, feminists appropriated the term feminicide for street demos that started spontaneously towards the end of 2014, protesting gender-related murders of women, the lack of political will to tackle the issue, and the media's sexist and re-victimizing reporting. These demos took place after every case throughout 2015 (they still do), reaching peak momentum and cohesion as they converged with regional protests against feminicide on the streets and in social media. In June that year, outraged by a gruesome series of cases of feminicide, feminist activists in Argentina called for mass action. Activists across the region answered and replicated this call, agglutinated in the refrain *Ni Una Menos*, recovering the words of Mexican poet and activist Susana Chávez Castillo, protesting feminicide in Ciudad Juárez: *Ni una mujer menos, ni una muerte más*[15] (Jay Friedman and Tabbush 2016). The feminist atmosphere then was, and continues to be, electric with pain, rage, and struggle.

But the affective atmosphere around feminicide emerges in the interaction of multiple discourses: the voices of a transnational feminist counter-public, on the one hand, and media and political voices on the other. The political atmosphere around feminicide in Uruguay a few years ago could be described as apathetic: delays in debating bills and failures at the judicial and police levels. As Haydée—who recorded cases of feminicide in Uruguay between 2001–2014—exclaimed in one of our dialogues: "they are so lukewarm here!" Just as legal change (potentially) "reverberates throughout society and can be a vehicle for education and emancipatory change" (Bueno-Hansen 2010, 291; citing Acosta Vargas 1999, 623), so media narratives also significantly influence how a particular (criminal) issue, and its context and possible solutions, are understood (Sacco 1995, 142). Contrasting political lethargy, the media atmosphere overflowed sensationalism—each case a hot story, as exemplified in Yamila's case then and others still today. Other feminist activists recording and mapping feminicide in their countries described the local atmospheres in similar ways. For example, through tracing cases in Ecuador, Gabriela's team found inadequacies in state institutions' processes and noticed most news media reproduce stereotypes or victim-blaming narratives. Alicia highlighted the challenges of recording and making visible gender-related murders of women that exceed existing legal frameworks, and also how the media's lack of training on gender perpetuated outmoded understandings of the issue. Charged with different affects (sensationalism and apathy), nevertheless the atmospheres of politics, law, and the media were attuned, at least, with regard to denial and indifference towards the *systemic* nature of gender-related violence.

Against this pressing atmosphere, feminist activists are attempting to make an "atmospheric correction" (Schuppli 2013) around feminicide.[16] Moved by a desire to "repair" an incomplete image of feminicide and "recuperate" the stories of women, feminist activists employ various activist subterfuges or strategies—such as recording and mapping data—to make feminicide visible and shift, or modulate, the affective atmosphere surrounding it.

Atmospheric Modulation

Ash (2012, 6) applies the term *affective design* to describe the techniques and technologies of *affective amplification* that interactive videogame designers deploy to modulate affect, generating "emotional responses [which emerge] in relation to the particular biographies, contexts and social position of those being affected" (Ash 2012, 13). These notions help us conceive interactive digital cartographies of feminicide as *affect amplifiers,* where amplification is a two-fold procedure of augmentation and clarification (Ash 2012, 12). On the one hand, by designing and making these artefacts, feminist activists attempt to augment and clarify feminicide, making the issue (more clearly) visible by revealing the "bigger picture," now showing each case of feminicide located as part of a systemic pattern. On the other hand, they are also moved to augment and clarify affective and emotional responses to feminicide, a process which implicates "attempting to generate and modulate between affective states [. . .] intimately linked to the types of attention generated" (Ash 2012, 12). Digital cartographies, as *feminist affect amplifiers,* call our attention to feminicide in an attempt to shift the atmosphere from indifference and denial, to awareness, concern, and action.

Feminist activists develop careful processes to establish, adjust, and refine such criteria through feminist readings, conversations with other feminist activists, and/or by consulting legal frameworks at national or regional level. For example, Alicia described how her team reached agreement on classifying criteria during a training session held before starting to populate the database of cases in Spain, but that they had to adapt it as new cases appeared that stretched or challenged the boundaries of these criteria. In contrast, Gabriela explained how, since activists in Ecuador were mapping feminicide reported in the media during a specific past period, they defined the criteria after sorting through all the data. As well as clarifying their own criteria, feminist activists must interpret the data we gather from media, government, and/or other sources, often reading against the grain. Alicia illustrated this by explaining how her team choose to record cases that currently fall outside the predominant (and legal) understanding of *machista violence* in Spain, for example, female sex workers or trans women murdered by someone who was not their partner or ex-partner. In this case, what is modulated is the dominant definition of gender-related violence, feminicide as an activist term used to press against its limits.

One of the issues raised in conversation with other activists was the importance, and desire, to make the criteria for mapping explicit, in order to be "responsible for what [the map] says and its effects on the world" (Pavlovskaya and St. Martin 2007, 589) and to facilitate the map's readability. Clarifying what is being mapped is a crucial component of the map's capacity to press on the atmosphere surrounding feminicide. And although, anonymity for this type of work can be a necessity,[17] it is also important to "credit the bodies that make visualization possible—the bodies that collect the data, that digitize them, that clean them, and that maintain them" (D'Ignazio and Klein 2016, sec. 3.6). The issue of authorship is important because the media and political spheres seem more likely to enter discussion with feminist accounts of feminicide when made by known institutions or NGOs. For example, although I know a few journalists consult *Feminicidio Uruguay*, it is rarely referenced in reports on the issue, whereas Haydée's webpage, openly credited as the work of Caminos NGO, has prompted public conversations, including, she remembered, a debate with a prominent journalist who vehemently critiqued her criteria.

When it comes to feminicide, disagreements are still live, not only between feminists and the media and political spheres, but also within/between regional and local feminist counterpublics. Alicia, for example, explained how several feminist organizations in Spain started recording cases, leading to a sense of competition or confusion. Similarly, several groups in Uruguay publish their own accounts, producing discordant data (Suárez Val 2020). Contestation and duplication can sometimes cause bad feeling,[18] exposing the feminist movement to a potential loss of perceived legitimacy and thus hindering their attempts to modulate the affective atmosphere. Nevertheless, these divergences can also lead to further refinement and development of feminist concepts and strategies. Feminist activists making digital cartographies and other records of feminicide are not staking a claim to a single truth but proposing to question the *status quo* and affect other actors to pay attention and participate in a conversation about feminicide and about ways to eradicate this form of violence.

And the *status quo*, as represented in the political and media atmospheres, has actually shifted (slightly). In 2016, the female caucus of the Uruguayan parliament presented two bills on gender-related violence against women: a proposal to typify *femicide* as an aggravating circumstance to homicide and a comprehensive law to guarantee women a life free of gender violence, both eventually approved in 2017 and 2018 respectively (Parlamento del Uruguay 2017; 2018). Nevertheless, the latter, more substantive project underwent several "compromise" adjustments that were heavily criticized by human rights and feminist organizations ("Comunicado por Ley Integral que Garantice a las Mujeres una Vida Libre de Violencia de Género" 2017), and even at the time of writing activists continue to demand an adequate budget be allocated to implement the law (Montevideo Portal 2019). Feminist activists' critique of media portrayals of feminicide also affected media coverage, as evidenced in more articles contextualizing cases by referring to a total number of cases and/or using the word *feminicide* (with headlines such as "In 50 days, seven women murdered" or "Fourth feminicide in the year under investigation"), the term *femicide* more frequent since the approval of the new laws.[19] The media's emphasis on figures echoes some of the refrains of feminist actions, including digital cartographies of feminicide. In the next section, I look at this quantitative orientation, as I examine the feminist feeling rules for activism on feminicide.

FEMINICIDE AND FEELING RULES

In recent years, feminicide has been expressed in refrains repeated and amplified, online and offline, at feminist demos across the region: *Ni Una Menos*, *Vivas Nos Queremos* or *Tocan a Una, Tocan a Todas*.[20] Feeling rules are recognized by "inspecting how we assess our feelings, how other people assess our emotional display, and by sanctions issuing from ourselves and from them" (Hochschild [1983] 2012, 57). Although these rules are rarely made explicit, the refrains on a digital flyer calling a march protesting feminicide (see figure 8.3) do just that: issuing a strong feeling rules reminder to feminist activists, including those making digital cartographies of feminicide. The flyer condenses into simple refrains, and circulates, feeling rules that prompt the community of activists against feminicide to be touched by each case; to perform emotion work to transform pain into rage; and to allow these private feelings to transmute into public struggle (see Hochschild [1983] 2012, 160). If feeling rules prompt feminist activists to transform pain into rage into action, how is this transmutation enacted through the practice of making digital cartographies of feminicide?

Transmuting Pain and Anger

A refrain "allows an individual (a group, a people, a nation, a sub-culture, a movement) to receive and project the world according to reproducible and communicable formats," it constitutes a form of sense-making that materializes as rhythm: the "relation of a subjective flow of signs (musical, poetic, gestural signs) with the environment" that can "trigger a process of agglutination, of sensitive and sensible communality" (Berardi 2012, 33, 35). In the atmosphere where digital records and cartographies of feminicide are being

Figure 8.3.   NOT ONE LESS / Let pain become RAGE, / Let rage become STRUGGLE, / And our voice, a SHOUT: / THEY TOUCH ONE, THEY TOUCH ALL OF US! Digital flyer for Ni Una Menos march—June 3, 2017, circulated by the Coordinadora de Feminismos del Uruguay.

made, antithetical refrains reverberate in terrible rhythm: feminicide as expressive violence (agglutinating violent masculinity) but, also, feminicide as empowered term (agglutinating feminists as counterpublic and as community of affect).

Recording and mapping cases of feminicide is the way in which some members of the feminist community of affect respond to the "affective charge of investment, of being 'touched'" (Cvetkovich 2003, 49; in Karatzogianni and Kuntsman 2012, 3) by gender-related murders of women. Feminist activists transmute emotions into affective arrangements of digital traces of feminicide. Their commitment fueled by pain about each case of feminicide and anger at the treatment of cases by the media, inaction by politicians, and society's attitudes. In fact, the apparent disattunement between society and feminist emotions about feminicide—i.e., pain and outrage and action against feminicide are not widespread—provides a significant motivation.

Ivonne, for example, recalls a TV image imprinted in her memory, where the display on the faces of Prosecutor's Office operatives, talking to each other as if nothing happened, smiling at each other, did not match her idea of a correct emotional public display in the presence of feminicide. Haydée transformed into commitment her indignation at the ineptitude and lack of interest she perceived in prosecutors and judges. Alicia and I also discussed our anger at below-the-line and social media comments that reveal society's failure to identify cases of feminicide as part of a systemic context of violence against women, choosing instead to imagine perpetrators as either monsters or, conversely, loving men turned mad by jealousy or alcohol.

Practice and Self-Care

Recording data about feminicide is painful. Gabriela, who worked on the map but chose not to participate in data collection while breastfeeding, described her colleagues as "beat" and needing "release," after scrutinizing two years of Ecuador's main newspapers for cases of feminicide. We all highlighted the importance of emotional self-care practices during the mapping process, to sidestep saturation and burnout. Haydée sought other interests, while Ivonne found emotional support in doing the map while sharing space with her intimate partner. The Ecuadorian collective decided to run a self-care workshop after the project ended—a space for the collective, especially those closer to the data, to relieve themselves, as Gabriela put it, so that all that pain they had stayed with could be pushed out. However, while pain is part of the process, sometimes other emotions are transmuted into action.

Over time, feminist activists recording cases of feminicide might adjust to the rhythm and, as Alicia and I confessed, get used to the violence. As we become more skilled in performing the actions of collecting data and placing markers on the map, "the repeated traversal of the same movement figure" might eventually help feminist activists attain "an ease and lightness that allows [us] to become detached from what [we] continue to do" (van Eikels 2008, 94). And detachment, when counting murder, "could become a means for ensuring a continuation of engagement" (Ruse 2016, 228). It is evident from our conversations and my own experience that feminist activists care about the women in the cases of feminicide we record and map. However, to ensure continued engagement, doing the work correctly—doing a good job— can be as important as doing it with the "correct" emotion. Even if pain or anger do not always materialize, feelings of duty, but also pride and enjoyment in applying skills to do a good job, "hook" feminist activists to the practice, as Alicia put it. Ivonne and I coincided that the self-imposed commitment can feel impossible to get out of, the hook too tight. However, pride and enjoyment in our practice—evident in Haydée's pride at pioneering of this work in Uruguay while mastering difficult technical skills, or Gabriela's joyful encounter with geographic thought that, she said, opened a whole universe—also surfaced as we discussed our works, enthusiastically sharing knowledges and experiences.

Since Haydée started recording cases in 2001, the number of women killed in gender-related murders in Uruguay has remained regular year to year. On average, one to three women will be murdered because of their gender each month. Recording cases of feminicide is a practice with a regular rhythm: feminicide and feminist protest repeating, alternating refrains. Here I return to Berardi's (2012, 33) characterization of *refrain* as both "a sound, a sign, a voice, a song that makes it possible to link our existence with the existence of other human beings" and a notion pointing to "the relation between the rhythm of singularity and the chaosmotic rhythm of the world." The refrains *Not One Less* and *Touch One, Touch All of Us* proclaim feminist activists' link to each murdered woman, and locate the singularity of a case within a broader rhythm of gender violence and inequality in the social world. And to know how many "less" women there are and where they are missing from, quantitative data must be gathered, and shouted out.[21] Markers on a map, or names on a webpage, become signs that reproduce in a communicable format the refrain of feminicide, revealing and amplifying its rhythmical, regular flow through affective data visualizations.

FEMINICIDE AND AFFECTIVE DATA VISUALIZATIONS

In feminist digital cartographies of feminicide, politics enters beyond/before the image's representational content "at the structural level of its information acquisition, processing, and transmission" (Schuppli 2013, 21). From data collection, through the arrangement of data into interactive visualizations for circulation, to the viewer's encounter with the resulting artefact: a vision, as

Figure 8.4. Black tears on Spain's map of gender violence (Blázquez Carpallo 2017). Screenshot altered by author.

Ivonne put it, of feminicide. The description on a map of feminicide in Spain helps reveal how recording and mapping feminicide is a practice of "being political affectively" (Anderson 2006, 749):

> *I attempt to create a tear for every victim of gender violence. Men murder them, citizens appear to ignore it, institutions are overwhelmed, and politicians do not intervene as they should. This will go on . . . (Blázquez Carpallo 2017)*

Markers on a map displaying cases can be conceived as the product of emotion work: visible (upside-down) tears (see figure 8.4). Filled with marker-tears, data visualizations are a form of "open grieving" (Butler 2009, 39): public displays of sadness, pain and outrage, public tears opposing the ruling feelings of indifference and inaction apparent amongst citizens, institutions, and politicians. Such public tears express feminist activists' emotions and they also express, in symbols and texts, data about feminicide. Thus, these arrangements of public tears constitute affective data visualizations, specific configurations of symbols and text designed to transform, intensify, or shape affective atmospheres by amplifying, producing, and modulating affective personal and political responses to violence against women.

Embarrassing Data

> *I'm talking about figures, statistics . . . that sort of thing . . . and I find it embarrassing, somehow, to do that. I know it's important to do it, I know I have the commitment . . . and at the same time, I feel totally ashamed of doing it that way [. . .] Because, well, in the end, it's a map that will not change the situation. Right? Other things must be done. So . . . I'm in a crisis.*

Ivonne's crisis—shared by me and the others—highlights two key tensions. On the one hand, there is the feeling that collecting data and creating visualizations might not change the situation. But there is also the embarrassment of using quantitative methods: figures, statistics, maps.

Feminist critiques of quantitative methods have deconstructed and challenged "quantitative methods' claims of objectivity and their assumed legitimacy" (McLafferty 1995, 436). Quantitative methods are not exempt from subjective bias, yet by "hid[ing] their subjectivity carry more weight and influence than qualitative methods" (McLafferty 1995, 437). Of particular relevance to mapping is the feminist critique of vision, which has "specifically linked the authority of vision and the practices of looking to the patriarchal nature of Western societies" (Pavlovskaya and St. Martin 2007, 587). While visualizations (might) seem to pull the "god-trick of seeing everywhere from nowhere" (Haraway 1991, 189), because of the developing nature of the term feminicide and the reliance on often partial or inaccurate reports, Alicia, Ivonne, Haydée, Gabriela and I agreed that the data and therefore the maps are imperfect and incomplete, as acknowledged in *Feminicidio Uruguay*. Moreover, the use of digital methods could be "discriminatory and complicit with state and military surveillance" (Leurs 2017, 133) and their redistribution to corporate platforms could lead to "the privatization of social research" (Marres 2012, 140). Against such critiques, visualizations of incomplete data, on Google-provided, military-developed geographic information systems (GIS), must certainly be problematized.

Notwithstanding, feminist scholars, especially feminist geographers, have advocated research methodologies that *rehabilitate quantification* (Oakley 1998), *reclaim vision* (Haraway 1991), and *re-envision GIS* (Kwan 2002). Quantitative methods can provide a means "to shed light on embeddedness, the texture of everyday lives [. . .] a texture comprising multiple mediations of 'facts', 'truths', 'conditions' and 'power'" (Moss 1995, 447, 446). Additionally, despite ongoing critiques of this, policy-making and public funding bodies consider quantification as a "gold standard" (Hughes and Cohen 2010, 190). Feminist activists can (and do) take advantage of this bias by strategically deploying quantitative methodologies "to get the attention of policy makers and to make visible and intelligible the differences that had been rendered invisible, misinterpreted or missed" (Rocheleau 1995, 461–62). While using methods and platforms associated with masculinist, colonial and neoliberal agendas, digital visualizations of feminicide critically contest extant common senses about violence against women, creating and re-orientating knowledge and feeling. By creatively experimenting with digital environments, feminist activists are producing "paradoxical and critical feminist science projects" (Haraway 1991, 188) that sensitively and affectively reclaim and mobilize quantitative methodologies against feminicide.

Despite some skepticism, activists see the potential of combining "traditional forms of mobilization with those typical of data activism" (Chenou and Cepeda-Másmela 2019, 397). By showing the frequency, location, and specific details about each case, the cultural embeddedness of feminicide is revealed, as well as gaping holes in the state's role as guarantor of women's right to a life free of violence. As Alicia affirmed, maps give us the power to do more, providing data to do advocacy on what is going on and how to face it. Similarly, Haydée explained how her webpage, by exposing the blunt fact of the number of cases, could be used to reproach sceptics and "opiniologists" who deny feminicide is a problem or claim it is a problem invented by feminists. Still, feminist activists make efforts to avoid the "decontextualization and the distorting simplification" of quantification (Sellar 2015, 132), using it instead as "a tool, not to strengthen a claim or privilege a position, but rather to contextualize and ascertain structural relations of everyday life" (Moss 1995, 445). Showing each case of violence against women is "not one woman's account of a singular act; rather, it is a singular woman's account of an experience many women have as part of their everyday lives" (Moss 1995, 447). In this way, the data provides an affecting, blunt vision of feminicide. And feminist activists are well aware of the power of the visual aspects. As Haydée emphasized, referring both to data visualizations *and* street demos: What matters isn't just counting, it's making visible.

Visible Tears

As research-creation projects, digital cartographies of feminicide are advocacy tools that make visible data about feminicide, but they also make visible, remember, and grieve the lives of women lost to gender-related violence. As Haydée expressed:

> *These women . . . women who had lived, fought, dreamed, who had trusted . . .*
> *who were killed by the persons who they believed loved them and who they*

forgave so many times . . . I thought they deserved a place in the collective memory . . . let's say, not to go unseen, right?

Knowing how powerful it is "to state and show the name, to put together some remnants of a life, to publicly display and avow the loss" (Butler 2009, 39), feminist activists pair quantitative symbols for location and date and qualitative narratives stating each woman's name, a brief account of her life, and of her murder into public tears: open displays of pain and anger strewn on a map. The resulting ensemble and arrangement of symbols and texts constitute an aesthetic object—a visual artefact and technical environment with its own affective atmosphere. In this interactive environment, cases of feminicide are juxtaposed against the instantly and intimately recognizable shape of a country, bluntly revealing, to the viewer, feminicide's embeddedness within the politics, values and traditions of their own community.

As Alicia emphasized, feminist activists take meticulous care in finding the smallest detail, becoming almost intimately familiar with each woman after reviewing absolutely every report. Care is also taken to modulate the narrative from the original data source, correcting sexist, discriminatory, or sensationalist slants. For example, in *Feminicidio Uruguay* the active voice narrative ("he killed her" rather than "she died"), reframes the media's "lifeless" women or women who have "lost their lives," redirecting focus towards the murderer's responsibility. Care is also taken to highlight women's agency, and avoid victim-blaming, for example, pointing out when a woman had actively ended or was trying to end the relationship. In these ways, feminist activists honor the memory of the women, as well as making visible data about feminicide.

Activists also make conscious decisions regarding visual design, within limits enforced by the platform. They might choose the "neutral" default marker (red dot, or upside-down tear); a black marker or a cross, mobilizing (region-specific) associations with death and mourning; a pink cross, associated with activism against feminicide in Ciudad Juárez (see Mendoza 2017); a woman's silhouette . . . The choice depends on local context and how feminist activists conceive the constitutive parts of feminicide (see Luján Pinelo 2018). For example, I find silhouettes problematic: do these universal "women" include Indigenous women? Black? trans? urban or rural? Does the scattering of female figures not obscure the male murderers? I discussed with Gabriela how I avoided crosses and their religious connotations, as a secular standpoint but also rejecting the rise of religious (and conservative) fundamentalisms in Latin America. Gabriela noted that, after evaluating the maps, her collective will review their use of iconography and legends to ensure feminist principles are upheld. Haydée also carefully considered visual aspects. For instance, to convey the historical dimensions of violence against women, she deployed a "vintage" look, which connects the

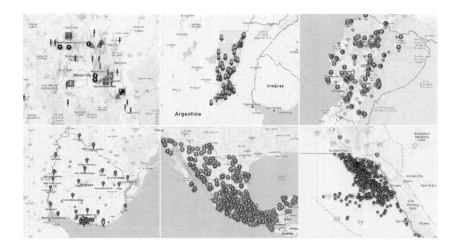

Figure 8.5.   Maps of feminicide in various countries and regions (screenshots taken July 31, 2017). Clockwise from top-left: Mexico state, Mexico 2004–2017 (ongoing) (Madrigal n.d.); Santa Fé, Argentina 2011–2017 (ongoing) (Piccolini 2016); Ecuador 2014–2016 (Colectivo de Geografía Crítica del Ecuador 2016); Uruguay 2017 (ongoing) (Suárez Val 2016); Mexico 2016–March 2017 (Salguero n.g.); 1985–2017 Ciudad Juárez, Mexico (ongoing) (Ramírez n.d.). Composite of screenshots by author.

murders of women today, understood through the feminist lens of feminicide, with the unexpressed suffering of "glamorous" women of the past who, exclaimed Haydée, must have gone through so much without knowing how to name it.

In this sense, digital cartographies of feminicide convey specifically feminist knowledge about the world (gender-related murders of women understood as feminicide) but also, hopefully, they generate an affective and emotional experience about feminicide that might move viewers from indifference and inaction, to pain, rage, and struggle. Through making specific compositions of signs and texts, feminist activists can "leverage affect in order to create an emotional bond with a story or issue, or to engage and impress" (D'Ignazio and Klein 2016, sec. 3.5). "Digital mapping interfaces invite us to touch, talk and move with them" (Lammes 2016, 1), conforming an environment where viewers can zoom in and out, click different markers, move from one location to the next, see different temporalities (cases in one year, two, more). The design and narrative choices made by feminist activists "anticipate and structure the user's interaction [. . .] construct[ing] the user as well as the interface" (Hayles 2002, 48). Such "techniques of affective modulation" (Ash 2012, 21) have the potential to impress viewers, generating affective responses that (hopefully) "construct users" into actively engaged participants in the struggle against violence against women.

GOING FORWARD

In this chapter, I have presented an exploration of the emotional and affective practices of feminist activists recording and mapping cases of feminicide in Ecuador, Mexico, Spain, and Uruguay. Through making digital cartographies of feminicide, feminist activists attempt to modulate, or "correct," the prevailing affective atmosphere of indifference and inaction. Guided by feeling rules that urge them to transmute pain and anger about feminicide into public actions, feminist activists gather, interpret, process, and circulate data about feminicide for a specific location, to make publicly visible the frequency, pattern, and range of the phenomenon, but also to publicly display grief and anger at the loss. Yet, the purpose is not solely to quantify feminicide or create memorials. Feminist activists understand the "enormous political potential" of public displays of grief bound with "outrage in the face of injustice and indeed of unbearable loss" (Butler 2009, 39). Through the affective practice of creating data visualizations marked with public tears, feminist activists engage in a politics that hopes to modulate the affective atmosphere surrounding violence against women. In this sense, digital cartographies of feminicide can be conceived as feminist affect amplifiers.

Returning to Ivonne's concern: can these affect amplifiers "change the situation"? It is evident from my own experience and conversations with other activists that viewers encountering these works do experience "effects at the level of affect or bodily intensity" (Sellar 2015, 131). Reacting to the map, Spanish-speaking viewers often call it *fuerte*,[22] which translates as "strong," but can also mean "affecting," "emotional," "upsetting," "shocking," "blunt." *Fuerte*, in another sense, also means "loud," invoking the notion of map as amplifier. Since "the success or failure of [modulating and amplifying affect] is often based on very small degrees of difference" (Ash 2012, 13), it could be argued that feminist activists should consider the under-explored affective dimensions of data visualization (D'Ignazio and Klein 2016, sec. 3.5), as these aspects impact the potential of digital cartographies of feminicide to provoke social change. Nevertheless, more work is needed to fully explore the political and affective impacts of digital cartographies of feminicide.

The concept of feminist affect amplifier is proposed in this work as an invitation and as a theoretical lens that could ground further research, to continue exploring how political affects reverberate in and through feminist data visualization practices that hope to help end feminicide and other forms of violence against women.

NOTES

1. Feminist authors in Latin America often adopt *murder*, in Spanish *asesinato*, as a gender-neutral alternative to *homicide*, disregarding its specific legal content (Toledo Vásquez 2009, 24 (footnotes); citing Monárrez Fragoso 2006, 361). In this work, I adopt this usage.
2. Some activists and authors include the murder of trans women as part of the definition of femicide/feminicide. New terms, such as *travesticide* or *transfemicide* (Bento, 2014; Berkins, 2015; Maffía, 2016), have also been developed to express the particularities of such violence.
3. *First Uruguayan Encounter of Feminisms.*
4. *Feminists on Alert and on the Streets.* This commission was part of the *Coordinadora de Feminismos UY,* an articulating/coordinating space for feminisms in Uruguay set up after the *Encuentro.*
5. Conversations were carried out in Spanish; all quotations are my own translations.
6. *They (female) Have Names.*
7. Gallego, Haydée. Interview by author. Skype, 22 May 2017.
8. *Who They Were*
9. Ramírez, Ivonne. Interview by author. Skype, 5 June 2017.
10. Since the works in this study share a conception of feminicide built on feminist theories, I use the term *feminist activists* as a catch-all. However, I want to note that other digital cartographers of feminicide self-identify in different terms. Ivonne, for example, self-identified as *feminist* when she started mapping, but now chooses *womanista*, to distance herself from white, Eurocentric feminisms. Haydée referred to her work as *militancia*, rather than *activismo.*
11. Ruales, Gabriela. Interview by author. Skype, 15 May 2017.
12. *Feminicidal Violence in Ecuador.*
13. "Alicia" (pseudonym). Interview by author. Skype, 5 March 2017.
14. Part of this ongoing conversation has resulted in a written reflection: *Monitoring, recording, and mapping feminicide—experiences from Mexico and Uruguay* (Madrigal et al. 2019).
15. Not one woman less, not one more death.
16. In satellite images, the term refers to the erasure of atmospheric phenomena or the correction of missing images due to technical glitches.
17. For example, María Salguero, who started mapping feminicide in Mexico in 2016 (Salguero n.d.), had originally worked under a pseudonym for safety reasons, although she eventually relinquished this anonymity (Personal communication).
18. Both Haydée and Alicia expressed negative feelings about "competition" between local feminist groups' often divergent records of feminicide. However, in Haydée's case this eventually provided an opportunity to step aside, when other NGOs started, as she put it, "doing it well."
19. Author's own media monitoring as part of Feminicidio Uruguay project.
20. Not One (woman) Less; We Want Ourselves Alive; Touch One, Touch All of Us.
21. Street demos in Uruguay in late 2014 and early 2015 often included shouting a count, a quantitative emphasis later abandoned because it was considered problematically reductive.
22. Personal communications.

REFERENCES

Acosta Vargas, Gladys. 1999. "La Mujer en los Códigos Penales de América Latina y el Caribe Hispano." In *Género y Derecho*, edited by Alda Facio and Lorena Fries, 424–72. Santiago de Chile: LOM.

Ahmed, Sara. 2014. *The Cultural Politics of Emotion.* 2nd edition. Edinburgh: Edinburgh University Press.

Anderson, Ben. 2006. "Becoming and Being Hopeful: Towards a Theory of Affect." *Environment and Planning D: Society and Space* 24(5): 733–52.

———. 2009. "Affective Atmospheres." *Emotion, Space and Society* 2(2): 77–81.

Ash, James. 2012. "Attention, Videogames and the Retentional Economies of Affective Amplification." *Theory, Culture & Society* 29(6): 3–26.

Bento, B. 2014. Brasil: Pais do Transfeminicídio [Brasil: Transfeminicide Country]. Centro Latino-Americano em Sexualidade e Direitos Humanos. June 4, 2014. http://www.clam.org.br/uploads/arquivo/Transfeminicidio_Berenice_Bento.pdf.

Berardi, Franco "Bifo." 2012. "Automation and Infinity of Language—Poetry versus Financial Semiocapital." In *Rehearsing Collectivity: Choreography Beyond Dance*, edited by Elena Basteri, Emanuele Guidi, and Elisa Ricci, 31–37. Argobooks.

Berkins, L. 2015. El travesticidio también es femicidio [Travesticide is also femicide]. *Página 12*. June 12, 2015.

Blázquez Carpallo, Puerto. 2017. "2017. Mapa de la violencia de género." Google Maps. España. https://www.google.com/maps/d/u/0/viewer?mid=1EX51lTKbhb5EQpENtrll3Ov pD0s.

Böhme, Gernot. 2006. "Atmosphere as the Subject Matter of Architecture." In *Herzog & de Meuron: Natural History*, edited by Philip Ursprung, 398–407. Thesis Eleven.

Bueno-Hansen, Pascha. 2010. "Feminicidio. Making the Most of an 'Empowered Term.'" In *Terrorizing Women: Feminicide in the Americas*, edited by Rosa-Linda Fregoso and Cynthia Bejarano. Durham; London: Duke University Press.

Butler, Judith. 2009. *Frames of War: When Is Life Grievable?* London: Verso.

Centro Interdisciplinario "Caminos." n.d. "Quiénes eran . . ." Who they were . . . Caminos. Accessed 22 November 2018. http://www.caminos.org.uy/quieneseraninicial.htm.

Chenou, Jean-Marie and Carolina Cepeda-Másmela. 2019. "#NiUnaMenos: Data Activism From the Global South." *Television & New Media* 20 (4): 396–411.

Clavero White, Carolina. 2012. "Mujeres Uruguayas Contra La Violencia Doméstica: Una Mirada Retrospectiva Sobre La Sociedad Civil Organizada." Uruguayan women against domestic violence: a retrospective look at organised civil society. *Distintas Latitudes*. May 8, 2012.

CNCLVD, and SIPIAV. 2013. "Primera Encuesta Nacional de Prevalencia Sobre Violencia Basada En Género y Generaciones." First National Survey on the Prevalence of Violence based on Gender and Generations. Uruguay: CNCLVD; SIPIAV. http://www.inmujeres.gub.uy/innovaportal/file/33876/1/resumen_de_encuesta_mides.pdf.

Colectivo de Geografía Crítica del Ecuador. 2016. "Violencia Feminicida Ecuador a 2016." Google Maps. Ecuador. https://www.google.com/maps/d/u/0/viewer?mid=1GJGbiCYk TZUS2ryoq3tCiW4txBs.

"Comunicado por Ley Integral que Garantice a las Mujeres una Vida Libre de Violencia de Género." 2017. Organizaciones de la sociedad civil uruguaya. http://www.cotidianomujer.org.uy/sitio/91-campanas/declaraciones/1726-comunicado-por-ley-integral-que-garantice-a-las-mujeres-una-vida-libre-de-violencia-de-genero.

Cvetkovich, Ann. 2003. *An Archive of Feelings: Trauma, Sexuality, and Lesbian Public Cultures*. Duke University Press.

D'Ignazio, Catherine, and Lauren F. Klein. 2016. "Feminist Data Visualization." In *IEEE VIS Conference 2016*. Baltimore.

Eikels, Kai van. 2008. "This Side of the Gathering. The Movement of Acting Collectively: Ligna's Radioballett." *Performance Research* 13 (1): 85–98.

Engle Merry, Sally, and Susan Bibler Coutin. 2014. "Technologies of Truth in the Anthropology of Conflict: Gender Violence and the Social Construction of Reality." *zeitschrift für menschenrechte*, Zeitschrift für Menschenrechte, no. Menschenrechte und Gewalt: 28–48.

"expresivo, va." 2019. In *Diccionario de la lengua española*, Edición del Tricentenario. Real Academia Española. http://dle.rae.es/?id=HL63TXL.

"Express." 2019. In *Lexico*. https://www.lexico.com/en/definition/express.

Fregoso, Rosa-Linda, and Cynthia Bejarano. 2010. *Terrorizing Women: Feminicide in the Americas*. Durham; London: Duke University Press.

Gambetta Sacías, Victoria, and Paula Coraza Ferrari. 2017. "Femicidios Íntimos En Uruguay. Homicidios a Mujeres A Manos de (Ex) Parejas." Intimate Femicides in Uruguay. Homicides of women at the hands of (ex)partners. Ministerio del Interior. https://www.minterior.gub.uy/images/2017/femicidios.pdf.

Goldsman, Florencia. 2018. "Datos y mapeo de feminicidios: visualizar y cuestionar." Feminicide data and mapping: visualising and questioning. *Conversaciones Feministas. GenderIT.org* (blog). 26 November 2018. https://www.genderit.org/es/feminist-talk/datos-y-mapeo-de-feminicidios-visualizar-y-cuestionar.

Haraway, Donna Jeanne. 1991. *Simians, Cyborgs, and Women: The Reinvention of Nature.* New York: Routledge.

Hayles, N. Katherine. 2002. *Writing Machines.* Mediawork Pamphlet. Cambridge; London: MIT Press.

Hemblade, Helen, Andrada Filip, Andrew Hunt, Marie Jasser, Fritz Kainz, Markus Gerz, Kathryn Platzer, and Michael Platzer, eds. 2017. *Femicide Volume VII: Establishing a Femicide Watch in Every Country.* Vol. VII. Femicide. Vienna: ACUNS Vienna Liaison Office.

Hochschild, Arlie R. (1983) 2012. *The Managed Heart: Commercialization of Human Feeling.* 3rd ed. University of California Press.

Hughes, Christina, and Rachel Lara Cohen. 2010. "Feminists Really Do Count: The Complexity of Feminist Methodologies." *International Journal of Social Research Methodology* 13 (3): 189–96.

Jay Friedman, Elisabeth, and Constanza Tabbush. 2016. "#NiUnaMenos: Not One Woman Less, Not One More Death!" *NACLA*, 11 January 2016. http://nacla.org/news/2016/11/01/niunamenos-not-one-woman-less-not-one-more-death

Karatzogianni, Athina, and Adi Kuntsman, eds. 2012. *Digital Cultures and the Politics of Emotion: Feelings, Affect and Technological Change.* London: Palgrave Macmillan UK.

Kwan, Mei-Po. 2002. "Feminist Visualization: Re-Envisioning GIS as a Method in Feminist Geographic Research." *Annals of the Association of American Geographers* 92 (4): 645–61.

Lagarde y de los Ríos, Marcela. 2008. "Antropología, Feminismo y Política: Violencia Feminicida y Derechos Humanos de Las Mujeres." Anthropology, feminism, and politics: Feminicidal violence and women's human rights. In *Retos Teóricos y Nuevas Prácticas*, edited by Margaret Bullen and Carmen Diez Mintegui, 209–39. España: Ankulegi.

———. 2010. "Preface: Feminist Keys for Understanding Feminicide: Theoretical, Political and Legal Construction." In *Terrorizing Women: Feminicide in the Americas*, edited by Rosa-Linda Fregoso and Cynthia Bejarano. Durham; London: Duke University Press.

Lammes, Sybille. 2016. "Digital Mapping Interfaces: From Immutable Mobiles to Mutable Images." *New Media & Society*, 1461444815625920.

Lan, Diana, Sabina Prado, and Sonia Vera. 2019. "Mapeo de los Espacios del Miedo de las mujeres en Tandil 8M 2018." Centro de Investigaciones Geográficas CIG; Instituto de Geografía, Historia y Ciencias Sociales IGEHCS.

Leurs, Koen. 2017. "Feminist Data Studies: Using Digital Methods for Ethical, Reflexive and Situated Socio-Cultural Research." *Feminist Review* 115 (1): 130–54.

Luján Pinelo, Aleida. 2018. "A Theoretical Approach to the Concept of Femi(Ni)Cide." *Philosophical Journal of Conflict and Violence* 2 (1): 41–63.

Madrigal, Sonia. n.d. "Distribución Geográfica de La Violencia Feminicida." La Muerte Sale Por El Oriente. Accessed 20 May 2017. http://soniamadrigal.com/lamuertesaleporeloriente/mapeo/.

Madrigal, Sonia, Ivonne Ramírez Ramírez, María Salguero, and Helena Suárez Val. 2019. "Monitoring, Recording, and Mapping Feminicide—Experiences from Mexico and Uruguay." In Femicide Volume XII: Living Victims of Femicide, edited by Helen Hemblade and Helena Gabriel, XII, 67–73. Vienna: Academic Council on the United Nations System (ACUNS). http://femicide-watch.org/sites/default/files/Femicide%20XII_0.pdf#page=73.

Maffía, D. H. (2016, November 30). Crímenes de género: Del femicidio al travesticidio / transfemicidio [Gender crimes: from femicide to travesticide/transfemicide]. Jornada de capacitación sobre "Crímenes de género: del femicidio al travesticidio / transfemicidio," Oficina de la Mujer de la Corte Suprema de Justicia de la Nación, Buenos Aires, Argentina. https://www.youtube.com/watch?v=b936aFQpWvg

Marchese, Giulia. 2019. "Mapeando feminicidios. Las ambigüedades cartográficas de los mapas de feminicidios." *colectivo ratio* (blog). 13 April 2019. https://www.colectivoratio.com/single-post/mapear-feminicidios-mexico-datos-violencia.

Marres, Noortje. 2012. "The Redistribution of Methods: On Intervention in Digital Social Research, Broadly Conceived." *The Sociological Review* 60 (December): 139–65.

McLafferty, Sara L. 1995. "Counting for Women." *The Professional Geographer* 47 (4): 436–42.

Mendoza, Elva F. Orozco. 2017. "Feminicide and the Funeralization of the City: On Thing Agency and Protest Politics in Ciudad Juárez." *Theory & Event* 20 (2): 351–80.

Monárrez Fragoso, Julia E. 2006. "Las diversas representaciones del feminicidio y los asesinatos de mujeres en Ciudad Juárez, 1993–2005." The diverse representations of feminicide and murders of women in Ciudad Juárez. In *Violencia infligida por la pareja y Feminicidio*, II:353–98. Sistema Socioeconómico y Geo-referencial sobre la Violencia de Género en Ciudad Juárez.

Montevideo Portal. 2019. "Intersocial Feminista: 'Desde que nacemos mujeres vivimos en Emergencia Nacional.'" *Montevideo Portal*, 16 September 2019. https://www.montevideo.com.uy/Mujer/Intersocial-Feminista--Desde-que-nacemos-mujeres-vivimos-en-Emergencia-Nacional--uc729975.

Moss, Pamela. 1995. "Embeddedness in Practice, Numbers in Context: The Politics of Knowing and Doing." *The Professional Geographer* 47 (4): 442–49.

Oakley, Ann. 1998. "Gender, Methodology and People's Ways of Knowing: Some Problems with Feminism and the Paradigm Debate in Social Science." *Sociology* 32 (4): 707–31.

Parlamento del Uruguay. 2017. "Actos de discriminación y femicidio." Acts of discrimination and femicide. Ley 19538. Parlamento del Uruguay. https://parlamento.gub.uy/documentosyleyes/leyes/ley/19538.

———. 2018. "Violencia Hacia Las Mujeres Basada En Género." Gender-based violence against women. Ley 19580. Parlamento del Uruguay. https://parlamento.gub.uy/documentosyleyes/leyes/ley/19580.

Pavlovskaya, Marianna, and Kevin St. Martin. 2007. "Feminism and Geographic Information Systems: From a Missing Object to a Mapping Subject." *Geography Compass* 1 (3): 583–606.

Piccolini, Alejandro. 2016. "Femicidios en provincia de Santa Fé (2011–)." Google Maps. Santa Fe, Argentina: Norma López. Concejala. https://www.google.com/maps/d/u/0/viewer?mid=17tuGTd-crQ3vCa429uh6W7D_eNk.

Radford, Jill, and Diana E. H. Russell. 1992. *Femicide: The Politics of Woman Killing*. New York: Twayne.

Ramírez, Ivonne. n.d. "Ellas Tienen Nombre." Accessed 23 April 2017. http://www.ellastienennombre.org/mapa-1.html.

Rocheleau, Dianne. 1995. "Maps, Numbers, Text, and Context: Mixing Methods in Feminist Political Ecology." *The Professional Geographer* 47 (4): 458–66.

Ruse, Jamie-Leigh. 2016. "Experiences of Engagement and Detachment When Counting the Dead for Menos Días Aquí, a Civilian-Led Count of the Dead of Mexico's Drugs War." *Journal of Latin American Cultural Studies* 25 (2): 215–36.

Sacco, Vincent F. 1995. "Media Constructions of Crime." *The ANNALS of the American Academy of Political and Social Science* 539 (1): 141–54.

Salguero, María. n.d. "Yo Te Nombro: El Mapa de Los Feminicidios En México." Web page. Accessed 2 August 2017. http://mapafeminicidios.blogspot.co.uk/p/inicio.html.

Schröer, Frederik. 2017. "Of Testimonios and Feeling Communities." Humboldt-Universität zu Berlin, Philosophische Fakultät III, Institut für Asien- und Afrikawissenschaften, Seminar für Südasien-Studien.

Schuppli, Susan. 2013. "Atmospheric Correction." In *On the Verge of Photography: Imaging Beyond Representation.*, edited by Daniel Rubinstein, Johnny Golding, and Andrew Fisher, 16–32. Birmingham: Birmingham Article Press.

Segato, Rita Laura. 2010. "Territory, Sovereignity, and Crimes of the Second State." In *Terrorizing Women: Feminicide in the Americas*, edited by Rosa-Linda Fregoso and Cynthia Bejarano, 70–92. Durham; London: Duke University Press.

Sellar, Sam. 2015. "A Feel for Numbers: Affect, Data and Education Policy." *Critical Studies in Education* 56 (1): 131–46.

Suárez Val, Helena. 2016. "Feminicidio Uruguay (2015–)." Google Maps. Uruguay. http://bit. ly/mapafeminicidioUY.

———. 2019. "Datos Discordantes. Información Pública Sobre Femicidio En Uruguay" (forthcoming). *Mundos Plurales*.

Toledo Vásquez, Patsilí. 2009. *Feminicidio: consultoría para la Oficina en México del Alto Comisionado de las Naciones Unidas para los Derechos Humanos*. México: OHCHR.

UN ECLAC. n.d. "Femicide or Feminicide." Gender Equality Observatory for Latin America and the Caribbean. Accessed 27 November 2018. http://oig.cepal.org/en/indicators/ femicide-or-feminicide.

Wetherell, Margaret, and David Beer. 2014. "The Future of Affect Theory: An Interview with Margaret Wetherell." *Theory, Culture & Society*, 15 October 2014. https://www. theoryculturesociety.org/the-future-of-affect-theory-an-interview-with-margaret-wetherall/.

Chapter Nine

Reproductive Justice and Activism Online

Digital Feminisms and Organizational/Activist Use of Social Networking Sites

Leandra H. Hernández and Sarah De Los Santos Upton

Over the past few years, news headlines have both highlighted and lamented reproductive and gendered violence in the United States and Latin America. As our research has analyzed (Hernández 2019; Hernández and De Los Santos Upton 2018; Hernández and De Los Santos Upton 2019), reproductive violence against women in recent years is evident in several problematic and gruesome case studies: lack of healthcare access for women throughout North America; women who were murdered because they were pregnant in Mexico (Knoll 2016); women who underwent "voluntary sterilization" in Venezuela because of economic, health, and political crises that prevented them from having more children under conditions free from political turmoil or poverty (Ulmer 2016); women who are being jailed in El Salvador for aborting a fetus conceived from a rape (Gies 2018); women throughout Latin America who are incarcerated because of miscarriage and stillbirth (Brice-Saddler 2019); and high maternal mortality rates for women of color throughout the United States, to name a few. These case studies are representative of what we term *reproductive feminicides*, issues of violence against women in reproductive contexts that range from lack of access to healthcare to the rape and murder of pregnant women in different contexts (Hernández and De Los Santos Upton 2018; Hernández and De Los Santos Upton 2019).

In response to reproductive feminicides more specifically and reproductive injustices more broadly, reproductive justice activists and organizations throughout the United States and Latin America have galvanized the use of social media and physical organizing to protest the 45th President of the United States' policies on immigration and reproductive control, among other reproductive and gendered types of violence (Mellen 2019). As research has illustrated, women of color reproductive justice activists have created a movement that redefines important reproductive activism terminologies (Price 2010), such as choice and justice; impacts reproductive justice activism and policy at local, community, state, and national advocacy levels (Price 2010; Ross, Gutierrez, Gerber, and Silliman 2016); and considers the transformative nature of intersectional feminist coalition building to inspire change (Chávez 2013; Cole 2008).

Within the context of intersectional feminist coalition building to fight for women's rights in reproductive spheres, we are interested in exploring the ways in which feminist reproductive justice organizations and activists utilize social media to communicate their organizational and activist goals and aims, to reach their target audiences, to organize and galvanize change, and to ultimately participate in larger efforts to eradicate reproductive and gender-based injustices. Thus, in this essay, building upon research from different theoretical areas such as intersectionality, coalitional politics, and social media/digital feminisms, we seek to explore the ways in which reproductive justice activist organizations, activists, and journalists utilize social media to advocate on behalf of women of color and reproductive justice. We examine digital feminisms and online assemblies via Twitter to explore different modes of collaboration to illustrate technologic reproductive justice feminism in real-time. We conduct a textual analysis of hashtag conversations surrounding #ProChoice, #ReproductiveRights, #ReproJustice, and #ReproductiveJustice. By examining digital reproductive justice feminisms via the official organizational Twitter accounts that emerged through our initial search of the outlined hashtags and extended hashtag conversations, we hope to provide a snapshot of current reproductive justice digital feminist activist efforts, persuasive appeals utilized to inspire change, and examine Twitter users' responses to these efforts, thus exploring the use of reproductive justice counterpublics' efforts in response to current misogynistic, patriarchal, and racist reproductive control legislation proliferating the United States and beyond.

REPRODUCTIVE JUSTICE, DIGITAL
INTERACTIVITY, AND FEMINIST ACTIVISM

As we mentioned above, this exploration of feminist reproductive justice organizational activism lies at the intersection of several different theoretical entry points. Thus, in this section, we first identify and describe the tenets of reproductive justice as are informed by and intertwined with intersectional coalitional politics. Then, we discuss digital interactivity and social media networking within feminist activism contexts.

Reproductive Justice & Intersectional Coalitional Politics

At its core, reproductive justice as a theoretical and activist framework seeks to ensure that women are supported in all reproductive endeavors, whether that includes a woman's right to have no children, a woman's right to have as many children as she pleases, and the fundamental right to parent and grow one's family in a context free from violence, institutional or governmental intrusion and intervention, and political or environmental hazards (Hernández and De Los Santos Upton 2018; Hernández and De Los Santos Upton, 2019; Ross 2017; Ross and Solinger 2017). Situated within a history of the larger debates between pro-choice and pro-rights activism, in the tradition of the Combahee River Collective, twelve Black women navigating the pro-choice movement in 1994 coined the term "reproductive justice" to "recognize the commonality of our experiences and, from the sharing and growing consciousness, to build a politics that will change our lives and inevitably end our oppression" (Ross 2017, 286).

Moreover, as reproductive feminist and anthropologist Lynn Morgan (2015) notes, SisterSong, a collective of US women of color, combined reproductive rights with social justice and coined the term "reproductive justice" to emphasize the need to support a spectrum of reproductive choices and contexts including parenting, not parenting, and having children in safe spaces, among other dynamics and rights. Furthermore, reproductive justice "decenters abortion and contraception to show how other issues—such as incarceration, immigration, racism, housing, and adoption policies—affect personal and social reproduction, including its non-procreative and eugenic forms" (Morgan 2015, 137). Ultimately, reproductive justice explains and interrogates intersections of race, class, and gender, thus eclipsing the pro-life/pro-choice binary, and examines the relationships between reproductive relations and forcefully imposed practices and policies as they shape women's reproductive experiences (Ross 2017). As Zavella notes, there are significant differences between reproductive health organizations and reproductive justice organizations:

Reproductive justice organizations distinguish themselves from those that only provide reproductive health services or those that focus on policy advocacy and instead strategically incorporate three foci in their movement building: (1) Base building includes organizing women of color and building constituencies that can be mobilized for demonstrations, organizational events, or lobbying; (2) Policy advocacy includes sponsoring legislation or advocating directly to local, state, or national legislators; and (3) Culture shift work involves critiquing negative representations of or discourses about people of color and promoting positive cultural images and messages. (2016 p. 39)

Ultimately, building upon the work of reproductive justice scholars and activists who have come before us (Ross, Gutierrez, Gerber, and Silliman 2016; Ross and Solinger 2017), we contend that reproductive justice is only achieved when "women of all backgrounds and walks of life have access to equitable and supportive health care throughout all phases of reproduction, ranging from the decision to reproduce or not have children through prenatal care, postnatal care, and more throughout the lifespan" (Hernández and De Los Santos Upton 2018, 42). Reproductive justice encompasses the following health contexts and considerations (and more): access to affordable fertility treatments, contraception, and abortion (Hernández and De Los Santos Upton 2018); the ability for women to make their own autonomous and empowered healthcare-related decisions without intervention or institutional legalities and restrictions (Hernández and De Los Santos Upton 2018; Ross, Roberts, Derkas, Peoples, and Bridgewater 2017; Ross and Solinger 2017); labor and birth free from problematic patient-provider interactions or unnecessary medical intervention (Craven 2010; Dubriwny and Ramadurai 2013; Mack 2016; Reed-Sandoval 2019); and the ability to mother and exist in spaces free from harassment, violence, or the threat of death (Hernández and De Los Santos Upton 2018, 2019).

Ultimately, reproductive justice is an intersectional enterprise precisely because of its focus on the interrelated identity categories (race, sex, gender, ethnicity, sexuality, physical ability, neurodiversity, and more) that converge to shape women's reproductive health experiences (Price 2011; Ross 2017; Zavella 2016). As Price notes, "The reproductive justice movement places intersectionality and human rights doctrine at the center of its mission and organizing activities with the belief that this new framework will encourage more women of color and other marginalized groups to become more involved in the political movement for reproductive freedom" (2011, S56). Research has found that women of color are dissatisfied with the contemporary pro-choice movement, do not identify with pro-choice messages and efforts, and ultimately assert that their reproductive needs have not been met (Price 2011). Thus, rooted in critical race theory and Black feminisms, intersectionality refers to the interrogation and analysis of women of color's experiences shaped by contexts like identity politics, anti/discrimination, social movement politics, and violence against women (Crenshaw 1990).

In an interview with Bello and Mancini, Crenshaw noted that "intersectionality attends to both the ways that categorization has facilitated and rationalized social hierarchy and to the institutional and societal structures that have come to reify and reproduce social power" (Bello and Mancini 2016, 15). In a reproductive health context, the institutional and societal structures that reify and reproduce social power refer to the cases we mentioned in the introduction, namely the stripping of women of color's reproductive rights; the impact of environmental toxins, pollutants, poverty, and crime on women's pregnancies and ability to raise children in safe and healthy contexts; and the ability to mother in contexts free from reproductive or gendered violence. As we have noted elsewhere, "In other words, women of color's experiences cannot be fully understood unless the outcomes of racism, sexism, and classism are seen as operating together" (Hernández 2019, 2). As Crenshaw's (1991) foundational research on intersectionality illustrated, women of color's subject positions are fundamentally different from those of white women, which necessitates an intersectional lens and frame to understand more thoroughly how women of color are "differently situated in the economic, social, and political worlds" (1250). In a reproductive justice context, we extend the functioning impact of interrelated outcomes of racism, sexism, and classism to also include gendered violence. Taken together, reproductive justice as an activist enterprise with organizations such as SisterSong and California Latinas for Reproductive Justice (to name a few) have utilized several frames, strategies, and tactics to achieve their goals in the interest of effective organizing, improve women's reproductive health experiences, and illustrate how reproductive justice activism can be deployed successfully at both local and national levels (Luna 2010, 2016).

Digital Interactivity & Feminist Activism

From a social media perspective, research has illustrated both the advantages and disadvantages associated with marginalized populations' use of social media. In this section, we utilize the term "marginalized populations" to refer to the fact that many reproductive justice organizations are both founded and operated by women of color. On the one hand, social media use can both enhance social support (Hanasono and Yang 2016) and activist efforts (Fritz and Gonzalez 2018; Jackson et al. 2017). On the other hand, marginalized individuals' use of social media can lead to harassment (Lawson 2018), discrimination (Jackson et al. 2017), and potentially death. The use of social media by marginalized populations, particularly in areas with high rates of violence against women, like Mexico and Venezuela, raises the concern of safety and security for women of color who speak out against the status quo surrounding violence against women. For example, Frida Guerrera, a Mexican journalist and activist, utilizes her journalistic and social media outlets to raise awareness about gendered violence against women in Mexico and throughout Latin America. However, the threat of retaliatory violence against her continually looms because of her activism.

In coalition building contexts, two key strategies to ensure the success of any campaign or organization include 1) the use of collaborative spaces where leaders and community members alike can work together to create equitable neighborhoods and decision-making contexts, and 2) an engagement in dialogue about how local and national policies impact reproductive health access, options, and experiences (Verbiest, Malin, Drummonds, and Kotelchuck 2016). When collaborative spaces and dialogue are coupled with the use of multi-modal technologies and approaches, then coalitional building efforts have the potential to reach larger audiences and communities in powerful ways. Although there is little research on reproductive justice organizational use of social media, existing research highlights the power of multi-modal technologies, which provides both promise for the use of social media specifically and concerns about its success concerning limitations about efficacy and safety. Zavella's (2017) research, for example, explored how the New Mexico Respect ABQ Women campaign shifted dominant narratives about pro-choice discourses, pro-life discourses, and reproductive justice through the use of an "innovative strengths-based, cross-sectoral approach" that included the use of storytelling and community mobilization. In this context, the New Mexico Respect ABQ Women campaign engaged with parties and multi-modal strategies not normally utilized in traditional feminist approaches, which was key to the campaign's success. Similarly, the California Latinas for Reproductive Justice (CLRJ) organization is "enmeshed in a national network of organizations working on reproductive justice for transnational subjects" (Zavella 2016, 42); more specifically, CLRJ collaborates with the National Latina Institute for Reproductive Health and the Center for Reproductive Rights, among others. CLRJ has also created El Instituto, a two-day event that is "part of a larger strategy to increase capacity for structurally vulnerable Latinas, informing them of their right to access health care" (Zavella 2016, 42). During El Instituto, women are provided educational trainings and health-related information from several partnering organizations through storytelling pedagogies, interactive exercises, self-reflexivity, and community advocacy (Zavella 2016). With an approach similar to the New Mexico Respect ABQ Women campaign, both CLRJ and the New Mexico Respect ABQ Women campaign utilize an applied reproductive justice praxis approach to community outreach, organizing, and activism that engages in and applies decolonial thinking, a process that transcends race, ethnicity and gender to understand human rights (particularly in health contexts) as a liberating praxis (Fregoso 2014; Zavella 2016). Thus, given the effective utilization of decolonial activist organizing strategies in reproductive justice contexts, the following research questions guide our analysis:

1. What sorts of frames can be utilized to understand reproductive justice social media discourses surrounding the hashtags #ProChoice, #ReproductiveRights, #ReproJustice, and #ReproductiveJustice?
2. In what ways do contemporary reproductive justice activism social media discourses replicate or diverge from earlier feminist debates and fissures between pro-choice and reproductive justice efforts?

Taken together, at a micro level we are interested in exploring the potentially transformative role of social media in feminist reproductive justice activist organizing; at a macro level, we are interested in uncovering trends associated with the deployment of technologic reproductive justice feminism in real-time as it is ensconced within larger feminist reproductive debates.

METHODS

Digital networks are important public spaces of assembly (Butler 2015), and for this study, we deliberately chose to explore the online, technologic digital assemblies of feminist reproductive justice organizations through a textual analysis of the Twitter hashtags #ProChoice, #ReproductiveRights, #ReproJustice, and #ReproductiveJustice, along with the hashtags associated discursive assemblies and groups of tweets. We argue that this collection of tweets serves as a digital archive of the ways in which digital feminisms serve as a "means of intervening in the world" (Taylor 2003, 15). As Price (2011) notes, intersectional and technologic approaches to reproductive justice are significant because of activist impacts on reproductive health policy and advocacy. In other words, as Price (2011) elaborates, an intersectional approach *must* move beyond the traditional reproductive health binaries illustrated by the pro-choice/pro-life dichotomy and its associated groups, such as Planned Parenthood, NARAL Prochoice America, and the National Right to Life Committee, among others. An intersectional approach:

> *Requires that we look at other groups that might not come immediately to mind, such as the Religious Coalition for Reproductive Choice, the National Black Women's Health Project, and the National Latina Institute for Reproductive Health* (emphasis ours). If a researcher is conducting a media analysis of reproductive health issues, it is important that she or he diversify the sources. The sample should not draw from mainstream newspapers and news magazines only (such as the New York Times and Time), but also from women's magazines and ethnic media sources (such as the Amsterdam News and Colorlines). Such sampling techniques can yield cultural, social, and political perspectives that might otherwise never surface. (S56)

Thus, the use of the hashtags #ProChoice, #ReproductiveRights, #ReproJustice, and #ReproductiveJustice allowed us to examine tweets from a broad range of organizational and individual accounts and diversify the tweets we draw from in our analysis. As we mentioned above, we recognize that public posts on social media sites can lead to harassment and discrimination, and we have therefore made strategic choices about how to refer to tweets and their authors in our analysis below. Because several Twitter users in our data set have Twitter handles that highlight their political stances on issues of reproductive health and reproductive justice, we made the conscious choice to refer to their tweets with these handles as we argue that they provide important context for uncovering and interpreting meaning in our analysis. In an effort to protect the identities of individuals, especially those who experience marginalization, we have removed Twitter handles that refer to authors using their first and/or last names. In order to better understand how these hashtags function in real-time, we chose to collect tweets for the three-month period immediately preceding the completion of this project. We recognize, however, that due to the dynamic nature of Twitter, these tweets represent only a snapshot of a larger textual body that is ever changing.

Analysis

A tweet by SisterSong (@SisterSong_WOC) on August 30, 2019 clearly summarizes the findings of our analysis. Sharing a graphic that lists differences between reproductive rights and reproductive justice, they state "Know the difference and do work that secures rights, but also ends oppression and advances justice and liberation." The graphic is set against a pink backdrop with an illustration of a red seedling at the bottom. Centered, in red capitalized letters, the word "reproductive" appears at the top, followed by "rights" on the left and "justice" on the right. Underneath rights is a list of characteristics in white letters including "started by white, middle-class women; narrow scope; focus on freedom to control own reproduction; seeks to secure legal rights; law-focused; and individualistic." Underneath justice the list of characteristics includes "started by POC and low-income folks; intersectional; focus on freedom to control own body and parent with dignity; seeks to dismantle barriers to access; and social justice focused, collectivist and structural." The bottom left hand corner of the graphic reads "learn more at sistersong.net" in all black, capitalized letters.

This graphic highlights what we overwhelmingly found in our analysis of the hashtags #ProChoice, #ReproductiveRights, #ReproJustice, and #ReproductiveJustice. Unsurprisingly, tweets with the hashtags #ProChoice and #ReproductiveRights largely focused only on the issue of abortion, and the hashtag #ProChoice especially was used to support both sides of the #ProChoice vs. #Prolife debate. #ReproJustice and #ReproductiveJustice on the other hand, covered a range of issues including abortion, respectful birth practices, obstetric violence, Black maternal health, gender inclusivity and trans rights, immigration detention, farmworkers rights, the climate, and violence against women living with HIV, to name a few.

Conversations surrounding the hashtags #ReproductiveRights and especially #ProChoice focused mainly on abortion, addressing the topic from several angles. Just as the graphic from SisterSong suggested, one of the most prevalent themes of tweets with these hashtags was a discussion of legal rights, policy, and law. For example, because our analysis focuses on the last three months, several #ProChoice tweets cover the defeat of Alabama's attempted ban on abortion. Sharing a story with the headline "Alabama abortion law temporarily blocked by federal judge" one user tweeted "#USDistrictCourt judge blocked #Alabama's #abortion ban—which didnt allow exceptions for victims of rape/incest & would make it a felony for doctors to perform procedure unless a woman's life was at risk. #news #WomenRights #politics #prochoice." The Twitter account W.H.O.R.E.S, which stands for "Women Helping Others Resist Exploitation & Sexism," shared the same headline with the hashtags #mybodymychoice and #prochoice. Another popular story during this time frame was testimony from the director of Missouri's State Health Department admitting to keeping a spreadsheet on the menstrual periods of Planned Parenthood patients. Sharing a story with the headline "Going Full Nazi: #Missouri Has Been Tracking the Menstrual Cycles of Planned Parenthood Patients," one user tweeted "WT Actual F, Missouri?!? This is real sick, authoritarian sh*t y'all are doing! Besides violating a person's privacy, it deprives the targeted patients their inherent right to bodily autonomy. I seriously hope those responsible for this face consequences. #ProChoice #HumanRights." These tweets focus on laws and policies that restrict abortion and police the bodies of individual women.

In addition to tweets celebrating advances in access to abortion, another prevalent use of the hashtag #ProChoice was from self-identified members of the "ProLife" movement who used the hashtag in two specific ways. First, #ProChoice was often used alongside #ProLife to draw attention to tweets from "ProLife" organizations and individuals. For example, on October 30, 2019 twitter user Rosary of the Unborn (@ProLifeRosary) posted:

> "#Abortion is a hideous sin against the Law of #Love" http://holylove.org #ChooseLife #ProLife #prochoice #AbortionIsMurder #AbortionIsA WomansRight (not) #AbortionIsNotaWomansRight #reprohealth #CatholicT-witter #Stand4Life #InformationIsTheBestProtection (not).

Here, the hashtag #ProChoice is deployed to draw attention to anti-choice messaging, in addition to other hashtags associated with the ProChoice movement such as #AbortionIsAWomansRight and #StandWithPP. The second way #ProChoice was used by "ProLife" individuals was to characterize members of the ProChoice movement as stupid, misinformed, and evil. For example, on October 31, 2019 the twitter user Exposing Corrupt Democrats #DeepState & "FakeNews" (@TrueNewsGlobal) posted:

> Common Sense FACT About Pro #Abortion Men: #ProChoice men just want to "USE" women for sex and forget them, thus damaging them psychologically. If they get women pregnant after using them for sex . . . they want to KILL their babies, thus further psychologically damaging the women.

Sharing an article about a man in the UK who screamed "You should have been aborted" to a girl in a wheelchair, the twitter user God is <3 (@UMatterToGod) posted "Sadly some people who are #prochoice would probably think this is okay: UK man screams at girl in wheelchair: 'You should have been aborted!'" One of the most widely shared "ProLife" uses of the hashtag #ProChoice during this time period was on Halloween, where several accounts shared the video of a person in drag with a pregnant belly who cut into the belly and pulled out a bloody doll. The user Red Pill Rooster (@RedPillRoosters) posted the video alongside the tweet "What in the actual hell . . . As disgusting as this may be, This might be a perfect portrayal of #ProChoice ppl. #ProLife" and user The Right News shared the video along with the message "They joke about it openly now, no longer with the safe, legal and rare lie, the more slaughter the better. . . . #prolife #prochoice."

Breaking away from the pro-life versus pro-choice paradigm, uses of the hashtags #ReproJustice and #ReproductiveJustice on Twitter covered a spectrum of reproductive justice topics, ranging from issues surrounding abortion and birth, to gender inclusivity, migrant rights, and HIV stigma. In addition to the SisterSong graphic described above, many tweets with the hashtags #ReproJustice and #ReproductiveJustice included definitions of reproductive justice and goals of the movement. For example, in a three-tweet thread on October 29, 2019, In Our Own Voice (@BlackWomensRJ) explained,

> #ReproJustice is the human right to control our bodies, our sexuality, our gender, our work, and our reproduction. That right can only be achieved when all womxn and girls have the complete economic, social, and political power and resources to make healthy decisions about our bodies, our families, and our communities. At the core of #ReproJustice is the belief that all people have: 1) the right to have children, 2) the right to not have children, and 3) the right to nurture the children we have in safe and healthy environments.

The hashtags #ReproJustice and #ReproductiveJustice were also used to discuss a spectrum of experiences surrounding birth, including respectful birth practices; obstetric violence, and Black maternal health. On October 30,

2019, a twitter user who identifies as a maternal and child health (MCH) analyst (@MCHequity) tweeted about respectful birth practices, "Defining #birthequity as part of #reproductivejustice" along with pictures of a presentation from the 2019 South Carolina Birth Outcomes Initiative Symposium, which included Joia Crear-Perry's definition of birth equity as "The assurance of the conditions of optimal births for all people with a willingness to address racial and social inequities in a sustained effort." This tweet exemplifies several reproductive justice tweets concerned with how to serve birthing people and families in respectful ways.

Obstetric violence was another common theme in #ReproJustice and #ReproductiveJustice tweets. Sharing a news story with the headline "Medical Students Regularly Practice Pelvic Exams on Unconscious Patients. Should They?" The Wrong Side of the Speculum (@twss_movement) tweeted "I don't even need to read this to know the answer is absolutely no, and if you can think of a single reason to justify this you're a peice of shit, and a rape apologist. End of story. #MeToo #wedeservebetter #ReproductiveJustice #reproductiverights." The twitter account identifies as "A community of people fighting to end abuse in the doctor's office and promote consent, accountability, patient dignity, and bodily autonomy for all." Other tweets about obstetric violence highlighted the intersections of maternal mortality and racism. For example, a "Black-owned anti-oppression consulting firm supporting organizations, schools, and communities in achieving intersectional equity and justice" shared the work of one of its co-founders in the following tweet "@JalessahJ explained how today's maternal mortality outcomes result in part from flaws in the medical system that can be linked to a history of eugenics and scientific racism." Another twitter user shared a March of Dimes tweet, again highlighting the intersections of maternal mortality, race, and class: "Sadly whether a woman like me who suffer from acute pre-eclampsia die as a result is often dependent on where she lives and the color of her skin. #MODMOMentum #MODAdvocacy #ReproductiveJustice." Rather than just calling attention to the systemic problems surrounding Black maternal health, several twitter accounts shared examples of resources and efforts to improve health outcomes. For example, ACTIONS (@ACTION-SIncubat1) shared the link to a podcast episode "In a @ScholarsStrategyNetwork Podcast, ACTIONS Intern, Rachel Applewhite, discusses the effectiveness of doulas and midwives and the promise they hold for improving Black maternal health: "https://scholars.org/podcast/promise-midwives?fbclid= IwAR3Ph8zoduviat5YAKotWxMnqwdBCM1soUBtQiRlMPgSVehc5E6O W9sWSn0 #reproductivejustice #doulas #midwives." Black Women's Health Imperative (@blkwomenshealth) shared a tweet about their reproductive justice delegation: "Late last week, BWHI hosted a dynamic delegation of African female leaders from 17 African countries, including Rwanda, Burkina Faso, Ghana and Kenya & focused on ways to empower black women and girls through lifestyle/wellness, reproductive justice & more. #BWHI #ReproJustice." These tweets serve as examples of how the hashtags #ReproJustice and #ReproductiveJustice are used to describe a spectrum of birth experiences through an intersectional lens and highlight problematic structures while also offering solutions.

The hashtags #ReproJustice and #ReproductiveJustice were also used to talk about gender inclusivity and trans rights. One author used the hashtag #ReproductiveJustice to share a news article about the need for gender inclusive reproductive health. Several accounts also discussed the need for gender inclusive discussions around menstruation. For example, sharing a story about Always's decision to remove the female symbol from their packaging, Equality Federation (@EqualityFed) tweeted "Yes, it's true, people of all genders menstruate. Thank you @Always! #Transgender #menstrualequity #ReproductiveJustice #lgbtq." Sharing a story about period poverty, URGE (@URGE_org) argued for the necessity of gender inclusivity: "A5: #ReproJustice means recognizing that abortion is not just a cissue. It means centering young Black and Latinx queer and trans people. It means knowing period poverty impacts trans & GNB ppl. https://teenvogue.com/story/period-poverty-transgender-and-gender-non-conforming-people . . . #EndVAWHIV." Finally, in a self-reflexive post, one Twitter user described the need for gender inclusivity in her own work: "Making an effort to catch myself when I use trans-exclusive language while talking about pelvic pain and reproductive disorders. I think this is a necessary step for me as a social worker and healthcare advocate. #ReproJustice." These tweets challenge followers to think past the gender binary that is problematically inherent to many conversations around reproductive health.

Several individuals and organizations used the hashtags #ReproJustice and #ReproductiveJustice to talk about the intersections of reproductive justice and migrant rights. For example, on October 29, 2019 a self-described single father, immigrant, lawyer, social fighter, and immigration consultant, tweeted a story entitled "Immigration Detention Is Dangerous for Women's Health and Rights" with the tweet "Immigration #detention is a barrier to #ReproductiveJustice. A new report from @CAPWomen examines abuses of the health & rights of women & girls in immigration custody and offers policy recommendations." In another tweet, one author shared a call to action, arguing for the need to pursue migrant justice as a reproductive justice issue: "Bioethics need to act about reproductive oppression at the US-Mexico Border @faBioethics @aBigMess #ASBH19 @AmerSocBioHum #bioethics #socialjustice #reproductivejustice." Several tweets went one step further to link other abuses against migrant bodies to reproductive justice efforts. On October 29, 2019, California Latinas for Reproductive Justice (@Latinas4RJ) used the hashtags #ReproJustice and #ThisIsRJ to share an article with the headline "Wine Country farmworkers stay in fields, can't avoid smoke — and face serious health risks." These tweets call attention to the intersections of migrant and reproductive justice, moving beyond a focus on birthing bodies to discuss the ramifications of injustice for whole communities.

Finally, on October 23, 2019, several reproductive justice organizations partnered for the Day of Action to End Violence Against Women Living with HIV. Using the hashtags #EndVAWHIV and #ReproJustice, Positive Women's Network USA (PWN-USA @uspwn) facilitated a Q&A session on Twitter. PWN-USA kicked off the conversation with the tweet: "This will be starting in about 90 minutes right here! Join PWN, @blkwomenshealth @WithoutViolence @ifwhenhow @thewellproject @URGE_org & more for a lively chat about the importance of #ReproJustice & #BodilyAutonomy in the fight to #EndVAWHIV! See you right here shortly!" and proceeded to tweet a series of 11 questions beginning with Q and a number. Responses then began with A and the corresponding number. For example, when PWN-USA (@uspwn) tweeted: "Q6. How is #ReproJustice for women & #TGNC folks living with #HIV impacted by #stigma & discrimination? What can we do to change that? #EndVAWHIV," If/When/How: Lawyering for Reproductive Justice (@ifwhenhow) responded: "A6) Increased policing, prosecution, and surveillance are real, especially for #TGNC folks living with HIV. Legal advocates must recognize this, push back, and make change while taking leadership from affected communities. #EndVAWHIV #ReproJustice." Centering the need for coalition building, PWN-USA (@uspwn) asked, "Q10. Working together is vital to #EndVAWHIV. What partnerships or coalitions have been formed or need to be formed to create lasting change? #ReproJustice." URGE (@URGE_org) responded, "A10) It's vital that there is cross-movement collaboration and education. We need all orgs dedicated to #reprojustice to continue to center and uplift the experience of folks living with HIV. #EndVAWHIV" and Futures w/o Violence (@WithoutViolence) added, "A10) #AntiVIolence #DomesticVIolence #SexualViolence folks need to be working in coalition with #ReproJustice #TransJustice #DecriminalizeHIV movements! #EndVAWHIV." Finally, when PWN-USA (@uspwn) asked if there was anything else their partners would like to share, URGE (@URGE_org) called attention to the importance of the chat itself: "A11. Culture shift!! We must continue to fight the stigma that is perpetuated both legally and culturally against folks living with HIV. That starts with education and advocacy. Tweetchats like this are one great way to shift culture. #EndVAWHIV #ReproJustice." These tweets demonstrate reproductive justice's emphasis on intersectional activism and coalition building.

Discussion

As we have discussed in earlier research (Hernández and De Los Santos Upton 2018), contemporary media and news discourses about women's reproductive health at a broader, macro level continue to spotlight debates about pro-choice and pro-life ideologies. For example, our earlier analysis of reproductive health news discourses spotlighted the prolonged silencing and erasure of reproductive justice voices, actions, and efforts in favor of a perpetuation of the larger pro-life/pro-choice binary (Hernández and De Los Santos Upton 2018). Similarly, our analysis here highlights that this trend persists in social media contexts, as well, between pro-choice women's rights organizations and specifically identified reproductive justice organizations. For example, we found that tweets with the hashtags #ProChoice and #ReproductiveRights emphasized the individual right to an abortion as the main issue, while the hashtags #ReproJustice and #ReproductiveJustice were used more broadly to highlight a spectrum of experiences for not only birthing people but also minoritized communities. The continued divide between pro-choice and reproductive justice activist approaches is reminiscent of earlier second wave feminist divides across color, race, and gender lines. In other words, the continued trend of separate, yet complementary reproductive health activist organizing efforts in digital social media contexts continues to highlight the fissures between pro-choice and reproductive efforts.

Furthermore, we contend that this discursive erasure of reproductive justice efforts in discursive and technologic spheres further contributes to fissures associated with reproductive coalition building, outreach, and praxis. As Ross illustrates so aptly:

> Praxis is a term most often used by oppressed groups to change their economic, social, and political realities through social justice actions based on theoretical reflections. Reproductive justice praxis puts the concept of reproductive justice into action by elaborating upon the connection between activism and intersectional feminist theory. Activists intentionally employ a complex intersectional approach because the theory of reproductive justice is inherently intersectional, based on the universality and indivisibility of its human rights foundation. (2017 p. 287)

If reproductive justice praxis is hinged upon the effective deployment of intersectionality and intersectionality as a theoretical and praxis lens is excluded from larger reproductive health technologic discourses, this is yet another limitation of the opportunity to create and extend coalitions among reproductive groups.

In previous work (Hernández and De Los Santos Upton 2018) we have discussed the possibilities of media literacy for responding to problematic news framing of reproductive injustices. One of the tenants of media literacy is the ability to create your own media (The Media Literacy Project 2014), meaning that "as audiences we can use media to both support and participate in activist efforts to combat violence against women in all its forms" (211). Social media platforms offer users vast opportunities to create their own media messaging, and this possibility carries with it possibilities for pushing back against the problematic framing of a number of issues surrounding reproductive health. Further, on Twitter specifically, the use of hashtags allows users to literally apply their own frames that then filter messages into searchable themes and in the cause of our study connect multiple users in conversations around the #ProChoice movement, #ReproductiveChoice, #ReproJustice, and #ReproductiveJustice. Through this self-filtering, followers are able to see the issues and arguments that cluster around these hashtags, offering a glimpse into what each movement is prioritizing at any given moment.

Ultimately, our present study extends earlier research (Hernández and De Los Santos Upton 2018) that found deep fissures associated with 1) pro-life and pro-choice discourses in the media, 2) the lack of a presence of discussions about reproductive justice efforts, and 3) reproductive justice efforts' continued focus on intersectional approaches to theory, methods, and praxis. Through our analysis of Twitter as a digital archive (Taylor 2003) of discourses about reproductive health and reproductive justice, we found that these fissures play out in similar ways in feminist digital assemblies (Butler 2015). The use of hashtags to group large numbers of tweets from both individuals and organizations demonstrates both the limits and the potential of #ProChoice and #ReproductiveJustice approaches. As exemplified in discussions of reproductive rights in the second wave of feminism (Hernández and De Los Santos Upton 2018), #ProChoice tweets limit the discussion to a debate around access to abortion. In addition, as we demonstrated in our analysis, the hashtag #ProChoice can be hijacked by anti-choice individuals and organizations to promote tweets that stand in direct opposition to the goals of reproductive rights activists. #ReproductiveJustice, on the other hand, creates a digital assembly that not only allows individuals and organizations to voice concerns, but also facilitates the sharing of information and resources in ways that have the potential to reach larger audiences and achieve reproductive justice in digital spaces and in communities on the ground. We argue that the hashtag #ReproductiveJustice not only emphasizes, but ultimately engages with, the need to have more inclusive conversations about reproductive justice, intersectionality, and coalition building.

REFERENCES

Bello, Barbara Giovanna and Letizia Mancini. 2016. "Talking about intersectionality. Interview with Kimberlé W. Crenshaw." *Sociologia del Diritto:* 11–21.

Brice-Saddler, Michael. 2019. "She was raped, had a stillbirth and almost died. Then she was charged with the murder of her fetus." *Washington Post*, August 19, 2019.

Butler, Judith. 2015. *Notes Toward a Theory of Performative Assembly*. Cambridge, Massachusetts. Harvard University Press.

Chávez, Karma R. 2013. *Queer Migration Politics: Activist Rhetoric and Coalitional Possibilities*. Champaign, IL: University of Illinois Press.

Cole, Elizabeth R. 2008. "Coalitions as a model for intersectionality: From practice to theory." *Sex Roles* 59(5–6): 443–53.

Craven, Christa. 2010. *Pushing for Midwives: Homebirth Mothers and the Reproductive Rights Movement*. Temple University Press.

Crenshaw, Kimberlé. 1990. "Mapping the margins: Intersectionality, identity politics, and violence against women of color." *Stan. L. Rev.* 43: 1241–300.

Dubriwny, Tasha N. and Vandhana Ramadurai. 2013. "Framing birth: Postfeminism in the delivery room." *Women's Studies in Communication* 36(3): 243–66.

Fregoso, Rosa-Linda. 2014. "For a pluriversal declaration of human rights." *American Quarterly* 66(3): 583–608.

Fritz, Niki and Amy Gonzales. 2018. "Privacy at the Margins| Not the Normal Trans Story: Negotiating Trans Narratives While Crowdfunding at the Margins." *International Journal of Communication* 12: 20.

Gies, H. 2018. The rape survivor facing 20 years in jail lays bare El Salvador's war on women. *New Statesman America*. November 14, 2018.

Hanasono, L. K., & Yang, F. 2016. "Computer-mediated coping: Exploring the quality of supportive communication in an online discussion forum for individuals who are coping with racial discrimination." *Communication Quarterly* 64(4): 369–89.

Hernández, Leandra Hinojosa. 2019. "Feminist Approaches to Border Studies and Gender Violence: Family Separation as Reproductive Injustice." *Women's Studies in Communication*: 1–5.

Hernández, Leandra Hinojosa and Sarah De Los Santos Upton. 2018. *Challenging Reproductive Control and Gendered Violence in the Américas: Intersectionality, Power, and Struggles for Rights*. Lanham: Lexington Books, 2018.

———. 2019. "Critical Health Communication Methods at the US-Mexico Border: Violence against Migrant Women and the Role of Health Activism." *Frontiers in Communication* 4: 1–12.

Jackson, Sarah J., Moya Bailey, and Brooke Foucault Welles. 2018. "#GirlsLikeUs: Trans Advocacy and community building online." *New Media & Society* 20(5): 1868–88.

Knoll, Andalusia. 2016. Pregnant women are the latest victims of Mexico's femicide crisis. *Vice News*. August 3, 2016.

Lawson, Caitlin E. 2018. "Platform vulnerabilities: Harassment and misogynoir in the digital attack on Leslie Jones." *Information, Communication & Society* 21(6): 818–33.

Luna, Zakiya T. 2010. "Marching Toward Reproductive Justice: Coalitional (Re)framing of the March for Women's Lives." *Sociological Inquiry* 80(4): 554–78.

———. 2016. "'Truly a Women of Color Organization' Negotiating Sameness and Difference in Pursuit of Intersectionality." *Gender & Society* 30(5): 769–90.

Mack, A. N. 2016. The Self-Made Mom: Neoliberalism and Masochistic Motherhood in Home-Birth Videos on YouTube. *Women's Studies in Communication* 39(1): 47–68.

Marrow, Jasmine. 2013. New Rankings Announced: Top 25 National Reproductive Health, Rights, and Justice Nonprofits. *GuideStar*. October 12, 2013.

Mellen, Ruby. 2019. "Mexico is taking steps toward legalizing abortion. But across Latin America, restrictions remain widespread." *Washington Post*. October 4, 2019.

Morgan, Lynn M. 2015. "Reproductive rights or reproductive justice? Lessons from Argentina." *Health and Human Rights* 17(1): 136–47.

Price, Kimala. 2010. "What is Reproductive Justice? How Women of Color Activists are Redefining the Pro-Choice Paradigm." *Meridians* 10(2): 42–65.

———. 2011. "It's not just about Abortion: Incorporating Intersectionality in Research about Women of Color and Reproduction." *Women's Health Issues* 21(3): S55–S57.

Reed-Sandoval, A. J. 2019. *Socially Undocumented: Identity and Immigration Justice*. Oxford, UK: Oxford University Press.

Ross, Loretta J. 2017. "Reproductive Justice as Intersectional Feminist Activism." *Souls* 19(3): 286–314.

Ross, Loretta, Elena Gutierrez, Marlene Gerber, and Jael Silliman. 2016. *Undivided Rights: Women of Color Organizing for Reproductive Justice*. Chicago, IL: Haymarket Books.

Ross, Loretta, Erika Derkas, Whitney Peoples, Lynn Roberts, and Pamela Bridgewater, eds. 2017. *Radical Reproductive Justice: Foundation, Theory, Practice, Critique*. New York, NY: Feminist Press at CUNY.

Ross, Loretta and Rickie Solinger. 2016. *Reproductive Justice: An Introduction*. Vol. 1. Berkeley, CA: University of California Press.

Taylor, Diana. 2003. *The Archive and the Repertoire: Performing Cultural Memory in the Americas*. Durham, NC: Duke University Press.

Ulmer, Alexandra. 2016. Venezuelan women reluctantly opt for sterilization amid economic crisis. *Huffington Post*. August 3, 2016.

Verbiest, Sarah, Christina Kiko Malin, Mario Drummonds, and Milton Kotelchuck. 2016. "Catalyzing a Reproductive Health and Social Justice Movement." *Maternal and Child Health Journal* 20(4): 741–48.

Zavella, Patricia. 2016. "Contesting Structural Vulnerability through Reproductive Justice Activism with Latina Immigrants in California." *North American Dialogue* 19(1): 36–45.

———. 2017. "Intersectional Praxis in the Movement for Reproductive Justice: The Respect ABQ Women Campaign." *Signs: Journal of Women in Culture and Society* 42(2): 509–33.

Chapter Ten

Racial Justice and Scholar-Activism

Angela Smith, Ihudiya Finda Williams, and Alexandra To

Increased conversations around Diversity, Equity, and Inclusion and racial inequality in higher education have made clear the need for racial justice in academia. Academic labor is often described through lenses of research, teaching, and service where efforts towards diversity are often relegated to service. In this chapter, we discuss the writing and reception of our 2020 paper, "Critical Race Theory for HCI," as a case study for research-based scholar-activism.

The writing of the "Critical Race Theory for HCI" paper continues a tradition of bringing in justice-oriented frameworks to human-computer interaction (HCI) research with novel focus on race, racism, and racial justice (Ogbonnaya-Ogburu et al. 2020). In 2010, Shaowen Bardzell outlined an agenda for "Feminist HCI" which identified points of entry for feminist ideals, values, and practices to integrate into interaction design research and practice. This and other work contributed to a movement towards "Social-Justice Oriented Interaction Design" in Human-Computer Interaction (HCI) research agendas (e.g., Dombrowski et al. 2016). Simultaneously, HCI has steadily increased conversations around diversity, equity, and inclusion in our professional organizations. For example, within the past decade, the Special Interest Group for HCI (SIGCHI) developed the Diversity and Inclusion Statement, noting how diversity and inclusion are not one in the same. They define diversity as individuals of varying backgrounds and experiences that can lead to a breadth of viewpoints, approaches, and reasoning. Whereas, inclusion references when the environment is characterized by behaviors that welcome and embrace diversity (ACM 2020). However, like many other organizations, SIGCHI fails in its definition of diversity, commonly citing accessibility, gender, sexuality, and more—but notably, ignoring race. Our paper contributed new lenses of reflection, autoethnography, and storytelling in order to personalize the necessity of justice and identity-frameworks not only to better our research, but to be seen and valued as individuals in HCI professional spaces.

In this writing, we seek to provide larger context for our work so that others might benefit from our processes and better see this work for what it is—a single step amongst many steps in bringing our full selves to our academic community through a labor of scholarly activism for racial justice. Throughout the chapter, we will interweave personal stories from the authors to further deeply illustrate our experiences in block quotations. As Huerta (2018) previously identified, being a "scholar-activist" can feel like a liminal space where academic colleagues may not have the tools to evaluate research-action contributions, while traditional activists may view our work as existing solely within the ivory tower. In this writing, we seek to unpack specific instances of scholar-activism through a case study of our collective experiences.

To provide tools for those who might come after, we discuss why and how we wrote this paper. To contextualize what scholar-activism might look like in practice, we reflect on how the paper has been received in the first year of its publication. Finally, we provide another call to action in response to the reception of our paper with specific recommendations for long-term engagement with critical race theory and racial justice for research and practice in technology more broadly.

WHY BRING CRITICAL RACE THEORY TO HCI?

There are many ways of being a scholar-activist—from organizing strikes (e.g., #ScholarStrike, Butler and Gannon 2020), advocating for diversity, equity, and inclusion (DEI) resources (e.g., creating funding and mentorship opportunities for underrepresented scholars), to conducting community-based participatory research to amplify the voices and empower underserved populations (e.g., Harrington et al. 2019), writing auto-ethnographically about one's own underserved identity-relevant experiences (e.g., Hammer, 2020), and more. Critical Race Theory (herein referred to as CRT) itself was borne from acts of protest, boycotting, and eventually organization and reformation by legal scholars (Delgado and Stefancic 2017). As researchers who work with returning citizens (e.g., Ogbonnaya-Ogburu et al. 2019), unhoused youth (e.g, Smith 2020), and adolescents who are underrepresented in STEM fields (e.g., To et al. 2018), it is necessary for us to incorporate an understanding of how racial identity impacts our participants and informs the systems within which we are designing. In doing our work, the absence of theory or engagement with important topics of race and racism actively harmed our ability to perform our research ethically and holistically, and harmed our ability to advocate for its value within the field of HCI:

> *In a recent racial justice organizer meeting, a colleague proposed the theme for an upcoming event as "Fighting for Justice as Radical Love." She wanted to reclaim the narrative around activism—that activists fight for justice not out of anger or disdain, but instead we pour our mental, emotional, and physical labor into agitation, reformation, and movement because of a deep love for our communities—to push them to live up to their full potential in values that prioritize our collective humanity. To me, scholar-activism achieves those same goals. It requires bringing my whole self to my work, giving of something personal, to center justice for those who come after me and those who are affected by the work that everyone in my academic discipline performs. It is not that I dislike HCI, or that HCI is not good enough—I deeply care about HCI and the HCI community and believe it can and should do better.*

The "Critical Race Theory for HCI" paper is a tool we created for researchers, like ourselves, who know that race was central to the research questions we investigate, but had no precedent for discussing as something that our discipline might prioritize as being central to how we construct research questions, collect and analyze data, or evaluate a research contribution. We wrote and published it in the form of a conference paper at the Association of Computer Machines Conference on Human Factors in Computing Systems (herein referred to as ACM CHI or CHI) as a way of formal legitimization by speaking in the language best read and understood by our disciplinary peers.

Critical race theory tells us about the ordinary, everyday nature of racism. As Black and Asian-American/Mixed women, we are each acutely aware of and consistently experience everyday racism. As residents of the United States, we are constantly in a context where racism, xenophobia, and other prejudiced and violent ideologies are increasingly publicly embraced. As HCI scholars, our academic computing spaces' conversations around race are typically at best entirely absent and at worst, microaggressive and systemically oppressive. Existing and thriving in computing spaces where both subtle and overt white supremacy live is a radical act (Rankin et al. 2020). As a result, we take matters into our own hands and do the labor for our colleagues—here is a framework so that you can see the white supremacy we are all surrounded by and engaging with. Here are tools so you can do the work of dismantling racial oppression in computing.

MAPPING CRITICAL RACE THEORY
TO HUMAN COMPUTER INTERACTION

Research commonly begins with philosophical assumptions and investigators' sets of beliefs that inform the manners in which studies are designed and implemented. Interpretive and theoretical frameworks further shape these investigations. As most of the researchers within the Human-Computer Interaction (HCI) discipline are white, typically cis-male individuals, we have witnessed these conscious and unconscious biases reflected in our academic institutions, built into the frameworks required for academic success (e.g., publications, research designs, review processes, etc. (Hammer et al. 2020), and designs created for *public* consumption. Issues of representation in HCI are most salient at events like CHI, where in 2019 over 3,800 professionals registered for the conference, though notably many were barred from attendance due to visa processing issues and other travel restrictions. Though introduction of Equity Chairs brought about many welcome changes to policy and resources for diverse needs ("Equity" 2020), more informal, coded networking brought us together through workshops (e.g., doing "sensitive" research) and paper sessions (e.g., "marginalized populations") that encoded a potential interest in race.

Being largely underrepresented in our academic space and finding a common need to engage more meaningfully with race, we discussed that critical race theory was a natural fit as a theoretical framework to bring to our community of HCI researchers. CRT situates race and racism within the structure of American society as well as acknowledges the complex relationships and intersections that reside within race, class, gender, and sexuality differences that feature in the world of ethnic minorities. Through collective learning and community building, critical race theory asks us how we can transform the relationship between race, racism, and power to work toward the liberation of people of color.

Early in our process, we met weekly to discuss our individual understandings of race and critical race theory. To center our conversations, we chose to read "Critical Race Theory: An Introduction" (Delgado and Stefancic 2017) and found that the majority of themes in the book aligned with our own understandings and needs and were appropriate as introductory material. From that we selected and adapted key tenets: 1) racism is ordinary, not aberrational; 2) race and racism are socially constructed; 3) identity is intersectional; 4) those with power rarely concede it without interest convergence; 5) liberalism itself can hinder anti-racist progress; and 6) there is a uniqueness to the voice of color, and storytelling is a means for it to be heard.

We submitted our work to the Association for Computing Machinery Conference on Human Factors in Computing Systems, self-described as "the premier international conference on Human-Computer Interaction" ("Welcome to CHI 2019 " 2020). CHI is a place where researchers and practitioners gather to discuss the latest in interactive technology. CHI considers itself a leader in its commitment towards diversity and inclusion, further describing, "we are a multicultural community from highly diverse backgrounds who together investigate new and creative ways for people to interact" ("Welcome to CHI 2019" 2020). The conference focuses on physical efforts, such as Diversity and Inclusion Statements, Women's Breakfast, accessible locations, and more. Seemingly, CHI creates opportunities to discuss issues that thwart innovation by impeding inclusion and collaborate on strategies to empower those who have been marginalized. However, these efforts were fruitless as CHI continues to exclude those individuals who have been marginalized most, persons of color:

> *I recall my first CHI, in which I was given a free ticket to the Diversity and Inclusion lunch. I was excited, I figured this was the opportunity to meet more individuals like myself in the HCI space. The time came, and I found my seat at a central table. I sat as I watched as a sea of white women walked by. Slowly, they filled my table. One other person of color sat down. The event began, in which the organizers wanted us to introduce ourselves to our table and spark discussion by asking questions about our experiences. The lunch continued by honoring different individuals on being a Trailblazer, Strong Ally, Star Mentor, and Emerging Voice. All of which were awarded to white women and one white male. This showed me how CHI was still not quite ready to welcome diverse individuals.*

This framing of the conference's goals and related shortcomings afforded us the opportunity to inject ourselves in more ways than one. We submitted our work to CHI, not only because of academic relevance, but also because if we could begin the discussion and illustrate the significance of incorporating race into our research practices and designs, we could carve out a haven for ourselves. We introduced HCI to a theory novel to the discipline in a way that challenged how they currently engage with race in their research and professions. Additionally, we started the conversation on various levels as to what was truly necessary from both a micro- and macro- level to be inclusive for persons of color, a marginalized group who had previously been left out of the conversation.

We wanted to be true to the theory; therefore, we leveraged the last of the tenets, *there is a uniqueness to the voice of color, and storytelling is a means for it to be heard.* We introduced storytelling to HCI in which we posited our own personal stories as autoethnographic, synthesized data. This *data* was used in three ways: 1) to emphasize the many ways in which the majority population has marginalized us, a smaller sample of persons of color representative of the HCI community; 2) to highlight the how impactful storytelling is as a methodology in drawing out the nuances of our voices and experiences and; 3) to illustrate the adaptation to HCI through concrete examples of the tenets.

Storytelling posed a unique experience for us as authors. We wrote honestly, thinking of the various ways in which technology and academia had excluded and minimized. We witnessed the *ordinariness of racism* and our numerous experiences with it. As we wrote the paper exposing ourselves, our fears emerged. We found ourselves questioning if these stories were truly anonymous and how the (lack of) reception may negatively impact our potential careers. We feared, more importantly, whether and how they would be received in a scholarly manner. As the academy is largely white, would they be able to step outside themselves to engage critically with the subject matter. The process of writing our personal stories was censored activism. We wrote story after story and connected it back to the tenet. However, if the story exposed us or possibly could make a reviewer too uncomfortable, we edited it away. We undercut our transparency to appease our audience. While critical race theory aspires to empower voices and perspectives that have been marginalized, unfortunately, it does not assuage the same fears and privilege that oppress them. Critical race theory encourages a problem to be placed in social, political, and historical context while simultaneously considering issues of power, privilege, racism, and other forms of oppression; but how do we accomplish such feats when our intended audience is ignorant to such contexts?

AMPLIFYING RACE THEORY ACROSS INSTITUTIONS

We have received a number of signals that this work has been positively received by our intended audience and community. The paper received a Best Paper Award at CHI, an award given to the top 1 percent of papers at the selected conference. Additionally, we have been invited by a number of academic and industry research organizations to present the paper. From May to October 2020, we presented the paper fourteen times at: four public universities, four private universities, five research and design groups at major tech companies, and one locally organized professional group. A subset of the authors attended informal Q&A discussions or gave shorter presentations at three additional universities.

CHI was canceled in 2020, initially denying our ability to share this work with our intended audience in a safe, structured environment. It was a tough decision, but the right decision. The cancellation also demonstrated a failure of waiting until a conference to present work. We had been relying on, but did not need, a conference to be connected. While serendipitous meetings and networking opportunities during the conference disappeared, the opportunity to join meetings that were formally closed due to accessibility, cost, disability, had increased.

For instance, we conducted a virtual workshop for CHI, "What's Race Got to Do With It? Engaging in Race in HCI" (Smith et al. 2020), and several people praised us for the reduced cost of joining a meeting and getting connected. Going virtual demonstrated the ease of getting experts into common spaces and the ability to facilitate collaborations across institutions. One of our goals was to lessen the burden to attend our workshop, so we secured funds to absolve registration fees. However, a virtual workshop posed new concerns. It highlighted the importance of consistent internet access and the difficulties of finding mutually agreed upon blocks of time across varying time zones. Virtual platforms raise concerns around the essentiality of lighting for persons of color, namely Black people, as well as the virtual room security to prevent Zoom bombing with this heightened topic. Zoom provided people access to our homes, our privileges and the lack thereof. While this cancelation of CHI opened many doors, it also closed down many:

> *Life was chaotic; I lived in the DC metropolitan area where there were protests occurring in Washington, DC, the COVID-19 pandemic, and life events happening in the background. But, an opportunity opened up. In a weird way, the stars aligned. The negligent murders of Black men and women highlighted the racial tension in America on both a micro- and macro-scale. People began to notice our paper, and we no longer had a CHI audience to expect. So, we used social media to make a simple request. We will present, if you will have us. Many were happy to receive this invitation. After receiving what seemed like an astronomical number of invitations, we used the summer to plan a calendar. We would be going on a world tour on the internet. Week by week, we connected digitally via Zoom or Google Hangout to one venue or two. Meeting new faces. Faces of individuals we otherwise would never have met. Our hope was they were allies. Concerned about our personal security, we often discussed measures to make sure we felt safe, e.g. not recording the meeting. We also thought about what was safe and comfortable for our participants, like asking questions anonymously. Discussing race, initially was scary, but as the weeks past, comfort set in, comfort in knowing you didn't need to have all the answers, and that everyone was learning, and it is ok to acknowledge wrong and do better . . . but you have to do better.*

Our talks began to have a cadence. Click join, introduce yourself, present your part, and stay for question and answer (Q&A) discussions. The most important part was to stay for Q&A. In question and answer discussions following our presentation, we received questions about our thoughts and opinions about the field of HCI, critical race theory, capitalism, the technology sector, and institutional racism. We became privy to personal thoughts, and micro- and macro-scale values from both academic institutions and professional organizations. We heard thoughtful reflections, as well as genuine surprise. For some, we heard surprise that racism is so pervasive in technology. For others, we heard surprise from our participants that our stories were so "tame"—racism in HCI can be far more visceral and violent.

We frequently encountered participants who wanted practical advice on how to apply CRT to their work. Questions ranged from individual everyday practice to the creation and management of our broader professional spaces including: "What is our obligation as researchers when race is not being brought up explicitly by our interviewees?," "Should we always be collecting race?," "What can CHI do to be better?," and "How can we institutionalize our values?" These are many of the questions we hoped readers would ask, and more importantly, put in the work to answer for themselves. Critical race theory is not a directive, but a lens to help us view and hopefully dismantle the racially oppressive systems we belong to, and we hope attendees of our talks continue to ask such questions.

Given the publicity around our talks, as the summer went on, participants wanted to hear how their peers had received our work. "Capitalism, attention, and money all seem like tensions that you highlight in your paper, it's a problem for all of us, how have other companies responded to your talk?" "Why, 40 years after the civil rights movement, are people in the tech community still so unaware of this?" One participant shared a personal experience: "What do you think is happening in design spaces today? I once had a research question years ago about gentrification and my colleagues looked at me like I was crazy." Participants were looking for other allies as well as leaders in the space. There seemed to be a strong desire to get a "temperature check" from others in HCI—is anti-racist sentiment in our community a passing fad or the sign of long-term and systemic change?

Finally, we found many participants who used that discussion space to air their personal anxieties and concerns about operating and benefitting from systems of oppression, white supremacy, capitalism, and interest convergence. One participant reflected on how managerial practice has roots in slave plantations in the United States, "I've been reflecting on the race blind approach, you can see how everything has roots in racist practice, do you have thoughts on the ability to do anti-racist work in systems that are inherently racist?" Another participant went as far as to say, "I've been thinking a lot the past couple weeks about capitalism and technology and how they're all so inherently bad . . . I guess I'm asking . . . Should we all just quit our jobs?" While we do not doubt the sincerity of this question, nor the likely accompanying grief and anxiety, we also want to acknowledge the kind of emotional labor this question puts on us, a group of complete strangers, in this individual's work space. This kind of question is neither productive nor does it signal true allyship and instead we urge readers with these questions to sit with these tensions and take stock of what they are and are not willing to risk or sacrifice (more on this in our call to action).

In the midst of these important but difficult conversations, the United States government issued an Executive Order banning the use of critical race theory in government agency spending and contracts ("Executive Order 13950" 2020). Though the order itself does not use the words "critical race theory," a memorandum accompanying the order insists that, "all agencies are directed to begin to identify all contracts or other agency spending related to any training on 'critical race theory,' 'white privilege,' or any other training or propaganda effort that teaches or suggests either (1) that the United States is an inherently racist or evil country or (2) that any race or ethnicity is inherently racist or evil" so that they might divert spending away from these initiatives (U.S. Executive Office of the President 2020). This description of critical race theory is fundamentally incorrect and actively seeks to shut down important, generative conversations on systemic and institutionalized racism by insisting that it does not exist. This order has had immediate ramifications for diversity, equity, and inclusion work as government organizations as well as schools that receive federal funding have begun cancelling critical race theory training sessions. E.O. 13950 presents not only professional threats to academics who work with and on critical race theory, but personal threats as well. Denying structural racism in the United States is an indignity to all people who operate within our systems. We call on allies and accomplices to affirm, protect, and uplift work on critical race theory. Below, we detail more direct action for allies and accomplices.

CALL TO ACTION

Within our CHI paper, we highlight several recommendations for researchers and practitioners in the field of HCI. Yet, we were often asked by participants of what they can do to center and include the topic of race and ethnicity in our research. Motivated by questions from attendees of our presentations, we have developed an additional set of recommendations to address both the practicality and the emotional implications of adapting and internalizing critical race theory.

Get Comfortable Being Uncomfortable

Gaining an awareness of how deeply entangled systemic racism and white supremacy are with our political, economic, social, and professional systems is painful. It hurts to find out you have had disadvantages outside of your own control, or that society does not value your life and your contributions because of your race. It hurts to find out that that has been happening to the people around you. More insidious than the pain is the discomfort when those of us with proportionally more privilege have to acknowledge that our accomplishments were not borne just of our own work ethic, contributions, or skill. It can be a natural instinct to fight that narrative. To argue back, "well nothing was handed to me" or "I earned my place" or to engage in the "Oppression Olympics" where intergroup competition emerges to assign weight to various experiences of marginalization (Hancock 2011). Instead, we urge accomplices to run towards that discomfort and sit with it. Get comfortable being uncomfortable. Recognize and identify how white supremacy hurts all of us, not just those of us from racial minority backgrounds.

Start Today

The urgency of performing racial diversifying in our research and communities cannot wait. Change starts now in the messiness of now, or we are bound to repeat the past. Perfectionism is just one more insidious, structurally reinforced tendency that holds us back from taking action.

While much of what needs to change involves long-term systemic restructuring, we have found that the magnitude of that need can be incapacitating. We encourage a balance of short-term action with long-term commitment to structural change. In the short-term you might: 1) audit your hiring and recruitment calls—are you taking advantage of best practice for inclusive hiring?; 2) develop and practice a short list of strategies for speaking up for yourself and others in the presence of both intended and unintended racism; 3) create small structures for ongoing conversations about race (e.g., reading groups). In the long term you might: 1) commit to lifelong learning about race; 2) plan ahead in budgeting to create long-term and sustainable funds for Black and Brown students to attend professional events such as conferences; 3) inventory the service work being performed at your workplace—are people of color taking on inordinate amounts of service? Are they being credited for that service work? (more on this below). As a more extended example in our paper, we discuss how imperative it is to identify the missing voices from our research so that we might avoid propagating and perpetuating the biases of majority group members. These are just a few concrete actions you might take, but of course we encourage you to use these as a starting point and to develop more that are contextually relevant to your own situation.

Redefine Success

From a systemic standpoint, we must acknowledge the current measures of success are incongruent for minoritized individuals and intersectional identities. Academic success is driven by the number of publications. Organizational success is driven by product development and sales. Top tier conferences and journals as well as industries' product deadlines often have looming timelines that do not allow for researchers to invest long term into populations in an effective way. In order to empower minoritized populations, we must focus on context to empower these populations.

We must take the time to consider the implications by moving to satisfy timelines, and the (un)intended erasure it causes. Research must take time to consider place, time, context, history, and intersectional identities in everything that we produce. By taking time to explore context, we can see how our research, tools, and designs alienate a few and examine ways to ensure our work is accessible and *inclusive* for all.

Assess Your Willingness and Ability to Take Risks

On the difference between being an ally and being an accomplice, Harden and Harden-Moore wrote (2019, para. 7), *"to be an accomplice, one must be willing to do more than listen; they must be willing to stand with those who are being attacked, excluded or otherwise mistreated, even if that means suffering personal or professional backlash. Being an accomplice means being willing to act with and for oppressed peoples and accepting the potential fallout from doing so."* Doing diversity, equity, and inclusion work requires moving beyond allyship to accompliceship. Speaking about racial disparity comes with very practical professional and personal risk ranging from harassment and social repercussions such as being seen as a "troublemaker" up to the loss of funding and speaking opportunities (e.g., as in the case of E.O. 13950). Is your racial justice work in words only, or are you willing to actually be in the fight alongside your colleagues? Where is there space in your personal and professional life to take on appropriate risks to individual success in favor of community and communal success?

Prioritize Minoritized Individuals

Race has and continues to make leaders uneasy. Organizations, systems, and structures incorporate diversity efforts that allow the majority to feel assuage guilt. Organizations commonly create cultures that dictate how individuals of color can contribute, often in ways that only comfort white people. In our writing, reviewers cite the desire for "objective" reports and case studies, while dismissing data from personal experiences as "too emotional." When horrific instances of racial injustice occurred, like the murders of Henry Dumas (1968), Rita Lloyd (1973), Margaret Laverne Mitchell (1999), Ronald Curtis Madison (2005), Oscar Grant III (2009) (Ater 2020), and more, white people carry on largely unaffected, with little acknowledgement or space for the emotions they triggered for individuals of color. These unjust murders have reverberated the racial injustices that have plagued Black and Brown minorities. Our work is not simply timely, it is about time that people are listening. Discussions about racism are discouraged as "divisive" or "unproductive." By not acknowledging these facets of minorities' identity, cultures, structures, and systems are unable to acknowledge the lives they live and value them for who they are.

It is clear that the existing diversity and inclusion tools and practices are insufficient for racial equity work. Instead of pushing for fundamental changes in organizations, they largely focus on patching new guidelines, practices, or programs onto the existing structures and culture in an attempt to help individuals of color better *fit in* and succeed. Instead, racial equity and inclusion efforts must radically change. We must focus on transforming the system to fit *all* people as opposed to focusing on how a select few can assimilate. To move toward racial equity, we must encourage structures to prioritize humanity. Individuals (of color) need the ability to work with the dignity of having their histories acknowledged and their life experience valued.

Be Explicit in Language

The horrific killings of Black and Brown individuals have sparked written correspondence from CEOs, COOs, college and university presidents, provosts, deans, and more that promise to understand and address the racism that Black people face. Individuals and organizations have been called out for their silence, how their lack of words rendered their compliance in a racist system. These situations have highlighted the importance of words and our language in our modern society. As an ally, it is not to use our words, but it is important to be using the *right* words.

For instance, when we think about the term, *underrepresented minority* (URM), it often refers to the lower participation rates of racial and ethnic groups in educational fields relative to their representation in the United States population. African Americans/Blacks, Hispanics/Latino(a), and Native Americans/Alaskan Natives are most commonly defined as URMs, which aligns with the National Science Foundation's definition (NSF, n.d.). URM is a well-established and accepted label in higher education, especially within the STEM fields. However, when we examine the term, "underrepresented," the Cambridge dictionary defines the prefix under- as "not enough," "not done as well," and "below." When we label individuals as underrepresented minorities, we need to think of the implications of this word on their identity and how implicit harm may be cast on their being.

Professor Tiffani Williams (2020) critiques the lack of representative identity in the label URM. One does not self-select this identity. There are no movements or television shows that celebrate the pride of being an underrepresented minority. More importantly, by "aggregating multiple groups together based upon their low levels of representation, the URM label becomes insensitive to the unique needs and circumstances of its group members"

(Williams 2020). One solution for Latinx student recruitment is increasing the number of Spanish-speaking professors at an institution. However, this solution does not solve the same concern for Black students. Using one label does not allow us to differentiate who is actually being discussed. We are hiding behind abstract language and convenient acronyms. We must be explicit in our language if we want to truly be inclusive and embrace and respect Black and Brown lives.

Redefine and Reward Service

It has been well documented that women of color in the academy are overburdened and underrecognized for their service work (Turner et al. 2008; Hammer et al. 2020). Service and administrative work are critical to the functioning of academic institutions but are not seriously considered in hiring, promotion, and tenure cases. Faculty and staff from racial and ethnic minority groups are often most safe for students of those same groups to seek out for formal and informal mentorship. And by intentionally or unintentionally avoiding this work, faculty and staff who do not belong to these groups have more time for research, increasing the time, resource, and productivity gap. However, mentorship relationships are often viewed as fulfilling by faculty and staff from racial or ethnic minority groups. Given this important role in the academic ecosystem, service has been proposed as the space where critical resistance and institutional change can occur to uplift faculty of color (Baez 2000).

We echo recommendations that push to prioritize service for faculty and staff—for departments and universities to formalize the documentation of service and to account for the time and labor that go into community-facing work (including but not limited to: informally and formally advising students, building community relationships outside the university, recommendation letter-writing, etc.). While these phenomena are best known to us through documentation in higher education research, we also urge accounting for gaps in administrative, service, and mentorship work in other professional spaces such as in the technology sector.

Move Back, Allow Minoritized Voices to Lead

Often in our work, we erase the representation that has been unduly called. Whites allow their pervasive privilege to occupy all roles and not seek other voices. We have seen this commonly in the media in which white actors are called upon for roles of varying ethnicities. This undermines the specificity of the experience of persons of color. Additionally, we see it in our research as we take from minoritized communities and rarely give credit.

Those in power must be comfortable taking a seat to let a minority shine. We cannot continue to glorify colorblind approaches and being ignorant to its implications. To truly embrace a diverse and accepting culture, those in power must be committed to creating opportunities and letting these voices have a place in our workplace and our scholarship for persons of color.

CONCLUSION

We close this chapter by stating, quite frankly, our goal of writing the "Critical Race Theory for HCI" paper was twofold—one, to make space for our research; two, to make space for ourselves. Critical race theory provides the opportunity to create safe spaces, center minoritized voices, and provide a framework to candidly discuss race—a discussion item often missed by our research community. Through storytelling, we demonstrate the many ways in which the majority population has marginalized us, a smaller sample of persons of color representative of the HCI community. We also highlight how impactful storytelling is as a methodology in drawing out the nuances of our voices and experiences, nuances that are missing from both academic and professional contexts. Our stories, both in this chapter as well as our paper show how racism is a persistent force in American society, and critical race theory illustrates racism's pervasiveness. CRT has allowed us to incorporate race and racism and its implications more broadly into our work and, more specifically, as Black and Asian-American/Mixed women. Broadly, CRT emphasizes the importance of examining and attempting to understand the socio-cultural forces that shape how we and others perceive, experience, and respond to racism. It has supplied a foundation upon which these discussions can begin and evolve. More specifically, critical race theory has illuminated the failures of varying diversity, equity, and inclusion practices and how some of these efforts continue to perpetuate racism in new forms. Critical race theory does not absolve white privilege or institutionalized racism, radical efforts are required for substantial and sustained change. While critical race theory is not the sole viable solution for the current divisive nature of our broken systems, it offers much more than a band aid on white supremacy.

REFERENCES

Association for Computing Machinery. N.d. *Diversity & Inclusion.*https://www.acm.org/diversity-inclusion.
Ater, R. 2020. "IN MEMORIAM: I CAN'T BREATHE." Retrieved fromhttps://www.reneeater.com/on-monuments-blog/tag/list+of+unarmed+black+people+killed+by+police.
Baez, Benjamin. 2000. "Race-Related Service and Faculty of Color: Conceptualizing Critical Agency in Ccademe." *Higher Education* 39(3): 363–91.
Bardzell, Shaowen. 2010. "Feminist HCI: Taking Stock and Outlining an Agenda for Design." In *Proceedings of the SIGCHI conference on human factors in computing systems*, 1301–10.

Butler, Anthea and Kevin Gannon. 2020. "Why we started the #ScholarStrike." *CNN*. September 8, 2020.

Delgado, Richard, and Jean Stefancic. 2017. *Critical Race Theory: An Introduction*. NYU Press.

Dombrowski, Lynn, Ellie Harmon, and Sarah Fox. 2016. "Social Justice-Oriented Interaction Design: Outlining Key Design Strategies and Commitments." In *Proceedings of the 2016 ACM Conference on Designing Interactive Systems*, 656–71.

"Equity." CHI 2019—Weaving the threads of CHI.https://chi2019.acm.org/for-attendees/equity/.

"Executive Order 13950 of September 22, 2020, Combating Race and Sex Stereotyping." 2020. *Code of Federal Regulations,* 60683–89.https://www.federalregister.gov/documents/2020/09/28/2020-21534/combating-race-and-sex-stereotyping.

Hammer, Jessica. 2020. "Envisioning Jewish HCI." In *Extended Abstracts of the 2020 CHI Conference on Human Factors in Computing Systems*, 1–10.

Hammer, Jessica, Alexandra To, and Erica Principe Cruz. "Lab Counterculture." In *Extended Abstracts of the 2020 CHI Conference on Human Factors in Computing Systems*, 1–14.

Hancock, Ange-Marie. 2011. *Solidarity Politics for Millennials: A Guide to Ending the Oppression Olympics*. Springer.

Harden, Kimberly, and Tai Harden-Moore. 2019. "Moving from Ally to Accomplice: How Far Are You Willing to Go to Disrupt Racism in the Workplace?." *Diverse*. March 7, 2019.

Harrington, Christina, Sheena Erete, and Anne Marie Piper. 2019. "Deconstructing Community-Based Collaborative Design: Towards More Equitable Participatory Design Engagements." *Proceedings of the ACM on Human-Computer Interaction 3*(CSCW): 1–25.

Huerta, Alvero. 2018. "Viva the Scholar Activist!." *Inside Higher Education*. March 30, 2018.

Ogbonnaya-Ogburu, Ihudiya Finda, Angela DR Smith, Alexandra To, and Kentaro Toyama. 2020. "Critical Race Theory for HCI." In *Proceedings of the 2020 CHI Conference on Human Factors in Computing Systems*, 1–16.

Ogbonnaya-Ogburu, Ihudiya Finda, Kentaro Toyama, and Tawanna R. Dillahunt. 2019. "Towards an Effective Digital Literacy Intervention to Assist Returning Citizens with Job Search." In *Proceedings of the 2019 CHI Conference on Human Factors in Computing Systems,* 1–12.

Rankin, Yolanda A., Jakita O. Thomas, and Nicole M. Joseph. 2020. "Intersectionality in HCI: Lost in Translation." *Interactions 27*(5): 68–71.

Smith, Angela D.R., Alex A. Ahmed, Adriana Alvarado Garcia, Bryan Dosono, Ihudiya Ogbonnaya-Ogburu, Yolanda Rankin, Alexandra To, and Kentaro Toyama. 2020. "What's Race Got To Do With It? Engaging in Race in HCI." In *Extended Abstracts of the 2020 CHI Conference on Human Factors in Computing Systems*, 1–8.

To, Alexandra, Jarrek Holmes, Eaine Fath, Eda Zhang, Geoff Kaufman, and Jessica Hammer. 2018. "Modeling and Designing for Key Elements of Curiosity: Risking Failure, Valuing Questions." *Transactions of the Digital Games Research Association 4*(2).

Turner, Caroline Sotello Viernes, Juan Carlos González, and J. Luke Wood. 2008. "Faculty of Color in Academe: What 20 Years of Literature Tells Us." *Journal of Diversity in Higher Education 1*(3): 139.

"URM definition" n.d. *U.S. National Science Foundation*.https://bit.ly/2BZx1ZO.

Vought, Russell, U.S. Executive Office of the President. 2020. "Memorandum for the Heads of Executive Departments and Agencies: Training in the Federal Government." Washington, D.C.https://www.whitehouse.gov/wp-content/uploads/2020/09/M-20-37.pdf.

"Welcome to CHI 2019." 2019. CHI 2021—Weaving the threads of CHI.https://chi2019.acm.org/for-attendees/equity/.

Williams, Tiffani L. 2020. "'Underrepresented Minority' Considered Harmful, Racist Language. *Communications of the ACM*. June 19, 2020.

Chapter Eleven

Hope Wears a White Collar

RBG Memes and Signifying Intergenerational Solidarity

Elizabeth Nathanson

While Ruth Bader Ginsburg was elected to the Supreme Court in 1993, she arguably did not become a household name until her image proliferated across digital platforms twenty years later. Around 2013, on the heels of Ginsburg's stances on landmark affirmative action and voting rights cases, a Tumblr was started by NYU Law student Shana Knizhnik. Then, following her scathing dissent in the Hobby Lobby case in 2014 her digital presence exploded.[1] By the summer of 2014, Rebecca Traister (2014) declared her "the most popular woman on the internet" (para 1). Local news outlets covered her mediated presence, remarking on her arrival as a "legit pop-culture icon" (Patrick Ryan 2018, para 1). And, she even has an entry on the website *Know Your Meme* which provides both her biographical information as well as a description of her online presence, noting that "Justices of the Supreme Court do not maintain online presences of their own, however, there is evidence online of a fandom of Ruth Bader Ginsburg" (Ruth Bader Ginsburg, *Know Your Meme*). From Tumblr to Instagram, Twitter to Facebook, Ginsburg's face, words, and statements appear with abandon. She is seen in private and public as an elderly justice and a youthful law student. The spread of her image across digital platforms appears to be produced by both activists and average "fans," and of course, critics of the liberal justice who respond to the proliferation of images with hatred and anger.

Ruth Bader Ginsburg's popular fame extends beyond mere online presence; she has been the subject of a documentary film (*RBG*), a biopic (*On the Basis of Sex*), has been a character played by Kate McKinnon in a variety of Saturday Night Live skits, appeared as the subject in coloring books, and her name has been invoked on television shows from *Parks and Rec* to *Law and*

Order: SVU (where a framed portrait of the justice sits behind the desk of the tireless, determined sex crimes detective Olivia Benson), just to name a few. When we compare the representations of the justice in these depictions, she is revered, explicitly and implicitly celebrated as a historical figure in the feminist movement. However, as Jill Lepore (2018) has written, Ginsburg's history as a judge does not have as straightforward a path as such popular culture accolades might imply. Lepore explains how the justice built a career in anti-discrimination cases, fighting for equal rights even while at times disagreeing with some women's groups. The apparent disjunct between Ginsburg's complex historical legacy as a lawyer and judge deftly mobilizing the power of the law in the interests of gender equality and the mainstream, popular representations leads the historian to bemoan the popular proliferation of Ginsburg's image; Lepore claims that it "is no kindness to flatten her into a paper doll and sell her as partisan merch" (Lepore 2018, para 9). Lepore's language of flatness implies a perspective on mass produced cultural reproductions that freeze figures into simplistic reductions of their actual complexity. These reductions can then be transformed into a "doll" dressed up to suit the needs of those reproducing the images to suit political ends, voiding the original woman of her supposedly authentic complexity.

When regarded through the lens of feminist media scholarship, however, we need not be dismissive of the meanings made of Ginsburg in and through popular culture. Rather than looking for historical verisimilitude, this chapter proposes that we consider her popular representations as indicative of shifts in feminine representations whereby older women are increasingly viewed as agents of change and liberal revolution rather than seen as symbolic of stasis or obsolescence. Furthermore, rather than implying that popular representations are empty, consumerist, or lacking authentic complexity, we might begin to think about the ways in which Ginsburg has not been "flattened" but rather has been circulated in ways that are unpredictable and distinctly energetic. Through the circulation of digital content that is positioned as in direct, critical dialogue with political events, Ginsburg's presumably staid image as both a supreme court justice and an older woman is granted pointed, resistant, affective vitality.

The unpredictability of Ginsburg's strident presence across the media landscape is particularly notable when considered in relation to its origins online. The combination of an octogenarian with digital content seems somehow like a contradiction. As digital media has also been known as "new media," the medium appears one bound to youth culture, an endlessly renewable resource which moves forward with seemingly limitless abandon. Popular and academic discourse abounds with articles about how the apparently symbiotic link of digital media and youth cultures has political effects. Teen Vogue celebrates how "The proliferation of social media platforms has brought young activists unprecedented opportunities and exposure" (Stauffer

2019, para 7). Media scholar Henry Jenkins (2017) similarly argues that youth cultures have changed political movements through their use of social media, claiming that "the internet may not have changed everything but it has definitely changed many things about the way politics operates in the twenty-first century, and youth have been on the front lines of this process" (para 14). Books such as *By Any Media Necessary: The New Youth Activism* similarly draw connections between youth cultures and digital activism that celebrate young activists' abilities to mobilize participatory cultures to work to transform democracy. Feminist media scholars, such as Jessalyn Keller (2019), similarly comfortably place social media activity in the hands of girls, arguing that girls are savvy, productive social media users who "not only understand social media platform affordances but actively use this knowledge when doing feminism online" (9). Across this discourse, assertions of the intimate ties between youth, digital media, and especially online activism reproduce definitions of the generation of digital natives who are not passive consumers but actively engaged with online content.

It is this point of seeming incongruity between Ginsburg as the elderly justice figure and "new," mobile digital platforms that is of interest in this chapter. This chapter will provide close, textual analysis of the ways in which RBG's circulation online performs a feminist politics through the body of the Supreme Court justice.[2] While the question of how Ginsburg became a popular presence is of course relevant, the larger concern here is what happens to conceptions of age, notably feminized age, when RBG became a meme. Her inscrutable face and powerful voice operate as a space through which digital communities can explore the concerns relevant to contemporary feminist politics. RBG memes vary from broad statements about anti-Trump "resistance," to addressing specific American liberal concerns such as abortion rights, gay marriage, and the Affordable Care Act. These memes articulate and capitalize on widespread assumptions and ideologies about older women, relying on popular understandings of older, grandmotherly women as fragile or anachronistic in order to generate some of the humor on which they depend. And yet, they also illustrate a more progressive idea of generational community building in which older women's bodies are positioned as vestibules of decades of experience and knowledge worthy of reverence. Similar to how other older white women like Betty White and Elizabeth Warren have been celebrated in popular and digital media, Ginsburg's digital circulation speaks of a continued desire to find hope in figures who otherwise might be discarded due to widespread ageism. As Susan Douglas (2020) recently argued, women like Ginsburg "are becoming role models for younger women. Bridges between generations are starting to be built" (173). These connections made across generations through the deployment of digital technologies promise to unsettle the disarticulations between generations of women that are all too present in postfeminist media culture.

As postfeminism fetishizes the illusory promise of youthful "can-do" girls, RBG memes reveal the future potentiality of figures like Ginsburg, who are celebrated, not denigrated for, their decades of work and experience. By revitalizing her image, again and again, the circulation of RBG memes bind a figure who might otherwise be relegated to the "past" to a very vital sense of present activism. Through deploying youthfulness in the memes themselves in the form of popular culture figures and texts used to construct new meaning around the figure of RBG, these memes express a desire for intergenerational solidarities. In doing so, the memes rewrite notions of aging femininity, grounding Ginsburg in corporal strength rather than merely wisdom, presenting her as affecting because of, not despite, the effects of time on her body.

DIGITAL FEMINISMS AND POSTFEMINISMS

Older women have been unsurprisingly historically subject to ageism in mainstream representations. From film to television, Twitter to Instagram "American youth culture" thrives, a concept Kathleen Woodward (2006) defines as an ideology where "youth is valued at virtually all costs over age and where age is largely deemed a matter for comedy or sentimental compassion" (164). Through systems of objectification, youthful feminized bodies are privileged under a male gaze, leaving older women unseen and unvalued. The fetishization of youth is further amplified when we consider the work of scholars addressing postfeminist discourses circulated in and through media; according to the logics of postfeminism, young women are granted agency, independence and freedom of choice, privileges bestowed upon them by the gains of feminist labors which have come before (Gill 2007; Harris 2004; McRobbie 2009). This discourse constructs a fantasy world in which women are expected to choose the shape their lives should take because of those gains, and yet continues to encode expectations of traditional, hegemonic femininity which underscore how these choices are not actually freely chosen, but pre-determined and internalized by feminine subjects. In the process, feminism as a politics is constructed as being of an earlier generation, distinct from, and unnecessary to the youthful contemporary one. Scholars have argued that this has resulted in young women disidentifying with feminism, as notably "one strategy in the disempowering of feminism includes it being historicized and generationalized and thus easily rendered out of date" (McRobbie 2019, 292). In effect, postfeminism is dependent upon a kind of ageism which situates the politics, and the women who signified it, in the historical past.

Recently, feminism does not appear as consistently repudiated in popular discourse and increasingly feminist media scholars turn to expressions of feminist online activity as indication of "emergent feminisms" (Ryan and Keller 2018). Scholars studying the increasing presence of feminist ideas in circulation often turn to digital media to understand how the platforms afford opportunities for such expression. This is often tied to fourth wave feminism whereby these technologies promise to give voice to discontent: "The internet has created a 'call-out' culture in which sexism or misogyny can be 'called out' and challenged" (Munro 2013, 23). Scholars have celebrated the possibilities of this space as offering users to freely articulate feminist discontent with the nature of gender politics (see for example, Thrift 2014; Rentschler and Thrift 2015; Barker-Plummer and Barker-Plummer 2017; Lawrence and Ringrose 2018). The digital promises what the IRL (In Real Life) lacks, notably it provides room for "consciousness raising and community building" (Rentschler and Thrift 2015, 329). Freed from some of the restrictions of time and place, digital technologies have been studied as offering opportunities to speak back to a culture that has historically denied women equal voice. Scholars who have studied the ways this is manifested through memes have found that "digital tools are allowing women to build a resilient, popular, reactive moment online" (Lawrence and Ringrose 2018, 213). In these ways, scholars demonstrate how the internet grants women opportunities to push back against anti-feminist discourses and resist having their views co-opted by postfeminist popular culture. As many of these scholars acknowledge, digital media are not free from the inequalities that exist IRL, for even while women may engage with feminism online, such engagement remains structured by privileges associated with race, class, ability, as well as age.

Of course, these digital cultures depend upon their publicness across digital platforms to render their power. In doing so, digital culture is not free from the ways such visibility renders feminist discourses vulnerable to backlash, misogyny, or objectification. As Sarah Banet-Weiser (2018) has charted, the increased presence of "popular feminisms" has been accompanied by a reactionary "popular misogyny," a trend marked by widespread vitriol and anger at women's so called-successes across media platforms including, and perhaps most particularly felt online. Feminist media scholars studying contemporary movements of popular culture have raised many concerns about the way postfeminist visibility operates to render women subjects of surveillance. And, as is articulated by Rosalind Gill (2016), just because we may see the presence of popular feminisms does not mean that we are free of the double, contradictory, dismissive binds of postfeminism. In fact, Gill has argued that while feminist goals and figures presume to become increasingly present in the media landscape, they are that much more vulnerable to hegemonic incorporation and the internalization of systems of surveillance that work to bind anti-sexist aims back into sexist practices, arguing that instead we consider "uneven feminist visibilities" (615) and how "feminist media storms arrive *always-already trivialized*" (616). Other scholars, such as Rosemary Clark-Parsons (2019), recognize the ways in which the risks associated with digital visibility are incorporated into the performative practices of hashtag activists. For Clark-Parsons, #MeToo activists specifically work by rendering visible the private on a global stage. While this makes the activists vulnerable to "re-traumatization, backlash, cooptation, complacency, and the exclusion of those most marginalized victims . . . "#MeToo participants developed *performance maintenance practices* to correct these erasures, maintain narrative control, and model actions audiences could take beyond tweeting" (16). Through careful attention to the needs, experiences and pains of others, hashtag activists demonstrate practices that both have limits and promises to advance the concerns of feminist politics.

For many of the scholars studying feminist digital activism, the question of impact haunts their scholarship; in other words, can the work being done online extend outside the networked feminism to mobilize a grounded sisterhood of activism? While scholars have cautioned against making grand claims about political productivity and online feminist activism (Fotopoulou 2014), questions of the efficacy of feminist digital engagement informs much of this scholarship. For example, Lawrence and Ringrose (2018) have studied how "social media platforms have produced new spaces for debates over feminism," and how humor and irony on Twitter "play a central role in increasing feminist audiences and mobilizing feminist collectivity" (212). Humor is another method through which Rentschler and Thrift (2015) analyze the feminist productivity of digital platforms, arguing that memes such as the "Binders Full of Women" facilitate online discourse. The seemingly unique affective capacities of digital content which spreads through networks and appears to generate relationships is the focus of Inger-Lise Kalviknes Bore, Anne Graefer, and Allaina Kilby's (2017) analysis of the online content stemming from the Women's March in 2017. They argue that "[t]he protesters' use of humor . . . can facilitate productive tensions between anger and joy, inviting others to feel a pleasurable sense of belonging to wider protest movements, to feel angry, energized and unruly together, and to laugh defiantly at patriarchy together" (535). In these ways, memes provide the opportunity to create, foster, and celebrate the feelings necessary to generate social, political, and cultural change.

While there has been much scholarship about how memes have played a role in articulating feminist concerns and promising a new kind of digital online activism, this work is largely attentive to social protest movements without attention to the way representations of age operate in the construction of such activism. While no doubt hashtag activism around rape culture and the concerns of #metoo speak across generations, scholarly analyses of online feminism have primarily attended to how hegemonic ideas about race and sexuality structure, constrain, and are negotiated by digital gender politics, without considering the role of ageism in online feminist practice. For example, Lawrence and Ringrose (2018) raise the point that digital "battles over feminism . . . provide space for discussing and debating differences between feminisms and debates over inclusive or 'intersectional' forms of feminism versus feminisms that are overly simplistic, reproducing forms of 'white' entitlement or gender binaries that limit the political potential" (214). Indeed, much scholarship about contemporary feminist concerns tends to disregard how such concerns are raised without attention to how youth is fetishized as the key site for future change (Gill 2016). For this reason, the RBG memes provide a provocative case study for considering how online activism and feminist politics are also invoking, explicitly or implicitly, the expectation that young people are at the forefront of change. Instead, these memes develop arguments about femininity and age, making connections between the figure of a notably older white woman and the promise of change that works towards a more feminist future. It is further notable that this work happens through digital culture, specifically memes which "have the capacity to move 'users' in new critical directions, encouraging them to challenge systems of inequality and oppression in contemporary society" (Bore, Graefer, and Kilby 2017, 529). In other words, while mobilizing the imagery of an older woman who is celebrated for her decades of work, the RBG memes deploy the imagined body of Ginsburg in order to create affective attachments to the promise of future feminist labors. In the process, they promise to reconstruct notions of the productivity of older feminine figures, affirming the legitimacy of her body, specifically, as a mode of strength rather than frailty.

CIRCULATING AND CELEBRATING RBG MEMES

While Ginsburg's image has proliferated across popular culture and consumer platforms, the production and circulation of the memes about Ginsburg powerfully reflect the needs and anxieties of popular feminist concerns. Bradley Wiggins (2019) argues that we consider memes in terms of their semiotic qualities, analyzing them not as a mode of "transmission" but as a "genre of communication" (3). In these ways, it is productive to explore the ways in which these digital artifacts function as an "ideological practice" in which they "propose or counter a discursive argument through visual and often verbal interplay" (1). Intertextuality is one of the primary ways in which memes work, creating meaning through the layering, juxtaposition and circulation of a range of texts simultaneously, and efficiently, combined. It is through semiotic and intertextual practices that memes generate meaning: "internet memes are not merely *content items* and thus simply replicators of culture but are rather visual arguments, which are semiotically constructed with intertextual references to reflect an ideological practice" (9). To this end, RBG memes draw heavily upon popular culture references in order to generate reflections about the agency and capacity of this older woman, creating textual connections across time, media platforms, and genre to generate new meanings. Many of the memes create juxtapositions between popular culture texts and historic photographic images of Ginsburg, invoking the kind of humorous combinations of associations that other scholars have argued is potentially generative of feminist political discourse. These images pose the seemingly staid image of the justice with text drawn from popular texts more often associated with youth culture and hip-hop music. In doing so, the jokes create generic contrasts that promise to raise consciousness about the agency and feminist politics of figures like Ginsburg, as well as popular celebrities. These contrasts are underpinned by ideologies of race, gender and age, resulting in a polyvalent, contradictory text upon which ageism is critiqued through the distinctly white body of RGB. The combination of Ginsburg's diminutive, increasingly aging image with text is startling because its generic juxtapositions highlight the age and whiteness of Ginsburg's body, with an ambivalence that threatens to reify hegemonic notions of gender and race.

The Notorious RBG feed on Instagram harkens back to the original Tumblr feed (https://notoriousrbg.tumblr.com/) and regularly features professional photographs of Ginsburg with text drawn from genres that seem antithetically positioned against the loftiness of the Supreme Court. For example, the June 25, 2015 meme published the day before the Supreme Court issued its decision on gay marriage features an image of Ginsburg, dressed in robes and positioned in a chair as if she were sitting for a professional portrait. Overlaid on top of this image is the directly confrontational text "come at me, bro." A meme like this presents the kind of masculine bravado epitomized by figures like Ronnie, from the show *The Jersey Shore*, who is credited for popularizing the term ("Come at Me Bro," *Know Your Meme*). The quote brings to this photograph a kind of straightforward, scrappy, physical confidence that is emblematic of the figure of white, youthful masculinity. The privilege of such confidence is transferred to the figure of the diminutive justice, granting her silent image a voice that is drawn from a reality TV show that bears little resemblance to the work, aesthetics, history, or politics of the Supreme Court. The text defies the poise and intellectualism expected of RBG, translating her judicial reasoning into direct confrontation with all of the boldness that is so easily displayed by figures like Ronnie. Such a meme, drawing upon such a popular culture form, grants RBG an implied deeply rooted masculinist physical strength, gesturing towards a steeliness that she apparently has within but is unable to fully embody when contained within the staid, traditional tropes of the Supreme Court or the expectations held for older women.

In memes like this, Ginsburg's status is augmented by rap and youth culture, depending upon them to speak for her in ways she apparently cannot when cloaked within her professional garb. The texts revise the image of the justice, using voices that capitalize on commodified notions of blackness as "urbanness" in order to render Ginsburg cool, and by extension, still relevant in the contemporary political moment. Sarah Banet-Weiser (2013) conceptualizes the way popular culture transforms civil rights and feminist activism in the interests of generating a "hip" style. As Banet-Weiser argues, "a market orientation toward the 'urban' has further consolidated the ways in which race is produced as a particular commodity more than a more traditional kind of engaged politics" (386). Thus, we must reconsider the ways in which the Notorious B.I.G. has historically been appropriated in the interests of granting Ginsburg strength and continued relevance. In these ways, memes that capitalize on black popular culture risk performing in acts of cultural appropriation that reify racist ideologies by presenting sensuality and physicality as invoking longstanding stereotypes about African Americans as threatening to white hegemony (Bogel 1990).

Such invocations of African American culture as commodified youthful coolness and strength harkens back to the original Notorious RBG memes. These memes remix Ginsburg with the rapper Christopher Wallace, aka Biggie Smalls or the "Notorious B.I.G." Wallace was murdered in a drive-by shooting when just 24 years old. Wallace was described by the New York Times as having "undeniable talent," "a reputation for preserving street-life authenticity in his lyrics" and who "described himself as a former neighborhood thug who found a better way to live" (Marriott 1997). Such mythos surrounding this rapper no doubt informs the invocation of his image and lyrics in relation to the Supreme Court justice. Both figures were born in Brooklyn and are painted as "self-made" through their own brilliance and determination, qualities that are apparently more significant than comparing the physical, ethnic, racial differences between the young African American rapper and the diminutive older Jewish woman. Instead, with memes like that tweeted by Justice Seeker on December 21, 2018, Ginsburg's face and collar is rendered in front of yellow rays evocative of sunshine.[3] With the crown and the hashtag #NotoriousRGB, the image is like many others which draw upon the iconic last photograph taken by Barron Claiborne in 1997 of Biggie Smalls before his death three days later. Wallace structures the figure of RBG, offering a "ghostly presence" that can be understood through Lauren Michele Jackson's discussion of memes in terms of blackness. For Jackson (2016), "memes in their emergence, development, transformation, and resurgence are imbued with a semantically Black mode of improvisation and revitalization" (para 20). Memes appropriate, rendering ghostly forms of Black popular culture, and yet, those forms remain, highlighting how "memes at their liveliness—that is, what allows them to keep living—is in

fact indebted to Black processes of cultural survival" (para 5). This ambivalent digital haunting appears in how the combination of imagery clearly
bestows the "kingly" status of Biggie on the tiny body of Ginsburg, a status
that depends upon the strength associated with the body of the rapper. Biggie
lives on through celebrations of his music and image, a kind of power that is
leveraged through the figure of the "Notorious RBG" but that power is deployed to serve the interests of celebrating the older white woman.

Perhaps, not surprisingly, figures from the world of feminized popular
culture also feature heavily as points of juxtaposition with Ginsburg. Beyoncé's lyrics from the song "Mine" appear written across images of Ginsburg in
another meme depicting a photograph of the justice sitting in her robes,
staring into the camera. The lyrics "All them fives need to listen when a ten
is talking" are superimposed in pink capital text surrounding her face, which
is thus safely framed within the words uttered by Beyoncé. The combination
of the image and the lyrics, which specifically associate Ginsburg with the
power of speech, work to call viewers to anticipatory attention, asking us to
attend to the voice of the justice. In this meme, the reference to Beyoncé's
prowess as a "ranked" woman, one who presumably competes with and beats
other women on the basis of both talent and sexualized beauty, is appropriated in the interests of establishing Ginsburg's distinctly feminized and thus
desirable authority. While Beyoncé's brand of "hip hop feminism" (Weidhase 2015, 130) has the potential to reclaim and grant agency to the historically denigrated, hypersexualized black feminized body, when Beyoncé's
lyrics are deployed in the interests of justifying the intellectual righteousness
of the older white justice, they reproduce a postracial colorblindness. Beyoncé, like Ronnie and Notorious BIG, grants Ginsburg assertive coolness, sum-
moning the aging figure into the contemporary present of youth culture. In
these ways, these memes operate in the "necessarily ambivalent" (Banet-
Weiser 2013, 391) way of postfeminist, postracial culture; rather than historicizing and rendering Ginsburg merely a figure of the past due to her status as
an older women, they affirm Ginsburg's continued strength and relevance,
however they do so through reproducing hegemonic norms associated with
the voices of younger bodies and of people of color.

While many of the memes draw upon lyrics and words and overlay them
on top of photographs of the justice, others create visual mosaics in which
Ginsburg's head is placed on the body of seemingly more virile bodies.
Feminized popular culture continues to offer sources of inspiration for such
digital circulations of RBG's image. Here, figures combine the youthful,
athletic bodies of women in powerful roles with the image of the justice.
Frequently, the face of Ginsburg is photoshopped onto the figures of Wonder
Woman or Black Widow. These images capitalize on the forward-looking
direct address of a feminized superhero to grant physical capacity to the
justice ordinarily seen seated, dressed in the gender neutral uniform of a

black robe. The aesthetic commonality between these posters notably presents the superhero bodies in similar ways whereby the heroine is directly facing the viewer, illustrating the body of a woman who appears to be emerging from battle. With a presumable phalanx of supporters behind her, wearing an outfit that invokes armor, these figures operate as shorthand for feminized bravery, leadership and strength. They also simultaneously reify traditional notions of feminized beauty as it is housed within a white, youthful slim yet buxom body with shiny hair flowing behind, and thus function as examples of postfemininity. In other words, autonomy and physical strength is only offered when it is contained within traditional femininity, offering the combination of feminist and anti-feminist values that scholars like Rosalind Gill have argued constitute postfeminist sensibilities.

The face of Ginsburg inserted into these celebrations of postfeminist strength somewhat changes the meanings of an older, cerebral figure. These images make uncanny the representation of strength coming from the bodies, and locate it more firmly with the reveal of the face of Ginsburg which is centralized and layered on top with the edges blurred to combine the figures of the two women. In their article about protest and affect through digital culture, Bore, Graefer, and Kilby (2017) argue digital media platforms "allow for the layering or intensification of affect that is a necessary ingredient for social change" (530). Similarly, these images of RBG's face are combined with the superhero bodies in such a way to contradict any claims that the particular body of Ginsburg might be read as having anything other than strength. The ordinarily stoic figure of the justice is transformed into one moving adamantly forward, mobilized and mobilizing the bodies of an assertive fantasy feminized figure to fight for an ethical future. Capitalizing on the fantasies of youth as forward-moving, the elderly justice herself becomes an idealized figure of righteous femininity. Wonder Woman and Black Widow are superheroines, figures who are granted otherworldly powers but use those powers to advocate for humanity. Through the aesthetic layering of these images, it appears as if RBG emerges from within these figures, representing her as someone who is secretly already aligned with the heroism of youthful femininity. These memes offer contradictory view of the aging justice. On the one hand the memes root RBG's strength in the idealized figure of the "can-do" postfeminist subject, literally propping her up on the body of youthful, conventionally feminine bodies. On the other hand, they work to affirm her strength as extending beyond her aging body, as grounded in the physical and metaphorical connections across generations of women.

The attention to the physicality of RBG speaks to a particular American historical context in which the desire for women's corporeal autonomy is under threat; for example, from the Hobby Lobby case to the rise in restrictions on abortions, to the public outcry about workplace sexual assault to efforts to defund Planned Parenthood, popular feminist discourse remains distinctly aware of how the battles associated with the dictum the "personal is political" are ongoing. If we work towards considering these memes as metaphorically operating to reflect the desires and anxieties of a cultural moment, the ambivalent status of Ginsburg's body takes on new meaning as we consider the nature of her corporeal age. The digital space of meme circulation becomes a way to grapple with the precarity of her body, and by extension, her embodied subjectivity as older woman who increasingly signifies the past, present and future potential of feminist aims. A variety of images circulated through Instagram or Twitter work through the concerns of her particular embodied status as an older woman with a body bound to uncertainty and precarity. For example, memes responding to her bouts with cancer and her fall in 2018 that caused her to break ribs deploy humor and the aesthetic affordances of digital media as "an affective strategy for self-defense" (Bore, Greafer, and Kilby 2017, 534). In one meme, an RBG figurine stands in front of a framed image of the "Notorious RBG." The figurine holds a gavel in one hand and a picket sign in the other on which is written "Bring Back the McRib." Posted on January 13, 2019, the post makes implicit reference to fall resulting in the fracture of Ginsburg's ribs, through intertextual linkages to fast food and the discourse surrounding the McDonald's McRib sandwich, one which is always in demand, and yet only sometimes offered. In another, Ginsburg's face is placed on a body covered in bubble wrap, with text overlaid "We've gotta keep Ruth Bader Ginsburg safe for at least 2 more years!" These memes articulate distinct anxiety about the tenuousness of Ginsburg's body, positioning the aging woman as at-risk, in need of community support as it stands in relation to the temporal deadlines established by the four year U.S. presidential election cycle; such awareness of a ticking clock as Trump is up for re-election in 2020 motivates concern about the effects of time Ginsburg's body as she is seen as a vital opponent to Trump policies. While working through anxiety about her body, and calling upon others to support figures like Ginsburg, such memes represent her as fragile, undermining the kind of authoritative strength which could be found in Ginsburg's faces emerging from Wonder Woman's body.

Ginsburg's body need not merely be depicted as weak because of precarity. In another series highlighting and working through anxieties about the security of her body, photographs of Ginsburg's face are given text which reads "I got another lung, bitches." Here, righteous anger emerges as the primary affect, one which promises to work in conjunction with her labors on the bench, not threaten it. In these memes, Ginsburg's corporeal form is granted a kind of anger that unifies and rallies with a confidence in her body's capabilities. In this way, these memes refute notions of aging female bodies as inherently frail and represent the aging woman as wise in her anger, not despite it; both her body and her feelings are strong. As Kathleen Woodward (2003) argues, age has long been associated with wisdom, which is "associated with thought and with the mastering of feeling" (56). Instead of presuming aged individuals should be celebrated for their decades of accumulated, impassively articulated wisdom, Woodward asks us to consider the way "wise anger" can be a feminist approach to the circulation of stories about the elderly. In other words, elderly women can embody wise anger as an honest, righteous critique of widespread sexism and ageism. Such wise anger appears through memes that defiantly speak back to those who would underestimate a justice who is all too easily assumed to be frail. Similarly, Hollis Griffin (2016) has argued that the circulation of the "Pussy Grabs Back" meme has the potential to speak of the "motivating" and "body phenomenon" of anger, thus pointing us to "an energy that harasses and persists." The Ginsburg memes which revel in righteous anger about Ginsburg's physical capacity speak to the feminist urgency of this contemporary moment, working against ageism specifically because the mobilize and glorify her body as a tool in feminist resistance, not as a hindrance to it. The memes can be taken up again and again, an endlessly reusable, rewritable resource to repeatedly, infinitely critique the infringements upon feminist and democratic values.

These RBG memes thus depend upon the promise of instability and how this instability specifically circulates around Ginsburg's body as well as how it speaks of the insecurity of the temporal present and future, in order to generate a feminist discourse about politics and aging. Her body is used to speak of the need for future change especially in the form of resisting the re-election of Trump and the threats he poses to gender equality, thus mobilizing the potential of her corporeal capacity to last, rather than decay, as is typically the narrative about aging women. These memes use the body of the individual justice to speak of collective hopes, reducing structural feminist concerns onto the body one person. In this way, such memes speak of the ways neoliberalism has co-opted digital protest and the feminist rhetoric of the personal as political. But, as Hester Baer (2016) argues, "digital feminisms are in a sense *redoing* feminism for a neoliberal age . . . by working through, making visible, and re-signifying central tensions in contemporary feminism, as well as the precarity of feminism itself in neoliberalism, these protests have begun to re-establish the grounds for a collective feminist politics beyond the realm of the self-styled individual" (19). For Baer, digital feminisms can expose the precarity in neoliberal dynamics as well as reveal how the feminine body is precarious "in a double sense as the insecure status of the female body within oppressive regimes of power but also as a site of ambivalence and potential resistance" (23). It is in this sense of precarity that righteous, angry feminist potential emerges forcefully through RBG memes, pointing to a way forward specifically through capitalizing on insecurity.

In a series focused on her workout, Ginsburg's strength is highlighted not through her work on the judicial bench so much as of the image of her as a persistent figure, one whose strength is built through time, not despite it. While representations of exercise could be interpreted as neoliberal fetishization of individual effort in the interests of appearance and "optimization" (Tolentino 2019), these memes are not the "end" of a perfect body but a "means to the end" of Republican rule; in other words, they represent an ongoing process towards a stronger, more progressive, future. In a close-up image of determination as she lifts weights, Ginsburg appears tough, and does not need words to articulate this strength. No doubt, the contrast between an older, unadorned woman wearing a sweatshirt stating, "super diva" is part of the humor of the piece, a comedic shot generated by the gap between conceptions of what constitutes a "diva" and the image of the justice. This gap is one of possibility, one which the self-determination of a diva can be transplanted into the intellectualized figure of an older woman. Furthermore, the solid security articulated by the "keep planking" memes epitomize the desire laden into these texts, speaking of how they function as what Lauren Berlant (2011) calls an "optimistic attachment" which "involves a sustaining inclination to return to the scene of fantasy that enables you to expect that this time, nearness to this thing will help you or a world to become different in just the right way" (2). The attachments forged with these images speak of persistence not obsolescence, promising to rewrite feminist politics and the ways that we see older women. Through this range of media texts, RBG offers an unruliness bound to her status as an older woman that depends upon a forward-looking sense of time, rather than positioning her as a figure of the past. Without text, which would imply the need to speak for the justice, the memes inspire through a performance of physical stability in unstable times. As they spread across a range of platforms, the RBG iconography disrupts the fetishization of youth central to postfeminist media culture and forging a feminist future of possibility. In the process, these memes aspire to construct intergenerational solidarities, presenting the figure of the justice as a part of and vital to youth cultures and forward-looking American politics. While scholars like Lepore have shown how Ginsburg's judicial actions exposed the different positions that can exist among feminists, all of whom may be arguing for equal rights but do so through different tactics, the digital image of Ginsburg is flexible, hailing a kind of feminist hopefulness that could be mobilized to create a digital community. Of course, as an "optimistic attachment" this promise of community despite difference is a fantasy limited to the digital world, but it is a powerful one that may point to promises of future change.

To this end, some of the most powerful uses of Ginsburg's digital images are thus those that are affectively evocative. Ginsburg's body is invoked as having and expressing a kind of mood, generating more connections between the justice and the affective experiences of generations of women. The wise anger that emerges here is motivating, pushing beyond the screen of the meme and onto the body of the user. For example, on January 11, 2019 Instagram user "saviem01" posted a screenshot of her iPhone lock screen with the image of Ginsburg as the wallpaper. Behind sunglasses, Ginsburg stares directly at the camera, jaw clenched, wearing a black collar top buttoned all the way to her neck and saviem01's caption reads: "my most recent lock screen gives me all the joy. #notoriousrbg #ruthbaderginsburg #rbg." Similarly, "louisianafeminist" circulates an image of Ginsburg at her confirmation hearing, holding up two fists as if getting ready to spar, and cites "Forever mood. #feminist #feminism #atheist #atheism #intersectionalfeminism #rbg #ruthbaderginsburg #rbg #notoriousrbg." Here, Ginsburg functions as inspiration, affectively binding her effectiveness to the current, and future mood of those striving to engage in "#intersectionalfeminsm." Posts like these make clear the ways her figure embodies the feelings required to live in the contemporary neoliberal focus on the self whereby the ability to personally customize one's technology to suit the feelings and moods of the user articulates the promise of late-consumer capitalism. And yet, the deployment of Ginbsurg's face in the interests of pushing the user forward into the precarious present also speaks of the ways digital media can be distinctly motivating. That they depend upon the body of this older woman to do so speaks of the feminist and anti-ageist potential of these memes. When user saviem01 cites the pleasure associated with channeling the undeniably assertive gaze of Ginsburg, the use of technology, and the larger world to which it promises access, is similarly given a kind of assertiveness. Ginsburg's gaze functions as a reminder, a sense of hopefulness of the political advances that may lay just beyond on the horizon. They offer a temporal discourse about the potential, not the limits, of this older woman's body, finding forward-looking hope, not discouragement, in the tough, wise, anger embodied by her gaze.

CONCLUSION: PLANKING OUR WAY TOWARDS INTERGENERATIONAL SOLIDARITY

Similar to the figure of Elizabeth Warren who achieved digital media fame through the proliferation of the "Nevertheless, she persisted" which became a hashtag "battle cry" in 2017 (Wang 2017), RBG gestures towards more widespread acceptance of the political efficacy of older women to inspire political change. The RBG memes promise to provide an alternative understanding of how older women are seen not as obsolete but in fact as containing a steely agency that comes through, not in spite of, their age. In the process, they rewrite the effects notions of time on women's bodies, associating the figure of RBG with desire for a feminist future that builds upon, but is not confined by her relationship to the past. Instead, it is Ginsburg's continued relevance that is highlighted here, and that ongoing sense of connection that binds these memes to the moods generated by the figure of the justice. They thus speak of aspirational feelings of intergenerational connection, feelings that appear through the intertextual quotes, the associations of her body with youthfulness and the circulation of her "mood." In doing so, they offer a view of feminism that is motivated by its relationship to older figures, not merely the youthful girl with agency so often celebrated by postfeminist media culture. In other words, while postfeminism may depict feminism as a "spent force" (McRobbie 2013, 289), feminist anger persists in the form of Ginsburg's wise body. In conclusion, the connections across generations that are signified by and aspired to through RBG memes is perhaps most poignantly embodied by the digital circulation of children dressed as Ginsburg. Across digital platforms, images of babies and children are clothed in the garb of this justice (presumably by their parents) wearing black robes, gavel, glasses, and the iconographic white collar donned by the justice when rendering decisions. These children illustrate the symbolic power of the RBG memes; young and old are united in these images, presenting a picture of intergenerational solidarity that draws upon the feelings and strength offered by generations of feminists committed to a more just future.

CODA

Ruth Bader Ginsburg died on September 18, 2020, a loss that threw into question her judicial record weeks before the U.S. presidential election. When Amy Coney Barrett, a conservative justice, was elected to replace Ginsburg on the U.S. Supreme Court, Barrett's image was quickly transformed into "notorious ACB" memes and merchandise, an act of appropriation which Dahlia Lithwik (2020) called "a turn to parody that is also cruelty" (para 3). And yet, this anti-feminist co-option underscored both the politically ambiguous flexibility and the ghostly power of the digital feminist iconography associated with Ginsburg. In the days and weeks following her death and leading up to the election Ginsburg's face appeared to haunt the Republican party. In what was perhaps the most uncanny example of the return of the RGB repressed, on October 8, 2020, immediately following the vice presidential debate, Twitter exploded with a meme featuring the poised justice seated in her robes and the words "I Sent the Fly."[4] Succinctly referring to the subtle yet unavoidable appearance of a fly that appeared on Vice President Mike Pence's snowy white hair mid-way through the debate, this meme sharply illustrates the enduring legacy of the justice; the authority of Pence, a staunch conservative with unyielding pro-life politics had his apparent authority undermined, without his knowledge, as the tiny fly distracted viewers of the debate from Pence's right wing positions. The "I Sent the Fly" meme reminded American voters that many figures can mobilize resistant disruption to patriarchal forces. While Ginsburg's Supreme Court rulings may be undermined by the harshly sharp turn to the right, the meanings she represented on digital media endure. We are left with the historic legacy of the elderly justice pressing onwards through the figures of the young feminists who poured into the streets and steps of the Supreme Court sporting RGB collars and gavels to pay their respects and signal their solidarity.

NOTES

1. The Hobby Lobby case ruled that family-owned corporations were protected by a U.S. law protecting religious freedom and thus could refuse to pay for health insurance that would provide contraception coverage to their employees. Seen as a major strike against women's reproductive justice, Ginsburg wrote the dissenting opinion to the 5–4 ruling. She critiqued the court's ruling which she wrote offered an "expansive notion of corporate personhood" that "invites for-profit entities to seek religion-based exemptions from regulations they deem offensive to their faiths" (Adam Liptak 2014).

2. I have deep gratitude for my research assistant Hallie Hoffman (Muhlenberg College) for tirelessly, enthusiastically digging into and collecting images representative of the digital circulation of RBG. Our collaboration represents the kind of intergenerational feminist community this chapter seeks to address.

3. https://twitter.com/tizzywoman/status/1076237015987965952?s=20.

4. https://twitter.com/Amy_Siskind/status/1314058961470001152.

REFERENCES

Know Your Meme. n.d. "Come at Me Bro." accessed November 16, 2019. https://knowyourmeme.com/memes/come-at-me-bro.

———. n.d. "Ruth Bader Ginsburg." accessed November 16, 2019. https://knowyourmeme.com/memes/people/ruth-bader-ginsburg.

Baer, Hester. 2016. "Redoing Feminism: Digital Activism, Body Politics, and Neoliberalism." *Feminist Media Studies* 16(1): 17–34.

Banet-Weiser, Sarah, 2013. "What's Your Flava? Race and Postfeminism Media Culture." In *The Media Studies Reader*, edited by Laurie Ouelette, 379–93. New York: Routledge.

———. 2018. *Empowered: Popular Feminism and Popular Misogyny.* Durham: Duke University Press.

Barker-Plummer, Bernadette and David Barker-Plummer. 2017. "Twitter as a Feminist Resource: #YesAllWomen, Digital Platforms, and Discursive Social Change." In *Social Movements and Media* edited by Deana A. Rholinger, and Jennifer Earl, J, 91–118. United Kingdom: Emerald Studies in Media and Communication.

Berlant, Lauren. 2011. *Cruel Optimism*. Durham: Duke University Press.

Bogel, Donald. 1990. *Toms, Coons, Mulattos, Mammies, & Bucks: An Interpretive History of Blacks in American Films*. New York: Continuum.

Bore, Inger-Lise Kalviknes, Anne Graefer, Allina Kilby. 2017. "This Pussy Grabs Back: Humor, Digital Affects and Women's Protest." *Open Cultural Studies* 1: 529–40.

Clark-Parsons, Rosemary. 2019. "'I See You, I Believe You, I Stand With You': #MeToo and the Performance of Networked Feminist Visibility." *Feminist Media Studies*: 1–19. DOI: 10.1080/14680777.2019.1628797.

Douglas, Susan J. 2020. *In Our Prime: How Older Women are Reinventing the Road Ahead.* New York: W.W. Norton & Company, Inc.

Fotopoloulou, Aristea. 2016. "Digital and Networked by Default? Women's Organisations and the Social Imaginary of Networked Feminism." *New Media and Society* 18(6): 989–1005.

Gill, Rosalind. 2007. "Postfeminist Media Culture: Elements of a Sensibility." *European Journal of Cultural Studies* 10(2): 147–66.

———. 2016. "Post-Postfeminism?: New Feminist Visibilities in Postfeminist Times." *Feminist Media Studies* 16 (4): 610–30.

Griffin, Hollis. 2016. "Biden Memes and 'Pussy Grabs Back': Gendered Anger After the Election." *Flow Journal*. December 19, 2016.

Harris, Anita. 2004. *Future Girl: Young Women in the Twenty-First Century*. New York: Routledge.

Jackson, Laura M. 2016. "The Blackness of Meme Movement." *Model View Culture*. March 28, 2016.

Jenkins, Henry. 2017. "How Young Activists Deploy Digital Tools for Social Change." *The Connected Learning Alliance*. September 25, 2017.

Jenkins, Henry, Sangita Shresthova, Liana Gamber-Thompson, Neta Kligler-Vilenchik, and Arely Zimmerman. 2016. *By Any Media Necessary: The New Youth Activism*. New York: New York University Press.

Keller, Jessalyn and Maureen Ryan. 2018. *Emergent Feminisms: Complicating a Postfeminist Media Culture*. New York: Routledge.

Keller, Jessalyn. 2019. "'Oh, She's a Tumblr Feminist': Exploring the Platform Vernacular of Girls' Social Media Feminisms." *Social Media and Society* (July-September 2019): 1–11.

Lawrence, Emilie, and Jessica Ringrose. 2018. "@Notofeminism, #Feministsareugly, and Misandry Memes: How Social Media Feminist Humor is Calling out Anitfeminsim." In *Emergent Feminisms: Complicating a Postfeminist Media Culture*, edited by Jessalyn Keller and Maureen Ryan, 211–32. New York: Routledge.

Lepore, Jill. 2018. "Ruth Bader Ginsburg's Unlikely Path to the Supreme Court." *New Yorker*, October 1, 2018.

Liptak, Adam. 2014. "Supreme Court Rejects Contraceptives Mandate for Some Corporations." *The New York Times*. June 30, 2014.

Lithwick, Dahlia. 2002. "The Contempt of 'Notorious ACB." *Slate*, September 27, 2002.

Marriott, Michel. 1997. "The Short Life of a Rap Star Shadowed by Many Troubles." *The New York Times*. March 17, 1997.

McRobbie, Angela. 2009. *The Aftermath of Feminism: Gender, Culture and Social Change*. London: Sage Publications Ltd.

———. 2019. "Postfeminism and Popular Culture." In *Cultural Theory and Popular Culture: A Reader*, edited by John Storey, 289–98. New York: Routledge.

Munro, Ealasaid. 2013. "Feminism: A Fourth Wave?" *Political Insight* (September 2013): 22–25.

Rentschler, Carrie and Thrift, Samantha. 2015. "Doing Feminism in the Network: Networked Laughter and the 'Binders Full of Women' Meme." *Feminist Theory* 16(3): 329–59.

Ryan, Patrick. 2018. "'RBG': How Ruth Bader Ginsburg Became a Legit Pop-Culture Icon" *NBC 5 News*. April 28, 2018 and December 22, 2018.

Stauffer, Rainesford. 2018. "Social media Transformed Teen's Ability to Build Activist Movements Online." *Teen Vogue*. December 19, 2018.

Thrift, Samantha C. 2014. "#YesAllWomen as Feminist Meme Event." *Feminist Media Studies*, 14(6): 1090–92.

Tolentino, Jia. 2019. "Athleisure, Barre and Kale: The Tyranny of the Ideal Woman." *Guardian*. August 2, 2019.

Traister, Rebecca. 2014."How Ruth Bader Ginsburg Became the Most Popular Woman on the Internet." *The New Republic*. July 10, 2014.

Wang, Amy. 2017. "'Nevertheless, She Persisted' Becomes New Battle Cry after McConnell Silences Elizabeth Warren." *Washington Post*. February 8, 2017.

Weidhase, Nathalie. 2015. "'Beyonce Feminism' and the Contestation of the Black Feminist Body." *Celebrity Studies* 6(1): 128–31.

Wiggins, Bradley E. 2019. *The Discursive Power of Memes in Digital Culture: Ideology, Semiotics, and Intertextuality*. New York: Routledge.

Woodward, Kathleen. 2003. "Against Wisdom: The Social Politics of Anger and Aging." *Journal of Aging Studies* 17: 55–67.

———. 2006. "Performing Age, Performing Gender." *NWSA Journal* 18(1): 162–89.

Index

150; and fiber crafting communities, 148, 155–158; hashtag activism in, 145, 149–150, 151, 156; intersectionality, 147, 148, 149–150, 154; nationalist origins of Indian women's movements, 153, 158; "people of color" as racial category, 146, 149–150; performativity, 151; queer and trans communities, 147; Savarna, 147, 159n1

Spence, Nadeen, 54

Spoonies, 74, 76

Spotted Eagle, Faith, 114, 115–116

Standing Rock Sioux Tribe, 114, 115–116

#StopAAPIHate, 1

#StopStreetHarassment, 33, 42

STS. *See* Science and Technology Studies (STS)

Super Bitch. *See* Cooper, Dreman

Survivor Healing Series, 1. *See also* #MeToo

#SurvivorPrivilege, 11

Take Back the Night marches, 99

Taylor, Breonna, 1

Taylor, Loris, 119

The Sims (video game), 74–76

#TheVajenda, 24

Thunder Hawk, Madonna, 114

TikTok, 6, 51

Tometi, Opal, 5

Topsy, 26

#TransCinayetleriPolitiktir, 24

transgender communities: and health care, 8; South Asian/Indian, 147; YouTube transition channels, 13, 129–142. *See also* Black LGBTQ+ communities

transmasculine YouTube community, 129–142; as alternative to mainstream media, 129–130; Black transmasculine communities, 140; and counterpublics theory, 131, 132–133, 138; expectations of medical transition, 139; Hispanic transmasculine communities, 140; and livability, 142; and norms of transmasculinity, 139–141; norms surrounding femininity, 140–141; and race, 139–140; reflexivity on gendered identities, 135–136, 137; rhetorics of openness, 135–136, 139; rhetorics of

passing, 135, 138–139; rhetorics of stealth, 135, 136–137; as space to connect, 129, 134; transmasculinity as intelligible subjectivity, 130, 133, 142; and trans women, 130; worldmaking in, 131–132

transnationalism, 123, 145, 146, 147, 149–150, 151, 169–170

trans women: and feminicide, 171; and transmasculine YouTube community, 130

Trump, Donald, 95–96

Tumblr, 6, 24, 51; and Ruth Bader Ginsburg, 221, 228

Turtle Island, 114, 115

Twitter, 6, 51; becomes publicly traded company, 30–32; Black Twitter, 10, 24; feminist Twitter, 11, 25; and reproductive justice activism, 188, 193–194; as site of resistance, 10; and South African student movements, 53; used by Black designers to expose theft of designs, 59

Twitter Feminism, 25

#TwitterFeminism, 32, 33

U+0023, 24, 40

Unicorn Riot, 117

United States Supreme Court: on *Burwell v. Hobby Lobby*, 33, 221, 231–232; death of Ruth Bader Ginsburg, 236; election of Amy Coney Barrett to, 236; gay marriage decision, 222; Kavanaugh hearings, 93; Marshall Trilogy, 111; Rio Yaqui dam ruling, 122

Valenti, Jessica, 40

video games as activism, 74–76

Vine, 30–32, 51, 57

virtual dwelling, 85–105

virtual sojourners, 49–62; Black and LGBTQ+ communities as, 50; comparison to nineteenth-century abolitionists, 51; Darnella Frazier as, 49–50; definition, 50; and internet law, 59–60

Walters, Maggie, 121

About the Contributors

Melissa Brown earned her PhD in Sociology at the University of Maryland, College Park. She currently works as a postdoctoral fellow at the Clayman Institute for Gender Research at Stanford University. Her research uses intersectionality as an analytical framework to examine big datasets harvested from social networking sites to study the formation of identity and community.

Tara L. Conley is an interdisciplinary Black feminist scholar, mediamaker, and assistant professor in the School of Communication and Media at Montclair State University. Her research centers Black life in the study and exploration of place, media histories, and technoculture. You can learn more about Dr. Conley's scholarship and multimedia projects by visiting www. taralconley.org.

Sarah De Los Santos Upton, PhD, is an assistant professor in the Department of Communication at the University of Texas at El Paso. Her research and teaching explore how communication can be used to create social change in the areas of community development, environmental conservation, and border activism. Her work also examines possibilities for greater civic engagement through service-learning on the Mexico/U.S. border where she lives and works.

Marisa Elena Duarte is an assistant professor in the School of Social Transformation, in the program of Justice and Social Inquiry. Duarte researches Native and Indigenous peoples' approaches to digital technologies—specifically Internet infrastructures, social media, and digitization of Indigenous Knowledge—toward decolonial resistance and imagination. Her

2017 book *Network Sovereignty* is about building Internet infrastructures across Indian Country. She currently teaches courses in digital activism, justice theory, Indigenous methodologies, and learning technologies for Native education.

Ace J. Eckstein received an MA in Rhetoric and Culture from the University of Colorado Boulder. He currently works as the teacher librarian at Peak to Peak Charter School in Lafayette, Colorado.

Sarah Ford is a PhD student at Bowling Green State University in American Culture Studies. Her primary field of interests are fan studies and digital environments and how these may impact a person's sociopolitical beliefs. Her twitter is @SarahEllenLou.

Radhika Gajjala (PhD, University of Pittsburgh, 1998) is professor of media and communication (dual appointed faculty in American Culture Studies) at Bowling Green State University, She is currently the managing editor of the *Fembot Collective*. Her books include: *Digital Diasporas* (2019); *Online Philanthropy in the Global North and South: Connecting, Microfinancing, and Gaming for Change* (2017), *Cyberculture and the Subaltern* (Lexington Press, 2012) and *Cyberselves: Feminist Ethnographies of South Asian Women* was published (Altamira, 2004). She has co-edited collections on *Cyberfeminism 2.0* (2012), *Global Media Culture and Identity* (2011), *South Asian Technospaces* (2008) and *Webbing Cyberfeminist Practice* (2008). She is currently working on a co-edited book on *Gender and Digital Labor*.

Leandra H. Hernández (PhD, Texas A&M University) is an assistant professor of communication at Utah Valley University. She enjoys teaching health communication, gender studies, and media studies courses. She utilizes Chicana feminist & qualitative approaches to explore Latina/o/x cultural health experiences, Latina/o/x journalism and media representations, and reproductive justice and gendered violence contexts. Her teaching philosophy is informed by social justice approaches, and she is passionate about mentoring undergraduate students through diverse and inclusive research projects. Her coauthored book *Challenging Reproductive Control and Gendered Violence in the Americas: Intersectionality, Power, and Struggles for Rights* was the recipient of the 2018 NCA FGSD Bonnie Ritter Book Award. Her other co-edited books—1) *This Bridge We Call Communication: Anzalduan Approaches to Theory, Method, and Praxis* and 2) *Military Spouses with Graduate Degrees: Interdisciplinary Approaches to Thriving* amidst Uncertainty—were the recipients of the 2020 OSCLG Outstanding Feminist Book Award and the 2019 Military Writers Society of America Bronze Medal Book Award respectively.

Adan Jerreat-Poole is a postdoctoral fellow in the School of Disability Studies at Ryerson University. They are a white settler in Canada who lives with chronic pain, depression, and anxiety. They study disability, digital media, and popular culture. Their work has appeared in Feminist Media Studies, a/b: Auto/Biography Studies, and Game Studies.

Vijetha Kumar teaches English and journalism at St. Joseph's College, Bangalore and has written about films, books, food, and caste.

Michelle MacArthur is assistant professor in the School of Dramatic Art at the University of Windsor. Her research focuses on four main areas: equity in theatre, theatre criticism, contemporary Canadian theatre, and feminism and performance. Dr. MacArthur's current project, "Gender, Genre, and Power in the Theatre Blogosphere," is mapping the demographics and generic characteristics of the theatre blogosphere and analyzing its findings in light of current studies and activism surrounding gender equity in theatre. Funded by the Social Sciences and Humanities Research Council of Canada, this project reflects her commitment to feminist scholarship with applications that extend beyond the academy.

Shana MacDonald is an associate professor in communication arts at the University of Waterloo and President of the Film Studies Association of Canada. Her interdisciplinary research is situated between film, media and performance studies, and examines feminist activism within social and digital media, popular culture, cinema, performance, and public art. Dr. MacDonald is co-director of the qcollaborative (qLab), a feminist design lab dedicated to developing new forms of relationality through technologies of public performance. Through the lab Shana co-runs the online archive *Feminists Do Media* (Instagram: @aesthetic.resistance) and the Feminist Think Tank. She has published in *Feminist Media Histories*, *Media Theory Journal*, *Feminist Media Studies.*

Elizabeth Nathanson is associate professor of media and communication at Muhlenberg College. She is the author of *Television and Postfeminist Housekeeping: No Time for Mother* (2013). Her scholarship on postfeminism, "women's work," and popular American media has appeared such journals as *Celebrity Studies* and *Television and New Media*, in the anthologies *Cupcakes, Pinterest and Ladyporn* (2015) and *Gendering the Recession* (2014) and in online publications, *Flow, Antenna* and *Cinema Journal/Teaching Media.*

Milena Radzikowska has over seventy-five coauthored publications and presentations on data visualization, HCI, and information design, including *Visual Interface Design for Digital Cultural Heritage* (2011) and upcoming in 2021, *Design + DH*, and *Prototyping Across the Disciplines: Designing Better Futures* (2021). She has designed 36+ interactive tools and interfaces and, in 2018, won the prestigious Design Educator of the Year Award from the Registered Graphic Designers of Canada. Dr. Radzikowska is a Full Professor in Information Design at Mount Royal University.

Angela Smith is a doctoral candidate in the Technology and Social Behavior program at Northwestern University, a joint degree in Communication and Computer Science. She is a designer and qualitative researcher who focuses on understanding and conceptualizing technology experiences that meet the information needs and practices of homeless emerging adults. Broadly, Angela's research leverages equitable design practices to give voice to vulnerable and marginalized populations. Her specific interests are finding ways to employ design as a catalyst to combat information poverty and provide socially responsible technology experiences. In her dissertation research, Angela conducts qualitative and exploratory design inquiries by leveraging cocreation and community-based participatory research methods to understand the technology needs and experiences of marginalized individuals. Angela believes that constructs of identity and social positioning impact our interactions with technology, including individual access to online information, the relevance of certain systems in our everyday lives, and the ways we accept certain interventions.

Helena Suárez Val is an activist, producer and researcher in the areas of human rights, feminism and digital cultures. She is on a PhD programme at the Centre for Interdisciplinary Methodologies, University of Warwick and is co-lead, with Catherine D'Ignazio (Data + Feminism Lab (MIT)) and Silvana Fumega (ILDA), on Data Against Feminicide, an action-research project. Since 2015, she maintains feminicidiouruguay.net, a database and map of gender-related violent deaths of women in Uruguay.

Sujatha Subramanian is a PhD Candidate at the Department of Women's, Gender and Sexuality Studies at the Ohio State University. Her research interests include girlhood and youth studies, juvenile justice and feminist media studies. She is also an editor with the Detention Solidarity Network.

Alexandra To is an assistant professor at Northeastern University in the Art + Design (Games) department and the Khoury College of Computer Science. Her core research interests are in designing social technologies to empower people in vulnerable and marginalized contexts using qualitative methods to gather stories and participatory methods to design for the future. Her recent research focuses on designing social technologies to empower support-seeking and coping with interpersonal racism. Alexandra is an activist, a critical race scholar, and award-winning game designer. She previously received her PhD in Human-Computer Interaction from Carnegie Mellon University and a BS and MS in Symbolic Systems from Stanford University. She has published work at CHI, CHI Play, DiGRA, FDG, UIST, CSCW, and DIS.

Brianna I. Wiens (she/her) is a postdoctoral fellow at the University of Waterloo and co-director of the qcollaborative, an intersectional feminist design lab. Her SSHRC-funded dissertation research, which was recently awarded a Provost Dissertation Scholarship, draws on her mixed-race queer activist-scholar experience to develop small-data feminist methodologies, considering the possibilities and constraints of feminist praxes for digital activism. Wiens's collaborative work has recently appeared in *Feminist Media Studies, Digital Studies/Le Champ Numériqe*, and *Leisure Sciences*.

Ihudiya Finda Williams is a doctoral candidate in the School of Information at University of Michigan. She is interested in understanding the impact of technology on low-income African Americans across the United States. She recognizes the great diversity of this community, and enjoys researching questions related to the intersection of race, class, and technology. Her current research focuses on the digital literacy development of individuals who were formerly incarcerated in the Detroit metropolitan area. She is an alumnus of Harvard Graduate School of Education and Rochester Institute of Technology. Prior to becoming a doctoral student, she has worked as a software engineer and product manager in various sectors of industry including Booz Allen Hamilton, Uplift Education Charter School, and the US. Department of State. Ihudiya Finda has published work at CHI and CSCW.

Byron and Scotland
Radical or Dandy?

Frontispiece: George Sanders' portrait of Byron landing from a boat,
reproduced by gracious permission of H.M. the Queen

Byron and Scotland

Radical or Dandy?

edited by
ANGUS CALDER

BARNES & NOBLE BOOKS
Totowa, New Jersey

Copyright © 1989 by Edinburgh University Press 1989

ALL RIGHTS RESERVED

No part of this publication may be reproduced or transmitted
in any form or by any means, without the written permission
of the publishers.

First published in the United States of America 1989 by
BARNES & NOBLE BOOKS
81 Adams Drive, Totowa, N.J. 07512
Printed in Great Britain

ISBN 0–389–20873–6

Library of Congress
    Cataloguing-in-Publication Data
Byron and Scotland : radical or dandy?/edited by Angus Calder.
176p. 19.8cm.
Includes index.
ISBN 0–389–20873–6
1. Byron. George Gordon Byron, Baron. 1788–1824—Knowledge—
Scotland—Congresses.   2. Scotland—Civilsation—19th century—
Congresses.   3. Scotland in Literature—Congresses.
4. Romanticism—Scotland—Congresses.   I. Calder, Angus.
PR4392.S34B9   1989
821'.7—dc20

Contents

v

Contributors

SHEENA BLACKHALL is a North East Scot, who has written four collections of poetry, and a volume of Scots short stories. A further collection of poems, and another short story volume, are due to be published later this year. A single parent, she is a ballad-singer, poet, writer, and illustrator.

DRUMMOND BONE is Lecturer in English Literature at the University of Glasgow. From 1979 until 1988 he was the academic editor of *The Byron Journal*, and he wrote the Romantic Poetry section of *The Years Work in English Studies* from 1985–87. He has lectured on Byron in most European countries, in the middle East and in India. Alongside his articles on Romanticism he also writes his own fiction.

NORMAN BUCHAN, a Labour MP since 1964, now represents Paisley South. He was a Minister in two Labour Governments and has held several 'shadow' positions including Arts and Media. He is a frequent contributor to political and cultural journals. He is the editor of 'The Scottish Folksinger' and '101 Scottish Songs.' He is Chairman of the Tribune Board.

DAVID CRAIG, born in Aberdeen, is Senior Lecturer in Creative Writing, School of Creative Arts, University of Lancaster. His latest published book *Native Stones* is being followed by one on oral memories of the Highland Clearances.

ANGUS CALDER is Reader in Literature and Cultural Studies and Staff Tutor in Arts, Open University in Scotland. He has written works of history (*The People's War: Britain 1939–45* and *Revolutionary Empire: The Rise of the English-Speaking Empires from the Fifteenth Century to the 1780s*) literary criticism

vii

(of African, Caribbean and Scottish Literature) and a short introduction to *Byron*. He co-edited *Journal of Commonwealth Literature* from 1981 to 1987 and frequently contributes to *London Review of Books, New Statesman, Cencrastus* and *Inter-Arts*.

JON CURT, born in Dundee, is doing a PhD on Byron for the University of Edinburgh and is currently a Lecturer in English at the University of Monastir, Tunisia.

WILLIAM J. DONNELLY obtained his doctorate from the School of Scottish Studies of Edinburgh University. He is a tutor in literature – including Romantic Poetry – with the Open University, author and co-author respectively of two study guides, *Scottish Society and Culture* and *Scottish Literature*, and co-editor, with Angus Calder, of a forthcoming Penguin *Selected Burns*.

DOUGLAS DUNN is a writer and poet. He contributes to *The Glasgow Herald, The New Yorker* and *TLS*. His most recent book of verse is *Northlight* (1988). He is an honorary professor in the University of Dundee.

MARGERY MCCULLOCH is a London graduate and was awarded her doctorate in Scottish Literature by the University of Glasgow. She has written extensively on twentieth-century Scottish Literature and is the author of a critical study of the novels of Neil M. Gunn. Her study of Edwin Muir is due to be published shortly. She is a literature tutor with the Open University and a visiting lecturer in Scottish Literature at the University of Glasgow.

ANDREW NOBLE is Head of the Literature Section in the Department of English Studies at Strathclyde University. He has written extensively on Scottish literature and is presently completing *Robert Burns and the Romantic Revolution* which expands some of the issues present in his Byron essay.

MICHAEL REES has been joint international secretary of The Byron Society since 1982, and was joint chairman 1975–78. Educated at Harrow and Cambridge, he studied in France and Italy and worked for the Wellcome Foundation 1968–88. He is currently translating Countess Teresa Guiccioli's *La Vie de Lord Byron en Italie*.

P. H. SCOTT is Rector of Dundee University. Born and educated in Edinburgh, and after many years abroad as a diplomat, he has been active in Scottish causes and has written extensively on historical, literary and political subjects. Publications include: *1707: The Union of Scotland and England, Walter Scott and Scotland, John Galt, In Bed with an Elephant*.

Preface

NORMAN BUCHAN MP

I was delighted to learn that the Open University and Glasgow University were organizing a symposium on the theme of Byron and Scotland. I was even more pleased to know that this book would emerge from the symposium, taking a closer look at the influences of Scotland and Byron, each upon the other. I declare an interest. I tend to agree with MacDiarmid that Byron was a great Scottish genius, and should be seen as such, despite his various anti-Caledonian strictures. (Perhaps indeed because of them. I have little doubt that he would have merrily helped to strangle the last Free Kirk Minister with the last copy of the *Sunday Post!*) In any case we cannot have better authority than Byron himself: he was, he said, half-Scot by birth and bred a whole one.

Nor is this is merely because of his origins, about which from time to time – notably at the Byron Bicentenary Dinner – we have had to give a mild corrective reminder. No, it is because there is a curiously Scottish hardheadedness, a permanent objective self-mockery that marks him out from almost all of his English literary contemporaries. His Romanticism for example is not that of Wordsworth. Imagine Wordsworth part of an Edinburgh scene – drinking in a contemporary howff in the High Street or in a pub in Rose Street in more recent times. For all his apparent English milordism, Byron, I suspect, would have been at home. And of course he was affected by his childhood. If the Jesuits can claim a seven-year sovereignty for a child, it is a nonsense to dismiss Byron's ten. And what an area to grow up in – singing Aberdeenshire no less.

I have always been struck by one enormous and significant difference between the Scottish (and Irish) poets of the Romantic Revolution and their English contemporaries. In Scotland, all of them, to a greater or lesser extent, were not

only poets but song-writers – Burns, Scott, Hogg, Tannahill . . .
and with a recent continuing ancestry, right back to Alan
Ramsay and the eighteenth century women writers ('Song-
stresses' the book calls them), Baillie and Nairn et al. But in
England, even when they essayed the odd lyric, not one of
them was a song craftsman, a song makar; not Coleridge,
Wordsworth, Southey . . . And to emphasise the point the only
other poet who came near it was, of course, the Irish Thomas
Moore. It is for this reason that I can never dismiss the
Aberdeenshire influence on Byron. Firstly, it clearly exists.
Not just his occasional echoes – 'Lachin Y Gair', for example –
but the more clearly derivative, like his 'So we'll go no more
a-roving.' Hamish Henderson pointed out, in a typically
generous and helpful letter to me, that the opening of that
song, in its ballad form as the chorus of *The Jolly Beggar*, had
already been printed several times before Byron included it
in a letter to Moore from Venice in 1817. It is not therefore
an oral, folk version of the Byron poem. (A phenomenon that
can and does happen.) On the contrary, it is the folk version,
pretty constant in its various forms, that is the origin of the
Byron:

> An' we'll gang nae mair a-rovin'
> Sae late intae the nicht
> We'll gang nae mair a'rovin'
> Though the moon shines ne'er sae bricht.

Hamish Henderson lists at least three examples printed before
1817: in Herd's *Ancient and Modern Scottish Songs* (1776),
Ritson's *Scottish Song* (1794) and, of course, *The Scots Musical
Museum*.

The primacy of the ballad is clear therefore. And it is equally
clear to anyone who knows how these things work, that the
echo is orally derived and not from the printed book. (Though,
as Henderson points out, Byron does make reference to his
awareness of printed balladry in the Introduction to *Childe
Harold*. 'The "Good Night" in the beginning of the first canto',
writes Byron, 'was suggested by "Lord Maxwell's Good
Night" in the Border Minstrelsy, edited by Mr. Scott.') He
liked the song form, as the ease of his lyrics show: 'She Walks

in Beauty', for example. His longer works are frequently punctuated by a lyric, *Childe Harold* and *Don Juan* itself. And in this characteristic the sense of Scottishness goes much deeper than a single remembered throwback to Aberdeenshire.

In the first place, the point about song – as compared to the printed poem – is its immediacy of contact and effect, its social rather than individual audience and therefore its deeply popular, indeed democratic, nature as an art form. And all of that is intensely Scottish – though, thank goodness, not exclusively so – in recent British cultural history. As in the form so, frequently, with the content. Despite its reference point presence in Scottish literary history, Ramsay's gentrified 'Gentle Shepherd' was alien to the surrounding song from which it was derived, which was inevitably popular, and therefore frequently radical.

That last quality is worth stressing. The picture of the gloom and doom Romantic is, of course, long exploded. Byron's hero was Pope. His style was eventually Augustan. He combined the radicalism of song with the brisk thinking-on-your-feet style of oral argument. He continually 'said' something. *Don Juan* is simply the best piece of hard-headed verse journalism we have in English, and at the same time a great and rollicking epic. He was one of the few who have combined coolness with passion. An agnostic who seemed to have respected the Calvinist zeal for argument. He said that as a boy he had read most of the Bible 'through and through – that is to say, the Old Testament, for the New struck me as a task, but the Old as a pleasure. I speak as a boy from the recollected impression of that period at Aberdeen in 1796.' He clearly revelled in the thundering of the language of the old Testament and the debate of the theologians – though he remained a sceptic, a common enough Scottish experience. Here he is in a letter to Edward Noel Long, quoted by Michael Foot in his *Politics of Paradise*: 'This much I will venture to affirm, that all the virtues and pious *Deeds* performed on Earth can never entitle a man to Everlasting happiness in a future State; nor on the other hand can such a Seat of eternal punishment exist . . .' And later, and with the customary wry sting in the tail, he comments: 'In Morality I prefer Confucius to the ten Commandments and

Socrates to St. Paul (even though the two latter agree in their opinion about marriage)' . . . But he liked the argument and the passion, and no man could use them better, and since he had no hope of a spiritual heaven in the hereafter he at least set about using these twin tools to get rid of a human hell in the present. In short, he became to his everlasting glory the poet of Revolution. He sang about it, he joked about it, he argued for it and ultimately he died for it.

In *Cain*, in *Manfred*, in *Marino Faliero*, he set out his stall. Man's noblest task was to revolt against tyranny whether of secular ruler, God or Devil. He seems to have equated the last two. And, rather like Milton, he gives the devil all the best lines. To Lucifer in *Cain* for instance:

> Souls who dare use their immortality,
> Souls who dare look the Omnipotent tyrant in
> His everlasting face and tell him that
> His evil is not good!

And here, as in *Manfred*, he hints at the argument that being given knowledge is not enough – man needs power too. An interesting and never properly examined shift from the Faust of Marlowe and perhaps also of Goethe. A comment on the insufficiency of the Liberal (bourgeois) revolution? I offer that idea freely to someone for a doctoral thesis! (preferably, I can hear Byron say, in Theology.) In *Marino Faliero*, above all, the theme of revolution is made clear. Indeed at times he theorises on the very nature of revolt. His analysis of the ambivalent position of the aristocrat Faliero himself, as he throws in his lot with the revolutionaries to destroy the Senate, must come close to Byron's own position. And it is the weakness of Faliero as opposed to the hardened revolutionaries, Bertuccio and Calendaro, which exposes their plot, leads to their common defeat and the ultimate execution of Faliero.

The last scene takes place in the centre of the courtyard of the Doge's Palace, with Faliero on the scaffold. The people, vaguely heard in the distance, are locked outside the gates watching the execution. He asks to speak and is told the people are too far away. They cannot hear him. The whole scene curiously echoes an earlier scene when one of the

conspirators asks: 'But if we fail . . .?' And Bertuccio replies: 'They never fail who die/In a great cause: the block may soak their gore;/. . . But still their spirit walks abroad. Though years/Elapse and others share as dark a doom,/They but augment the deep and sweeping thoughts/Which overpower all others, and conduct/The world at last to freedom.' In the same way, Faliero speaks unheard and says: 'I speak to time and to eternity . . .' Then, in an extraordinarily modernist way, the entire action of the execution scene is repeated once more, with a different perspective both of location and mood. This time the scene is enacted from outside the locked gates and from among the crowd. As they watch the execution, the people murmur of continuing revolt. (In modern Venetian terms,' la lotta continua!') 'Then they have murdered him who have freed us . . . we would have brought/Weapons, and forced them! 'and as the gates open and the crowd rush in, the curtain falls. I know of no comparable technical experiment like this until our own century, in the Japanese film *Rashomon*, for example. Regrettably, Byron, who described the play, in self-comparison with Napoleon, as his 'Leipzig', forbade its production in his own time. (Though it was, I believe, performed once in the Green Room at Drury Lane, against his wishes.)

We need not wonder at Byron's identification with revolution. His Scottish mother (one of the 'mad Gordons of Gight') was herself a fairly outspoken radical. Writing in 1792, in the middle of the wildest anti-French Revolution hysteria, she remarked calmly: '. . . I do not think the King. [i.e. King Louis XVI] after his treachery and perjury, deserves to be restored. To be sure there has been horrid things done by the People, but if the other party had been successful, there would have been as great cruelty committed by them . . .' And she goes on to refer with some interest to the formation of the Friends of the People in the principal towns of Scotland.

And not for nothing was Byron often identified in the public mind with King Ludd himself. Writing at a time when we have witnessed the nobility of the realm summoned in their hundreds to pass a Poll Tax that benefitted all of them at the expense of the mass of the people, we can guess at how Byron

might have flayed them. Indeed we have evidence. He spoke three times in the House of Lords, each time with courage and each time for freedom. The nobleman speaking for the poor mechanic. The agnostic speaking for Catholic freedom. Even allowing for the rhetoric of the time his speech attacking the Bill to bring in the death penalty against the machine-breakers rings tough and hard. And it remains worth the quoting. He had not long returned to England from his travels in the Balkans, which he compared with what he was now witnessing in England:

I have been in some of the most oppressed provinces of Turkey. But never under the most despotic of infidel governments did I behold such squalid wretchedness as I have seen since my return in the very heart of a Christian country. And what are your remedies? After months of inaction, and months of action worse than inactivity, at length comes forth the grand specific, the never failing nostrum of all state physicians . . . the warm water of your mawkish police and the lancets of your military. . . . How will you carry your Bill into effect? Can you commit a whole county to their own prisons? Will you erect a gibbet in every field and hang up men like scarecrows? . . . Are these the remedies for a starving and desperate people? When a proposal is made to emancipate or relieve, you hesitate, you deliberate for years; but a Death Bill must be passed off-hand without a thought for the consequences.

And even if it were passed, he said, even with all the battery of powers in the Bill, 'It would still need two things more – Twelve butchers for a jury, and a Jeffreys for a judge!'

Then all of this – the direct commitment to political involvement, the humour and the passion, the directness of song, his glorying in rhyme and rhythm – came together. He was working on a new poem, he told Moore. It was a bit facetious, but even so it might be 'too free for these very modest times'. Gone the occasional posturing, the mock romantic, the overstrained language of his dramas. In its place the marvellous stanza form of *Don Juan*, carrying what Michael Foot calls the 'Politics of Paradise' in a seemingly unending flow of invention, witty, sharp, confident, his passion for

liberty, sharpened by his love for Italy (and Teresa Guiccioli), his hatred of tyrants, now clearer as he moved away from his ambivalence about Napoleon. And through it all a happier and easier narrative line than exists anywhere else in English-language poetry – not Pope, not Dryden, not Scott. It has a kind of artlessness at variance with the enormous skill even of the rhyming techniques. Only in the very different verse form of the ballad do we find this same colloquial ease. (Perhaps also in the Burns of 'Tam o' Shanter' and he, of course, was steeped in ballad and song.) Perhaps 'singing Aberdeenshire' influenced him more than he knew – or we can know. (He remembered enough to call the Grammar School the 'squeel'.)

Don Juan was an enormous achievement. It remains a masterpiece. All the old targets all there, but his stance is cool and his aim sharpened. God or evil man together:

> 'Let there be light!' said God and there was light,
> 'Let there be blood!' said man, and there's a sea!

Through all the fun and the sex and the vitriol of *Don Juan*, there remains a conscious combining purpose. He is comment-ing on the world, as a kind of merry didactic journalist – and *that's* certainly a Scottish trait. But so is the radical heart of the message:

> For I will teach, if possible, the stones
> To rise against Earth's tyrants. Never let it
> Be said that we still truckle unto thrones –
> But ye, our children's children! think how we
> Showed what things were before the world was free!

I wish this book well. It is timely, following Byron's bicenten-ary year. And even more timely, perhaps, because of the moral and intellectual climate we presently endure. When even the work of Lord Elgin has still to be dealt with, and restitution made to Greece.

For the spoken and the written word is under attack again as it was with Byron. We are facing an indirect but comprehensive and increasing censorship. His publisher, Murray, seemed

timid enough to Byron. His correspondence is full of com-
plaints about it. But would similar work, taking on latter-day
Castlereaghs or Prince Regents, find any readier a publisher
today? In Byron's time and after, seven hundred men, women
and children went to gaol for distributing the unstamped press
– The Poor Men's Guardians. Today the popular press is in the
hands of only three men: Maxwell, Murdoch and Stevens. The
government pursues across five continents, and with all the
dented majesty of the law, an old man's book on his
disreputable trade of spying. We have restored a *de facto* Lord
Chamberlain in the shape of Rees-Mogg. The freedom of the
broadcaster is under attack and badly dented after 'Real Lives',
'Zircon' and the invasion of Scottish BBC by the Special Branch.
Self-censorship has lowered the threshold at which censorship
begins. Clause 28 would endanger a contemporary Byron. A
new Obscenity Act is threatened and with it all the joyous life
of a modern rhyming *Don Juan* would be in jeopardy.

What red meat all this would be to Byron! How he would
have scarified the visit of a premier to the General Assembly!
The evocation of St Paul, of Christ, of God! All that and *The
Sunday Times*, too – how could he have resisted it? The memory
of his *English Bards and Scotch Reviewers* . . . perhaps an English
Bird and Scottish Prelates? The combination would have been
irresistible!

No. He would certainly have written it. But in today's
developing climate, would he, could he, have found a
publisher brave enough to publish him?

Introduction

ANGUS CALDER

Of the dozen contributors to this volume, all but one are Scots. The occasion which brought us together was a Conference on 'Byron and Scotland' organised in January 1988 to celebrate the bicentenary of Byron's birth, by myself and Peter Gilmour on behalf of the Open University and Ann Karkalas for the Department of Adult and Continuing Education of Glasgow University, which provided the venue.

As organisers, we aimed to put on something which would involve well-equipped specialists in Byron and in Scottish literature but would be 'open' to any of the enquiring adults who provide a clientèle for our month-in month-out activities. The atmosphere was deliberately informal. A session addressed by David Craig and Andrew Noble generated particularly vigorous debate, in which Norman Buchan MP intervened on lines suggested by the Preface which he has kindly contributed to this volume. Proceedings concluded with a reading by Douglas Dunn, an award-winning Scottish poet who has edited Byron, and Sheena Blackhall, who not only writes verse in the Aberdeenshire Doric which Byron heard around him in his childhood, but sings unaccompanied in the North-Eastern tinker tradition.

So, the vitality of the occasion was drawn not only from its topic, but from the current resurgence of Scottish culture, involved as it has been with passionate politics. When Saintsbury and Grierson, to whom Jon Curt refers in his contribution, were professors at Edinburgh University, Scottish literature was regarded as an annexe to English. Very few well-read people in Scotland see it that way now. The independent Scottish tradition, which T. S. Eliot, writing on Byron in the 1930s, perceived, albeit rather dimly, is now recognised as something which from Barbour to the present

1

day has created its own characteristic modes of address to the reader, its own complex of intertextualities. No one in their senses would claim that this tradition wholly determined Byron's writing, but only a very obtuse person could miss the signs that his Scottish childhood left a strange mark on him, or fail to see some significance in his eager appreciation of Burns and Scott, poets who demonstrably influenced him.

His work, of course, has to be seen in relationship to other traditions. Jerome McGann's wonderful notes to his new definitive edition of the *Poetical Works* demonstrate Byron's saturation in the heritage of classical literature which he shared with every well-educated Scottish, English or European contemporary. Italy gave him the stanza form which he used in his finest work. His veneration of Pope and his commitment to Popeian satire help to explain certain scathing passages about Scotland and Scots to which Douglas Dunn draws attention in this volume. Pope's most notable successor had been Charles Churchill. In the phase of intense anti-Scottishness directed at the person of Lord Bute, George III's Prime Minister, Churchill's *Prophecy of Famine* (1763) had been lavish with invective. The Scottish landscape, for instance, had been described thus:

> Far as the eye could reach, no tree was seen,
> Earth, clad in russet, scorn'd the lively green:
> The plague of locusts they secure defy,
> For in three hours a grasshopper must die . . .

Innovative though he was elsewhere in form and feeling, Byron wrote couplet-satire as a follower of Pope and Churchill, and I think *English Bards and Scotch Reviewers* and *The Curse of Minerva* reveal no serious anti-Scottish animus, rather a young poet's emulation of his predecessor: he drew from a bank of anti-Caledonian jibes.

Likewise, it can be argued that his famous House of Lords speeches exhibit a young Whig striving to imitate the admired effects of Fox, Sheridan, Erskine. I see no reason to dispute Malcolm Kelsall's assertion in his recent *Byron's Politics* (1987, p.2) that Byron inherited the tradition of 'the patrician Whigs'

and that, like his friends Hobhouse and Kinnaird, he was self-consciously a would-be successor to the late, great Charles James Fox. He was not a democrat. He had a Whiggish horror of mob rule, though he shared the Whig belief that men had a right to use arms to preserve themselves and their liberty. Hence his admiration for Washington, the Republican slave-holder, was not self-contradictory, and hence Pushkin and his aristocratic Decembrist friends were right to sense a fellow-spirit in him. But does his aristocratical Whiggery account fully for the power of his best political verse? Was he, in spite of his cooler judgement, a revolutionary writer of democratical tendency? Our volume opens with Norman Buchan's impassioned claims for Byron's radicalism, and the intriguing link which he makes between this and the Scottish song tradition. David Craig, on the other hand, views him in the context of English radicalism, and attributes the 'shallowness' of all but his best verse to his remoteness from the popular movement. Andrew Noble goes further to accuse Byron of 'collusion' with his purported audience, of creating fluent trash, and of exemplifying, with his 'displaced nationalism', the capitulation of Scottish culture to Unionism and Imperialism, – and yet acknowledges the 'archetypal Scottish Calvinist' who wrote a masterpiece in *Don Juan*.

The interesting – indeed, all-important – question of Byron's religious perspectives proved harder to expose at our conference than problems to do with politics, class and nationality. For far too long the Scottish literary intelligentsia have tended to assume that our nation's Christian inheritance is a Bad Thing, typified by bigotry, joylessness, betrayal (the supposed role of the Church in the Clearances) and, latterly, the prudery of Roman priests. Dr Donnelly's contribution on 'Byron and Catholicism' is especially welcome in that it takes seriously the weight of Calvinism's perfectly logical presentation of the unsaved individual, and humanity in general, as 'uncatered for, unreconciled, irredeemable', and the pressure which it put on Byron, so helping to explain what is otherwise a puzzle, this pre-eminently 'Freethinking' poet's attraction to Catholic ritual and the Catholic concept of Purgatory.

The traditions of the Gordons of Gight were Catholic. From

Jenny Wormald's admirable *Court, Kirk and Community: Scot-land 1470–1625* (1981, pp133–4) we learn that in 1610 the Protestant Bishop of Moray 'wrote a moving letter to King James pleading that the Catholic Laird of Gight should be left in peace because he was ill . . . "the papists, I perceive, are not universally of ane corrupt disposition".' The surprising toler-ance which characterised Scottish society in the first few decades after the Reformation has been overlaid by images of the Covenanting period which followed – yet it helps to explain the persistence of a folk culture in which James' great Catholic court-poet, Montgomerie, was read and revered while gentry and peasantry alike preserved and extended the national heritage of song. The supposed meanness of the Scot, denounced by Churchill and others and still part of our image abroad, is belied by the generosity of feeling in that heritage.

Byron could be intensely snobbish: the posture went with his romanticisation of his ancient lineage, his elevation of Pope and contempt for Wordsworth, and his dislike (not merely paranoid, in view of the extreme penalties which his homosex-uality might have incurred) of the Evangelical morality of the English middle and, in some part, lower classes. But the claim that he was a democrat in spite of himself could (perhaps) be based on the generosity, the human fellow feeling, which he extends in *Don Juan* to people of all nations and classes. A similar quality was present in the writings of Walter Scott. Paul Scott's very thorough investigation here of his great name-sake's relations with Byron shows both men in an excellent light.

Margery McCulloch's study of the dealings of another major Scottish novelist, John Galt, with 'the Noble Lord', illuminates both writers. It also brings out the point – obvious enough, when one thinks of it – that in that heyday of British expansion, in which Scots joined so avidly, one did not have to visit Scotland in order to meet plenty Scots. Douglas Kinnaird, Byron's intimate friend, banker and financial adviser was like himself of noble Scottish antecedents. Douglas Dunn's review of Byron's concern over the 'Marbles' riven from the Parth-enon by another wandering Scot, Lord Elgin, illuminates European vistas in which Scots intruded themselves every-

where and raises again the question of 'displaced nationalism', a matter which I address, though not directly, in my own contribution on *The Island*.

The popular verse narratives of the Romantic heyday cry out to be decoded in terms of political ideology. Scottish practitioners of the genre were paramount – and Walter Scott, Thomas Campbell and Byron were all intensely 'political' people. It is my own conviction that the specific characteristics of *Scottish* romanticism could bear a lot more investigation, and that Byron's now unfashionable verse tales might be found to share in them. Meanwhile, Drummond Bone takes further than most have done the interesting but painful process of probing minutely into the presences and absences in Romantic literature which accompanied the evolution of a nostalgic, tartan-patterned image of Scotland for home and foreign consumption.

Mercifully, Scottish culture has had enough vitality to defeat the stereotypers and nostalgia-mongers. Sheena Blackhall writes from a North-Eastern milieu of literature and song which, more than any regional culture in Britain, has retained continuity and self-sufficiency. The fact that a poem by local-boy Byron is sung in Aberdeen as if it had no author but the people points to an enduring characteristic of Scottish literature: the closeness of our greatest writers to the everyday life around them. MacDiarmid's posture as a self-confessed élitist is *not* an exception that proves the rule, since his élitism was expressed, paradoxically, in demotic terms. I don't myself care very much whether Englishpersons or other foreigners choose to regard Byron as a Scottish writer: I do heartily hope that many more Scots will come to feel, as I do, that the author of *Don Juan* is as close to us as 'Davie' Lyndsay and 'Rabbie' Burns. As Scotland seeks to express its identity within the structure of the European Community, to be strengthened in 1992, the Aberdeen boy who loved Italy and died in Greece might seem an especially apt culture-hero.

Byron the Radical

DAVID CRAIG

In Aberdeen forty or fifty years ago Byron was very much 'our poet'. My grandfather had had a bronze statue of him placed on a granite tower in front of the Grammar School. At that school I was in the 'house' named after him, I played rugby and cricket for 'Byron', and cheered on the athletes wearing the blue silk ribbon on Sports Day with a splendid illusion of taking part in something rather heroic. His lyric to Lochnagar ('Lachin Y Gair', 1807), for all its quality of a blurred and florid painting in oils, resonated for me because Lochnagar was for us *the* mountain: we walked up it from Glen Muick by Fox's Well, climbed its thousand-foot cliffs, and it was even visible from the city limits fifty miles away as a peaked rampart darkened by rainstorms, standing up like a blue banner striped with white, announcing summer.

The point of all this (apart from sheer self-indulgence) is to suggest what a heady multiple charge we got from the Highlands in their most local form, so much so that I have spent the years since trying to sort out what is valid and useful from the merely attitudinising. Byron's sense of mountainous wilderness was not all rhetoric. One crucial snatch of *Childe Harold* rings true, even though we may feel he has needed Wordsworth's help to say

> I live not in myself, but I become
> Portion of that around me; and to me,
> High mountains are a feeling, but the hum
> Of human cities torture: I can see
> Nothing to loathe in nature . . .[1]

Byron is here creating a *persona* very much of its time, a member of the same family as Rousseau, from the Swiss Alps, and Wordsworth, from the dales and fells of Cumberland.

Writers (mainly of the middle and upper classes) were now getting charges of energy and whole wells of material from beyond the pale – from the not-yet-colonised moors and mountains and deserts, and from the lives of peasants and hill-farmers, factory workers, the peoples of remote countries. A habit of identifying with the more turbulent and rough, the less genteel, biddable, schooled, or affluent, lies near the marrow of Romantic art during those years, from Blake's *Songs of Experience* and the storming of the Bastille at one end of the epoch to Delacroix's *Liberty on the Barricades* and the overthrow of the Bourbons in 1830: 'the definitive defeat of aristocratic by bourgeois power in western Europe', as it has been called, and 'the emergence of the working class as an open and self-conscious force'.[2]

Byron was born the year before the French Revolution and died the year the Combination Laws were repealed. Such were the events which challenged a person of that time to understand, to commit himself or not. Now, Byron was a vehement reviler of despots; 'king' for him was one of the dirtiest of words. His maiden speech in the House of Lords (on 27 February 1812) was against a Bill to make lawful the hanging of Luddites. One of the most sustained, least shallow or fitful passages in his poetry is the ten verses at the start of *Don Juan* Canto IX (1822), in which he analyses with implacably critical scorn the career of Wellington, great military hero and Prime-Minister-to-be. How deep did all this run in him? Was it from the centre of his poetic self? When he wrote most originally, how far was he drawing on social-radicalism, by which I mean solidarity with the forces making for change and against the ruling class?

Let us remind ourselves of the poetry concerned here. It consists of that page from *Don Juan* and many scathing asides from the same poem, for example on Castlereagh, the 'intellectual eunuch' –

> The vulgarest tool that tyranny could want,
> With just enough of talent, and no more,
> To lengthen fetters by another fix'd,
> And offer poison long already mix'd.[3]

Or on George IV –

> Gaunt Famine never shall approach the throne –
> Though Ireland starve, great George weighs twenty stone.[4]

These are epigrams, strung loosely on the picaresque tale. As such they belong with his other political two-liners, for example:

> So *He* has cut his throat at last. – *He? Who?*
> The Man who cut his Country's long ago.[5]

This is crunching stuff, and fearlessly outspoken, if you consider that it was aimed at a senior member of the Cabinet in an age when it had been made illegal to agitate against the monarchy and for a republic. True, it smacks of the squib, and specifically of a young man's naughty wish *pour épater les bourgeois*. He knew this himself, and when he wrote his 'Song for the Luddites' in the safety of Venice and sent it to Tom Moore, he lightlied it as 'an amiable *chanson* . . . I have written it principally to shock your neighbour . . . who is all clergy and loyalty'.[6] How far was Byron able to found a sustained poem on such materials, or rise to as convincing a political vision as Dryden's conservative ideal of 'solid Pow'r' in *Absalom and Achitophel*, say, or Pope's more liberal conception of enlightened landownership in his fourth *Moral Essay*? Twice, it seems to me, he nearly did, in *The Vision of Judgment* and *Don Juan* Canto IX, both written in 1822, at the juncture when the long and dreadful abeyance of libertarian politics under the war economy and the Gagging Acts was breaking out into the revolutionary ferment that produced the Reform Act of 1832.

At the start of *The Vision of Judgment*, before he has turned to his main business, which is the lampooning of the conformist Southey – his favourite butt and incarnation of mediocrity, as Shadwell was for Dryden or Dennis for Pope – Byron delivers himself of an impressively considered judgement on those last twenty-five years of George III's reign, which had seen the heyday of police spies, the hounding and parliamentary suppression of the early trade unions, and the frequent hanging of political prisoners:

In the first year of Freedom's second dawn
    Died George the Third; although no tyrant, one
Who shielded tyrants, till each sense withdrawn
    Left him nor mental nor external sun;
A better farmer ne'er brushed dew from lawn,
    A worse king never left a realm undone!
He died – but left his subjects still behind,
One half as mad – and t'other no less blind.

A wonderfully black image is then created of the funerary
pomp, but it does not rest there: a traditional line of thought
about mortality is turned to insist on all the monarch has in
common with his subjects:

So mix his body with the dust! It might
    Return to what it *must* far sooner, were
The natural compound left alone to fight
    Its way back into earth, and fire, and air;
But the unnatural balsams merely blight   ·
    What Nature made him at his birth, as bare
As the mere million's base unmummied clay –
Yet all his spices but prolong decay.

This is a levelling wit, and the republicanism it implies is given
full voice when Satan makes his case for carrying King George
down to the nether world:

'Tis true, he was a tool from first to last
    (I have the workmen safe); but as a tool
So let him be consumed . . .

'He ever warred with freedom and the free:
    Nations as men, home subjects, foreign foes,
So that they utter'd the word "Liberty!"
    Found George the Third their first opponent . . .

'The New World shook him off; the Old yet groans
    Beneath what he and his prepared, if not
Completed: he leaves heirs on many thrones
    To all his vices, without what begot
Compassion for him – his tame virtues; drones

> Who sleep, or despots who have now forgot
> A lesson which shall be re-taught them, wake
> Upon the throne of Earth; but let them quake!⁷

The strength of this is not only in the idiomatic verve of the
disparagements but also in the plain candour with which
principles are invoked and tyranny confronted: 'he leaves
heirs on many thrones'. To speak out so ringingly for
republicanism was as bold, and personally as dangerous, as it
has been in other places and times to declare for socialism or
communism (or against Stalinism). For a generation it had
been illegal 'to advocate the dethronement of the King, the
establishment of a Republic, or the destruction of the Church,'
or 'to organise bodies to achieve any or all of these things.' But
by now, as a result of their hardships during the French Wars,
the organised workers, for example the wool-combers and
cotton-weavers, who had been 'Church and King' men, had
declared for reform under a republic and Byron, in theory at
least, was with them.⁸

   This is also the tendency of that supreme passage in *Don
Juan* Canto IX and the masterly poetry of it is sheerly Byronic,
not in the best-seller sense of the word (the world-weary
misanthrope cloaked and brooding in his exotic retreat), but in
the aristocratic sophistication with which he turns and points
and angles and inflects what at first seems to be a courtly
tribute until it becomes an indictment:

> Oh, Wellington! (or 'Vilainton' – for Fame
>   Sounds the heroic syllables both ways;
> France could not even conquer your great name,
>   But punned it down to this facetious phrase –
> Beating or beaten she will laugh the same) . . .

Already the conservative reader might be feeling a little
disconcerted as disrespect steals in on the heels of the
honorific address. Within ten lines it is unmistakable:

> Though Britain owes (and pays you too) so much,
>   Yet Europe doubtless owes you greatly more:
> You have repaired Legitimacy's crutch, –
>   A prop not quite so certain as before:

>The Spanish, and the French, as well as Dutch,
>   Have seen, and felt, how strongly you *restore* . . .
>
>You are 'the best of cut-throats:' – do not start;
>   The phrase is Shakespeare's, and not misapplied: –
>War's a brain-spattering, windpipe-slitting art,
>   Unless her cause by Right be sanctified.
>If you have acted once a generous part,
>   The World, not the World's masters, will decide,
>And I shall be delighted to learn who,
>   Save you and yours, have gained by Waterloo?

So it unfolds, the apparently civil surface more and more rough-
ened and darkened by cutting innuendoes, the formal address
more and more loosened to let direct challenges break through:

>I am no flatterer – you've supped full of flattery:
>   They say you like it too – 'tis no great wonder:
>He whose whole life has been assault and battery,
>   At last may get a little tired of thunder . . .
>I've done. Now go and dine from off the plate
>   Presented by the Prince of the Brazils,
>And send the sentinel before your gate
>   A slice or two from your luxurious meals . . .
>
>I don't mean to reflect – a man so great as
>   You, my Lord Duke! is far above reflection . . .
>Though as an Irishman you love potatoes,
>   You need not take them under your direction;
>And half a million for your Sabine farm
>   Is rather dear! – I'm sure I mean no harm.

The charade of courtesy and tribute has now played itself out:
mention of past leaders (Epaminondas, Washington, Pitt)
brings recent European history crunching in and the poem
(virtually a self-contained anti-ode) raises itself to deliver an
explicit and absolutely serious judgement:

>Never had mortal Man such opportunity,
>   Except Napoleon, or abused it more:
>You might have freed fall'n Europe from the Unity

Of Tyrants, and been blest from shore to shore:
And *now* – What *is* your fame? Shall the Muse tune it ye?
   Now – that the rabble's first vain shouts are o'er?
Go, hear it in your famished Country's cries!
Behold the World! and curse your victories![9]

How supply Byron has toyed with his subject – how he has entered demure disclaimers; promised to stop, then carried on; appealed to virtues (while implying they are signally missing); offered (impossible) advice with perfectly mimicked mock-humility; kept his most pungent points for (apparently) throw-away asides – until at last he draws himself up for that clinching judgement. The passage is as fine as Mark Antony's key speech in *Julius Caesar* (and twice as long) in its command of speaking tone used right across the gamut, from transparent insolence to naked challenge, and the subject it takes on so frontally is the most powerful British person of the time, the lynch-pin of its ruling class.

How disappointing it is, then, that the one movement of the time with which Byron might have made common cause – the Left Radicalism of Cobbett and Henry Hunt – found him recoiling in a dislike which he overstates with the huffing-and-puffing vehemence of any true-blue. In 1820 the Cato Street conspiracy was unmasked. *Agents provocateurs* had pretended that the Cabinet were to be together at a banquet and then egged on a few desperate ultra-Radicals to plot their assassination. Secret policemen shopped them, five were hanged, and their heads were held up for the London crowd to see – 'barricaded at a safe distance from the scaffold so that no rescue could be attempted and no dying speeches heard.'[10] They had planned, on a fairly fantastic plane, to seize the arsenals, burn down the barracks, and set up a provisional government in the Mansion House. But there was no rescue attempt when the arrested men passed through London; the leading Left Radicals, in the aftermath of the Peterloo Massacre, were studiously avoiding any rash provocations.[11] Yet here is Byron's reaction to the affair, in a long political letter sent from Ravenna to his closest friend in England, the MP John Cam Hobhouse:

What a set of desperate fools these Utican Conspirators
seem to have been. – As if in London after the disarming
acts, or indeed at any time a secret could have been kept
among thirty or forty. – And if they had killed poor
Harrowby – in whose house I have been five hundred times
– at dinners and parties – his wife is one of "the Exquisites" –
and t'other fellows – what end would it have answered? –
"They understand these things better in France" as Yorick
says – but really if these sort of awkward butchers are to get
the upper hand – *I* for one will declare *off*, I have always been
(*before you* were – as you well know) a well-wisher to and
voter for reform in Parliament – but "such fellows as these
will never go to the Gallows with any credit" – such
infamous Scoundrels as Hunt and Cobbett – in short the
whole gang (always excepting you B. and D.) disgust and
make one doubt of the virtue of any principle or politics
which can be embraced by similar ragamuffins. – I know that
revolutions are not to be made with rose-water, but though
some blood may & must be shed on such occasions, there is
no reasons it should be *clotted* – in short the Radicals seem to
be no better than Jack Cade, or Wat Tyler – and to be dealt
with accordingly.[12]

To take the measure of Byron's capitulation here to the merest
reactionary panic, contrast the response of Cobbett himself to
the labourers' revolt ten years later, when 'the fires were
blazing, more or less, in SIXTEEN *of the counties of England*'. In
the *Tuppenny Trash* for November 1830 he writes:
. . . our first feeling is that of *resentment against the parties*; but,
when we have had a little time to reflect, we are, if we be not
devourers of the fruit of the people's labours, led to ask,
What can have been *the cause* of a state of things so unnatural
as that in which crimes of this horrid sort are committed by
hundreds of men going in a body, and deemed by them to be
a sort of *duty* instead of *crimes*?

And in the *Political Register* for 4 December he answers his own
question, ironically quoting
a great landholder, in Wiltshire, named BENNETT, who,
upon being asked how much a labourer and his family ought
to have to live upon, answered, "We calculate, that every

1960s; it also achieved a formidable unison. The full-throated judgmental voice Byron levelled at the nation's leaders just after 1820 was closely anticipated by Shelley's treatment of the Cabinet at the start of *The Mask of Anarchy*:

> Next came Fraud, and he had on,
> Like Eldon, an ermined gown;
> His big tears, for he wept well,
> Turned to mill-stones as they fell.

> And the little children, who
> Round his feet played to and fro,
> Thinking every tear a gem,
> Had their brains knocked out by them.[20]

Here is that same voice of barely-contained righteous scorn, gathered up into implacable accusation. Again, however, it is not sustained; the allegory of the *Mask*, after its biting specification of bishops, lawyers, peers, and spies, tails off into abstraction and fails to become a narrative. The fact is that both men were writing from a distance – from Italy. Both had been absent from Britain when the quickening undercurrents of revolution desperately needed spokesmen and rallying-points (and funds). Both, as a result, were lacking in a density of first-hand social materials such as would have been needed to nourish a major poetry and both, within a few years, were dead – avoidably, it could be said – the one kicking his heels in the malarial swamps of the Peloponnese while yearning for a soldier's part in the Greek wars of independence which was in fact closed to him, the other sailing a small boat into a storm in the Gulf of Spezia.

The two men's career's were remarkably alike: they were maverick aristocrats, people at a loss for function, tremendously eloquent in their attacks on the bugbears of the time – this king, that Secretary of State – but uncertain how to embody their sense of a better way or a way forward. E. H. Coleridge remarks that Byron was 'more of a king-hater than a people-lover. He was against the oppressors, but he despised and disliked the oppressed.'[21] Neither could come out with any positive value as rooted as Burns's

The Man's the gowd for a' that . . .

The pith o' Sense, and pride o' Worth,
    Are higher rank than a' that,

which gives lyric force to fundamental egalitarian thoughts
uttered by Paine in *The Rights of Man* a few years before (for
example, France has 'put down the dwarf, to set up the man').
And both Shelley and Byron came out *against* the most
humane and honest politician of the time, and its most vivid
and informative writer, William Cobbett. In an epigram of 1820
Byron damns him to hell along with Paine.[22] In *Peter Bell the
Third*, written a few months after Peterloo, Shelley lumps
Cobbett with Castlereagh and Canning as one of the 'caitiff
corpses' who inhabit the contemporary inferno ('Hell is a city
much like London'), and then gets him quite wickedly wrong
in making out (as he also does in the swipe at 'vulgar agitators'
in his pamphlet on Reform) that Cobbett's politics were
vengeful or retributive:

Sometimes the poor are damned indeed
    To take – not means for being blessed, –
But Cobbett's snuff, revenge; that weed
From which the worms that it doth feed
    Squeeze less than they before possessed.[23]

Such ideas strike me as no more valid or thought-through than
the cant of the official spokesmen, in all countries, who try to
slight the outcry of the prisoners of starvation by smearing it as
a thing fomented by troublemakers from elsewhere.

To conclude by putting my argument dialectically: the
Radical movement in Scotland and England (as historians
have argued[24]) failed to find the leadership it needed, from
thinkers and poets among others, and the poets failed to
engage in the practical work, the engagement at close quarters
with native social experience, which might have saved them
from being what Arnold called 'beautiful and ineffectual
angels, beating their luminous wings in the void in vain'.
Furthermore, their more radical poems were alienated: they
were actually rejected or cast out by the society of their time.

John Murray would not publish *The Vision of Judgment* because Byron's *Cain* was being prosecuted and Longmans refused it because it might have spoiled the sales of Southey's encomium on the dead king. John Hunt published it and was prosecuted for his pains – as a result of which, in part, his brother Leigh withheld *The Mask of Anarchy* till 1832: it was thus denied any active role in the movement which had inspired it in the first place, and the rest of Shelley's poems in the revolutionary 1819 group stayed unpublished till 1839. Byron's Luddite song was unpublished until six years after his death. In the meantime a lotus-eating trance, a prolonged ascendancy of 'second-rate sensitive minds' (as Tennyson called himself), settled over literature. It was not replaced by something more talented and potent until Chartism and the Year of Revolutions in 1848 helped to engender the great age of realism in the novel.

NOTES

1. Byron, *The Complete Poetical Works*, ed. Jerome J. McGann (Oxford, 1980), II, 103.
2. E. J. Hobsbawm, *The Age of Revolution* (1962), 111.
3. *Poetical Works*, V, 7.
4. Ibid., 403.
5. *Byron's Letters and Journals*, ed. Leslie A. Marchand (1973–82), vol. 7, 210.
6. Ibid., vol.5, 149.
7. Byron *Works*, ed. E. H. Coleridge, (1901), IV, 489–91, 501–2.
8. G. D. H. Cole and Raymond Postgate, *The Common People* (1956 ed.), 181.
9. *Poetical Works*, V, 409–11.
10. E. P. Thompson, *The Making of the English Working Class* (1963), 705.
11. Ibid., 700.
12. *BLJ*, vol. 7, 62–3.
13. Quoted from *The Opinions of William Cobbett*, ed. G. D. H. and M. Cole (1944), 306, 308–9.
14. *Poetical Works*, IV, 48.
15. Quoted from A. L. Morton, *A People's History of England* (1948 ed.), 236; *A Handbook of Freedom*, ed. Jack Lindsay and Edgell Rickword (1939), 236–7.
16. *BLJ*, vol. 6, 226.

17. *Byron: A Self-portrait*, ed. Peter Quennell (1950), 752.
18. T. C. Smout, *A History of the Scottish People 1560–1830* (1972 ed.), 417.
19. Allan Rodway, *The Romantic Conflict* (1963), 13–22; the following wave he sees as a phase of 'unattained liberty' in which poets of 'European vision', Byron and Shelley, supplanted Wordsworth and Coleridge as the chief innovative talents.
20. Shelley, *The Complete Poetical Works*, ed. Thomas Hutchinson (Oxford, 1934), 338.
21. *Encyclopaedia Britannica* (1929), IV, 281.
22. *Poetical Works*, IV, 345.
23. *Poetical Works*, 350, 352; 'A Philosophical View of Reform': *Shelley's Prose*, ed. David Lee Clark (1988 ed.), 261.
24. Morton, *People's History of England*, 371.

Byron: Radical, Scottish Aristocrat

ANDREW NOBLE

The first generation of English Romantic poets was, initially, intensely radical. Whatever rejection or their aspirations followed, the young manhood of Wordsworth, Coleridge and Southey was in theory and, indeed, in self-endangering practice, committed to revolutionary change as the cause of the common people. Shelley and Byron, the aristocrats of the following generation, saw in their middle-class, middle-aged predecessors little but craven apostasy expressed in sycophantic, ambitious conduct. Southey, as Laureate, was the easiest target. Wordsworth, however, secreting himself in the decaying body of Lowther's aristocratic patronage, was almost equally heinous.

Neither Byron nor Shelley gave any credit to the fact that Wordsworth's conservatism was partly the result of direct experience of the Terror, with its manifestation of what Octavio Paz has defined as 'critical reason'.[1] More complex, however, is that Byron in particular seems not to have been aware that the prosody of Wordsworth's early, radical period and of his great poetry not only spoke of the suffering of the common people but, to a degree, employed the language of the common people. He shared Shelley's satirical sentiments, expressed in the latter's fine parody of Wordsworth, *Peter Bell the Third*, that Wordsworth's language was not only banal and pietistic but sententiously designed to make the people quiescent, rather than provoke them to action. As Shelley wrote of Peter Bell, standing as a servant in a demonic London behind Wordsworth's chair:

> And these obscure remembrances
> Stirred such harmony in Peter,
> That, whatsoever he should please,

23

He could speak of rocks and trees
    In poetic metre.

For though it was without a sense
    Of memory, yet he remembered well
Many a ditch and quick-set fence;
    Of lakes he had intelligence,
    He knew something of heath and fell.

He had also dim recollections
    Of pedlars tramping on their rounds;
Milk-pans and pails; and odd collections
    Of saws and proverbs; and reflections
Old parsons make in burying grounds.

But Peter's verse was clear, and came
Announcing from the frozen hearth
Of a cold age, that none might tame
The soul of that diviner flame
It augured to the Earth . . .[2]

Paul Foot has seen Shelley as such a divine, radical flame, burning with a pure intensity in the people's cause.[3] In 1988 Michael Foot published *The Politics of Paradise: A Vindication of Byron* in which he seeks to extend similar radical honours to Byron. Certainly Byron, in his later poetry, displays a virile, 'flying', satirical destruction of the cant that glued together the British establishment. A major part of Byron's enormous poetic output is, however, neither of this quality nor nature. Nor, though it might overwhelm a less credulous soul than that of Michael Foot, is the splendid savaging of his apostate elders irrefutable evidence as to the complete integrity of Byron's own radicalism. Though Byron's Preface to *Don Juan*, the companion piece to Shelley's *Peter Bell*, does display satiric verve and analytic acumen it also, unintentionally, raises difficult problems of Byron's language and prosody and, hence, the authenticity of his politics:

The poem, or production, to which I allude, is that which begins with – 'There is a thorn, it is so old' – and then the

Poet informs all who are willing to be informed, that its age was such as to leave great difficulty in the conception of its ever having been young at all – which is as much as to say, either that it was Coeval with the Creator of all things, or that it had been *born Old*, and was thus appropriately by antithesis devoted to the Commemoration of a child that died young. The pond near it is described, according to mensuration,

'I measured it from side to side:
'Tis three feet long, and two feet wide.'

Let me be excused from being particular in the detail of such things, as this is the Sort of writing which has superseded and degraded Pope in the eyes of the discerning British Public; and this Man is the kind of Poet, who, in the same manner that Joanna Southcote found many thousand people to take her to Dropsy for God Almighty re-impregnated, has found some hundreds of persons to misbelieve in his insanities, and hold him out as a kind of poetical Emanuel Swedenborg – a Richard Brothers, a Parson Tozer – half Enthusiast and half Imposter.

This rustic Gongora and vulgar Marini of his Country's taste has long abandoned a mind capable of better things to the production of such trash as may support the reveries which he would reduce into a System of prosaic raving, that is to supersede all that hitherto by the best and wisest of our fathers has been deemed poetry, and for his success – and what montebank will not find proselytes? (from Count Cagliostro to Madame Krudener) – he may partly thank his absurdity, and partly his having lent his more downright and unmeasured prose to the aid of a political party, which acknowledges its real weakness, though fenced with the whole armour of artificial Power, and defended by all the ingenuity of purchased Talent, in liberally rewarding with praise and pay even the meanest of its advocates. Amongst these last in self-degradation, this Thraso of poetry has long been a Gnatho in Politics, and may be met in print at some booksellers and several trunk-makers, and in person at dinner at Lord Lonsdale's.[4]

There is no little truth in this. Wordsworth's language can

descend from the prosaic to the disconcertingly banal. His ultimate adherence to the Tory cause was politically dire and is accurately reflected in the almost complete sterility of his late poetry. Yet the early Wordsworth was not a decadent poet, the betrayer of Pope's bequest. When radical in poetry and politics, his plain speech had both borne historical witness to the turbulence and sufferings of the age and brought into human consciousness hitherto undiscerned states of memory and being. If 'this Thraso of poetry has long been a Gnatho in Politics' Hazlitt, perhaps echoing the phrase, pointed to an anomaly in Byron at least as great. 'Lord Byron,' he wrote, 'who in his politics is a *liberal*, in his genius is haughty and aristocratic.'[5]

Hazlitt perceived Byron as decadent aristocrat in both politics and language. He had highly qualified opinions of Pope's powers, and was less than impressed by Byron's enthusiasm for him. For reasons we will later discuss, Hazlitt was confused by the mock-heroics of Byron's great achievement, *Don Juan*. The tantrums of behaviour and language in Byron's earlier poetic tales Hazlitt saw as largely vacuous and self-indulgent. While Hazlitt could be as incisively bitter as Byron about the apostate Wordsworth, he saw as the traitor to poetry, poetry as the conserver of our common humanity, not Wordsworth but Byron:

I do not recollect, in all Lord Byron's writings, a single recurrence to a feeling or object that had ever excited an interest before; there is no display of natural affection – no twining of the heart round any object: all is the restless and disjointed affect of first impressions, or novelty, contrast, surprise, grotesque costume, or sullen grandeur. *His* beauties are the *houris* of Paradise, the favourites of a seraglio, the changing visions of a feverish dream. His poetry, it is true is stately and dazzling, arched like a rainbow, of bright and lovely hues, painted on the cloud of his own gloomy temper – perhaps to disappear as soon! It is easy to account for the antipathy between him and Mr Wordsworth. Mr Wordsworth's poetical mistress is a Pamela; Lord Byron's an Eastern princess or a Moorish maid. It is the extrinsic, the uncommon that captivates him, and all the rest he holds in

sovereign contempt. This is the obvious result of pampered luxury and high born sentiments. The mind, like the palace in which it has been brought up, admits none but new and costly furniture. From a scorn of homely simplicity, and a surfeit of the artificial, it has but one resource left in exotic manners and preternatural effect. So we see in novels, written by ladies of quality, all the marvellous allurements of a fairy tale, jewels, quarries of diamonds, giants, magicians, condors and ogres. The author of the Lyrical Ballads described the lichen on the rock, the withered fern, with some peculiar feeling that he has about them: the author of Childe Harold describes the stately cypress, or the fallen column, with the feeling that every schoolboy has about them. The world is a grown schoolboy, and relishes the latter most.[6]

While Byron, in bad faith, might attack Wordsworth for finding 'some hundreds of persons to misbelieve in his insanities' so that he was 'half Enthusiast and half Imposter', it can be argued that imposture was of the essence of his own relationship to the hundreds of thousands who formed his credulous, immature audience. Hazlitt discerned that, paradoxically, it was Byron and his heroes' projection of aristocratic disdain which aroused mass enthusiasm and fed Byron's need to dominate his poetical contemporaries:

Whatever he does, he must do in a more decided and daring manner than anyone else – he lounges with extravagance, and yawns so as to alarm the reader! Self-will, passion, the love of singularity, a disdain of himself and of others (with a conscious sense that this is among the ways and means of procuring admiration) are the proper categories of his mind: he is a lordly writer, he is above his own reputation, and condescends to the Muses with a scornful grace![7]

Almost as seminal a literary sociologist as Coleridge, Hazlitt grasped that this was not merely regressive, *literary* excitement. It pointed in Byron's mass audience to an innate predisposition for aristocratic authority even, or indeed especially, in degenerate form. Byron as hero in life and letters was, if not a rebel in the cause of entrenched sycophancy, then at least the beneficiary of it:

Lord Byron complains that Horace Walpole was not properly appreciated, 'first, because he was a gentleman, and
secondly, because he was a nobleman.' His Lordship stands
in one, at least, of the predicaments here mentioned and yet
he has had justice, or somewhat more done to him. He
towers above his fellows by all the height of the peerage. If
the poet lends a grace to the nobleman, the nobleman pays it
back to the poet with interest. What a fine addition is ten
thousand a year and a title to the flaunting pretensions of a
modern rhapsodist! His name so accompanied becomes the
mouth well: it is repeated thousands of times, instead of
hundreds, because the reader in being familiar with the
Poet's work seems to claim acquaintance with the Lord.

> 'Let but a lord once own the happy lines:
> How the wit brightens, and the style refines!'

He smiles at the high-flown praise or petty cavils of little
men. Does he make a slip in decorum, which Milton
declares to be the principal things? His proud crest and
armorial bearings support him: – no bend-sinister slurs his
poetical escutcheon! Is he dull, or does he put of some trashy
production on the public? Is it not charged to his account, as
a deficiency which he must make good at the peril of his
admirers. His Lordship is not answerable for the negligence
or extravagances of his Muse.[8]
Hazlitt found negligence and extravagance in abundance in
Byron's poetry. Without the dramatic objectivity and realism
of Walter Scott, far less of Shakespeare, with his unsurpassed
empathetic power, Byron's work was foetidly subjective to the
point of unreality. Hazlitt saw Byron, then, not as the inheritor
of eighteenth century satirical wit but as an aristocratic
vulgarian, who transmuted the golden rhetoric of the
seventeenth century into strident and sensational dross. His
poetry was the verbal equivalent of Beckford or Walpole's
Gothic architectural tastes, where historical forms reappeared
in unstable structures. Byron, for Hazlitt, was perpetually
shoring up his ruined poetic worlds with other men's fragments.
This may account for the charges of plagiarism which have

been repeatedly brought against the Noble Poet – if he can borrow an image or a sentiment from another, and heighten it by an epithet or allusion of greater force or beauty than is to be found in the original passage, he thinks he shows his superiority of execution in this in a more marked manner than if the first suggestion had been his own. It is not the value of the observation itself he is solicitous about; but he wishes to shine by contrast – even nature only serves as a foil to set off his style. He therefore takes the thoughts of others (whether contemporaries or not) out of their mouths, and is content to make them his own, to set his stamp upon them, by imparting to them a more meretricious gloss, a higher relief, a greater loftiness of tone, and a characteristic inveteracy of purpose.[9]

Edwin Muir has cogently argued that the secret of the Victorian bestseller's success was that the middle-brow audience could indulge and, indeed, endorse, their fantasies (erotic, martial, historical and exotic) in terms of literary allusiveness based on their still existent, if deteriorating, grasp of classic texts.[10] Their world was almost as densely intertextual as that of the post-modernist literary critic. If this is true, Byron is the progenitor of the nineteenth century bestseller. He certainly begat Disraeli.

The over-blown and sloppily second-hand defined by Hazlitt arguably derives from ethnic as well as social causes. Writing in 1937, in what is arguably still the finest contribution to an understanding of Byron's poetry, T. S. Eliot remarked that 'I therefore suggest considering Byron as a Scottish poet – I say "Scottish", not "Scots", since he wrote in English.'[11] Eliot's distinction is an important one, to which he returns. For the moment, however, let us not discuss the strengths which the Scottish tradition gave, albeit largely unconsciously, to Byron but the weaknesses so characteristic of the mainly unhappy history of the anglicisation of Scottish writing in the eighteenth and early nineteenth centuries. On the face of it they are a wildly mismatched couple, but as Eliot is subtly aware, there is considerable affinity between his own predicament and Byron's. Both were expatriates who aspired to manufacture a personality with which to advance themselves

in their adopted land, and subsequently satirized that land in heroic-couplets. If for Eliot *Don Juan* marked the success of Byron's satire, it was a splendid summit resting atop a disconcerting mountain of verbiage:

> Of Byron one can say, as of no other English poet of his eminence, that he added nothing to the language, that he discovered nothing in the sounds, and developed nothing in the meaning of individual words. I cannot think of any other poet of his distinction who might so easily have been an accomplished foreigner writing English. The ordinary person talks English, but only a few people in every generation can write it; and upon this undeliberate collaboration between a great many people talking a living language and a very few people writing it, the continuance and maintenance of a language depends. Just as an artisan who can talk English beautifully while about his work or in a public bar, may compose a letter painfully written in a dead language bearing some resemblance to a newspaper leader, and decorated with words like 'maelstrom' and 'pandemonium': so does Byron write a dead or dying language.[12]

Unlike Hazlitt, Eliot thought Byron's true weakness lay not in platitudinous schoolboy thought but in his 'schoolboy command' of the language. He compared Byron's treatment of Waterloo unfavourably to Stendhal's in its lack of 'minute particulars' but, disconcertingly, felt that poetic good had been achieved in this stanza from *Childe Harold*:

> And wild and high the 'Cameron's gathering' rose!
> The war-note of Lochiel, which Albyn's hills
> Have heard, and heard, too, have her Saxon foes; –
> How in the noon of night that pibroch thrills,
> Savage and shrill! But with the breath which fills
> Their mountain-pipe, so fill the mountaineers
> With the fierce native daring which instils
> The stirring memory of a thousand years,
> And Evan's, Donald's fame rings in each clansman's ears![13]

This spirited burst, Eliot believed, came from his 'mother's people'. It seems rather to have come from Walter Scott, whom

Byron displaced as the best-selling narrative poet of the age and whose enormous prestige and commercial success was in major part due to his rendition of 'romantic' Scottish themes.

The Enlightenment manifested itself in Scotland principally in an extraordinary evolution of economics, the social sciences, mental philosophy of a sceptical toughness and technology. Simultaneously, Scotland, due to the accident of Highland geography and history, became the favoured location for the fantasies of educated nineteenth-century man. Overtly given to progress, European man was suffused with a profound nostalgia for the world he was leaving, indeed often destroying, and so much of eighteenth century bestselling writings are compensatory fantasies for this passing world. James MacPherson was the primary, knowing benefactor of these appetites for the militaristic, the melancholic, for feudal and aristocratic trappings, all tinged with the fey and supernatural. Scott perfected this by creating forms which permitted fantasy, yet assured the bourgeois reader of the ultimate success and security-of his ordered, respectable commercial world. There may be a law which states that bad literature abhors a partial national vacuum and rushes in to fill it. In part a synthetic, literary personality, Byron derived his fiscally rewarding insubstantiality from the synthetic Scottishness apparent in his early reworkings of Ossian and other Highland themes.

The principal cause of Scott's enormous literary and commercial success was his sensational variations on a theme wherein a martial insurrectionary hero of primitive, feudal energies *temporarily* endangers an increasingly prosperous, homogenizing commercial society. Thus, the failed Jacobite threat to the Union created a perfect historical paradigm for his fictions. Less obvious was the fact that Scott was more concerned with the Jacobins than the Jacobites. Even the sceptical Hazlitt did not read in Scott's fiction the conservative elements he discovered in his personality. Hazlitt, in fact, provides perhaps unsurpassed testimony to the power Scott's rebarbative fantasies had for prudent, rational man in the early nineteenth century:

Protestants and Papists do not now burn one another at the

stake: but we subscribe to new editions of *Fox's Book of Martyrs*; and the secret of the success of the *Scotch Novels* is much the same – they carry us back to the feuds, the heart-burnings, the havoc, the dismay, the wrongs and the revenge of a barbarous age and people – to the rooted prejudices and deadly animosities of sects and parties in politics and religion, and of contending chiefs and clans in war and intrigue. We feel the full force of the spirit of hatred with all of them in turn. As we read, we throw aside the trammels of civilisation, the flimsy veil of humanity. 'Off, you lendings!' The wild beast resumes its sway within us, we feel like hunting-animals, and as the hound starts in his sleep and rushes on the chase in fancy, the heart rouses itself in its native lair, and utters a wild cry of joy, at being restored once more to freedom and lawless, unrestrained impulses. Every one has his full swing, or goes to the Devil in his own way. Here are not Jeremy Bentham's Panopticons, none of Mr Owen's impassable Parallelograms, (Rob Roy would have spurned and poured a thousand curses on them), no long calculations of self-interest – the will takes its instant way to its object; as the mountain-torrent flings itself over the precipice, the greatest possible good of each individual consists in doing all the mischief he can to his neighbour: that is charming, and finds a sure and sympathetic chord in every breast! So Mr Irving, the celebrated preacher, has rekindled the old, original, almost exploded hell-fire in the aisles of the Caledonian Chapel, as they introduce the real water of the New River at Sadler's Wells, to the delight and astonishment of his fair audience. *'Tis pretty, though a plague,* to sit and peep into the pit of Tophet, to play at *snap-dragon* with flames and brimstone (it gives a smart electrical shock, a lively fillip to delicate constitutions), and to see Mr Irving, like a huge Titan, looking as grim and swarthy as if he had to forge tortures for all the damned![14]

The last thing, of course, that Scott wanted was a phenomenon anything akin to Rob Roy let loose in contemporary society. Byron's eruption on the scene caused him, consequently, far more alarm than the loss of his role as the age's chief narrative poet. His fictional success was more than adequate consola-

tion for that. What did alarm him was that Byron, as Napoleon, appeared unholily like a terrible dream made flesh; that a rough, slouching beast had, indeed, been born. Scott saw on Byron and his associate of his early European travels, John Hobhouse, distinct marks of the revolutionary beast. As Malcolm Kelsall has noted:

> Scott categorized Hobhouse's commentary as a 'frenzy' improper to 'individuals of birth and education' (patricians), and Byron and his commentator showed that they were 'trained in the school of revolutionary France' (Jacobins). It ill became Byron to complain of British liberty when it was only as a result of the freedom which Britain has won that the poet might visit Europe (thus Scott is directly in the tradition of Addison's *Letter from Italy*).[15]

Scott was suffused with desire to fill his own life with all the trappings of aristocracy. Hence Hazlitt's acerbic aside that 'Sir Walter Scott (when all's said and done) is an inspired butler.' Consequently, Scott found Byron's treachery to his own class incomprehensible. In discussing Canto IV of *Childe Harold*, Scott saw the threat to social order and relative prosperity of the new individualism personified in Byron:

> This moral truth appears to us to afford, in great measure, a key to the peculiar tone of Lord Byron. How then, will the reader ask, is our proposition to be reconciled to that which preceded it? If the necessary result of an enquiry into our own thoughts be the conviction that all is vanity and vexation of spirit, why should we object to a style of writing, whatever its consequences may be, which involves in it truths as certain as they are melancholy? If the study of our own enjoyments leads us to doubt the reality of all except the indisputable pleasures of sense, and inclines us therefore towards the Epicurean system – it is nature, it may be said, and not the poet which urges us upon the fatal conclusion. But this is not so. Nature, when she created man a social being, gave him the capacity of drawing that happiness from his relations with the rest of the race, which he is doomed to seek in vain in his own bosom. These relations cannot be the source of happiness to us if we despise or hate the kind with whom it is their office to unite

us more closely. If the earth be a den of fools and knaves, from whom the man of genius differs by the more mercurial and exalted character of his intellect, it is natural that he should look down with pitiless scorn on creatures so inferior. But if, as we believe, each man, in his own degreee, possesses a portion of the ethereal flame, however smothered by unfavourable circumstances, it is or should be enough to secure the most mean from the scorn of genius as well as from the oppression of power, and such being the case, the relations which we hold with society through all their graduations are channels through which the better affections of the loftiest may, without degradation, extend themselves to the lowest. Farther, it is not only our social connections which are assigned to us in order to qualify that contempt of mankind, which too deeply indulged tends only to intense selfishness; we have other and higher motives for enduring the lot of humanity – sorrow, and pain, and trouble – with patience of our own griefs and commiseration for those of others.[16]

As with a host of other contemporary phenomena, there seems good cause to believe that Scott vastly overestimated the revolutionary potential present in Byron. It took Dostoevsky in *The Possessed* to see how aristocratic *ennui* might become the medium through which psychotic revolutionary violence could discharge itself. Marilyn Butler is much more accurate in discerning the manner in which the Byronic here did not invert Scott's morality but, in fact, subtly reinforced it.

Byron now succeeded Scott as the most fashionable author of the day, and he did not because his appeal, which was superficially rebellious and hence exciting, remained at a deeper level bi-partisan. In *Childe Harold, The Giaour, The Bride of Abydos, The Corsair* and *Lara* he developed the Byronic hero from prototypes such as Schiller's Karl Moor and Scott's Marmion. Masterful, moody outlaws, haunted by some secret consciousness of guilt, these heroes act as a focus for contemporary fantasies. Not the least element of guilty complicity about them is that they echo the French cult of Napoleon: they are fictional equivalents of Géricault's handsome idealized portrait of the French emperor on a

white charger surmounting the Alps. By this daring hint, and by translating his hero from Scott's historical setting to a present-day theatre of war, Byron implies the possibility of effective action in the real world. Even so, his rebellious Corsair is sanitized, as far as the English public is concerned, by wielding his sword well away from the French proponents of liberty and equality, and still further from the machine-breakers and petitioners of the English provinces. Nor has his rebellion any hint of a philosophic dimension. It is drained of ideological content, to a degree actually remarkable in the literature of the period. An image potentially of revolution is presented in terms sufficiently unintellectual to allay the fears of the propertied public.[17]

The collusion between Byron and his propertied audience is perhaps even more subtle. Byron purveyed melodramatically the moody-browed man of will, a cardboard cut-out of Hegel's world spirit. This figure, however, was already beyond action, in a state of vacuity and emotional paralysis. Perhaps a celebrant of a love that did not dare speak its name, he was certainly a possessor of a guilt that refused to define its nature and origin. In a brilliant new essay, 'What's Eating Lara?', Dan Jacobson notes that on this core issue of unrevealed guilt Byron 'like any other bad writer, is trying to get out of the reader (amazement, fear, admiration) on tick, as it were – without delivering the goods.' Jacobson continues:

> It is not just that the readers of Byron's age, like readers today, enjoyed trash; and the more portentious it was, the more fluent, the more titillatingly knowledgeable about evil, the more 'literary' in certain obvious ways, the more they admired it. (Again, just like readers today – though the styles of vulgarity we respond to are different.) There is another and more interesting sense in which one can account for their success. What made them so effective in reaching into the minds of their readers is that they are *about* hollowness, they are about bad faith and insincerity. Indeed one can go further and say that the central tormenting secret which the heroes of the poems (and their creator) try to guard so jealously from prying eyes is their own suspicion that they are fakers.[18]

For Jacobson, Byron's poetry, other than *Don Juan* with its comic, multi-voiced scepticism, is to be read as revealing or, indeed, betraying the reverse of its intentions:

> His narrative poems, with that fixed demonic scowl on their foreheads, were the literary record of his attempt to overcome by an effort of will what I have called the torment of insincerity which had always haunted him. Every clenched denial and fierce assertation in these poems, every dark hint of what had better been left unspoken, had been intended to silence not only the doubters and deriders without – though on the whole they had shown themselves only too eager to be convinced – but the many sceptical voices within.[19]

The true revolutionary, for good or ill, is by definition a man of sincerity; for him desire and action are wholly congruent. In a recent book, *Byron's Politics*, Malcolm Kelsall explains Byron's vacuity and near paralysis of political will, despite his manifest political involvement in the Lords and elsewhere, as stemming from his belonging to a Whig tradition and establishment that had not only lost power but could not foresee its restitution. Kelsall very successfully repudiates the concept of Byron's 'iron revolutionary will' by demonstrating the relationship between his private correspondence and his dramatization of the dilemma faced by both himself and his party. From the outset, of course, Whig radicalism had always been brought up short when the claims of social citizenship infringed on the rights of property. Byron, the Nottinghamshire mine-owner, was actually aware of this predicament:

> I look upon [convulsion] as inevitable, though not revolutionist: I wish to see the English constitution restored, and not destroyed. Born an aristocrat, and naturally one by temper, with the greater part of my property in the funds, what have *I* to gain by a revolution?[20]

If revolution for the Whigs was to stop well short of redistribution of property, taking arms with the people was also far beyond the pale. Peterloo and Cato Street evoked in him responses akin to the more comprehensible paranoia present in Scott.

The reverse of the Byron, who at a safe distance from England donned the garb of a revolutionary leader in Italy and

Albania and Greece, and who saw such a leadership as a pure manifestation of the radical, nationalist popular will, this native Byron saw in revolution at home merely the manipulation of the common people by psychotic demagogues:

> I am convinced – that Robespierre was a Child – and Marat a quaker in comparison of what they would be [Hunt and Cobbett] could they throttle their way to power.
>
> I can understand and enter into the feelings of Mirabeau and La Fayette – but I have no sympathy with Robespierre – and Marat – whom I look upon as in no respect worse than those two English ruffians.[21]

Kelsall shows cogently how this moral dilemma, a product of the practical impotence and theoretical confusion of Whig politicians, runs through Byron's work. *Marino Faliero*, a bad play showing, as so often, Byron's lack of proper anxiety concerning Shakespeare's influence, is however perhaps the purest paradigm of this dilemma. Historical Venice, a 'sea-Sodom', mirrors the decay Byron felt present in another naval empire, contemporary England. Falierio, the Doge cannot bring himself to make common cause with the people against his peers, and cleanse the corruption of the state in a bloody flux of aristocratic blood. It is Falierio himself who ends in melodramatic execution. Hazlitt, eager for the overthrow of a corrupt establishment, liked the play's artistry little and its dramatization of vacillating politics even less.[22]

If, as Kelsall suggests, Byron was rendered impotent as a revolutionary by his bonds of birth and ideology to the Whig élitist cause, it is also true that this problem of class was intensified by his Scottishness. The revolutionary period into which Byron entered was not only that of republicanism but of nationalism; it was the age of the birth and rebirth of nations. In his excellent *The Break-up of Britain*, Tom Nairn has pointed to the peculiar paradox that in an age of national revolutions and heightened national consciousness throughout Europe, Scotland, peculiarly, intensified its links with England. Nairn cogently argues that we can account for this paradox because, uncharacteristic of other national movements, the Scottish élite had been so advantageously educated in the eighteenth century that they did not form an alienated group when

revolution broke in America and France, and were commer-
cially and professionally integrated into British society to a
degree that a commitment to a Scottish cause would have
seemed socially and fiscally self-endangering. The dispos-
sessed Burns was certainly fired by General Washington and
events in France, but his exception proves Nairn's rule. Sir
James MacIntosh in *Vindice Gallicae* did envisage the importa-
tion of Scotland's awareness of its historical roots, but this was
an idea which found little favour among his peers and one
which he himself was soon to deny. Sir Walter Scott, notor-
iously for Hugh MacDiarmid and other modern nationalist
intellectuals, solved the problem by creating massive enthusiasm for inessential, 'romantic' symbols of Scottish nationalism while in reality being terrified of the resurrection of what
he considered a rebarbative Scotland, and consequently passionately committed himself to social and economic salvation
through the Union. These lines, however, are not by the
pro-Unionist, Tory Scott but the allegedly radical Byron:

> Who hath not glowed above the page where Fame
> Hath fixed High Caledon's unconquered name;
> The mountain-land which spurned the Roman chain,
> And baffled back the fiery-crested Dane,
> Whose bright claymore and hardihood of hand
> No foe could tame – no tyrant could command?
> That race is gone – but still their children breathe,
> And Glory crowns them with redoubled wreath:
> O'er Gael and Saxon mingling banners shine,
> And England! add their stubborn strength to thine.
> The blood which flowed with Wallace flows as free,
> But now 'tis only shed for Fame and thee![23]

It was not the least of the successes of the imperial enterprise:
to convert their most feared eighteenth-century adversaries
into the sharpest cutting edge of its nineteenth century
overseas conquests. It is extraordinary to find Byron lending
his name to the cause. It is absurd when we consider that in
Italy, Albania and Greece he lent his name, money and
energies to succouring European national movements not his
own. A similar pattern can be observed in Boswell earlier and

R. L. Stevenson later. Such Scottish writers seem to be subconsciously driven into a displaced nationalism, which asserts its inauthenticity by degenerating into a kind of costume melodrama. Byron's theatricality had a source in an even less likely element of his Scottishness. Byron, Edwin Muir has remarked, 'appeared assiduously in society as an *âme damnée'*.[24] Given its long history of sermonising monologues against the dialectical nature of the theatre, it is no little irony that by the late eighteenth century, Scottish Calvinism appears in theatrical form in British culture's burgeoning Gothic sensibility. While it never had the obsessive force of Italian Roman Catholicism, with its orgiastic fantasies polluted by sadistic monks and erotic nuns, Scottish Calvinism did breathe sexually titillating tongues of fire. Hazlitt has a marvellous, comic description of London's Caledonian Chapel with the Rev Edward Irving as celeberity preacher at that unfallen stage in his career when Christ's sexual nature was still only a matter of rhetoric.[25]

A major part of Byron's appeal, his ability to 'express before anybody else what a great number of people wanted to have expressed for them', seems to have stemmed from his capacity to appear as simultaneously the rebel against and victim of sexually repressive religion.[26] It is, of course, of the nature of radical politics to see in unexploited sexuality an essential element of the democratic freedom it seeks. Byron follows on the heels of William Blake and Mary Wollstonecraft. Yet the cleansing purity of Blake's vision seems quite lost in Byron. At worst he seems to have made himself master of ceremonies for the children of the middle-classes to indulge in promiscuity in the name of liberty – as in the 1960's – and subsequently retreat back into comfortable conformity. We also feel that Byron's constant posing as 'something demonic rather than human, a Miltonic Satan or fallen angel' is not like Blake's radical transvaluation of values but a kind of theatricality. I have argued elsewhere, comparing Blake with Burns, that we can perceive a similar self-dramatising demonism in Burns who, like Byron, often referred to his sibling relationship to Milton's Satan.[27] These Scottish poets, at the end of the day, always imply that their manic rebellion will end, like Milton's Satan,

in depressed defeat at the hands of institutional authority. It
lends veracity to these sentiments that this is how Burns and
Byron did end. Calvinism, at least as late as Stevenson, seems
to have politically unmanned the Scottish creative writer.
Since terrible paternal authority can be attacked but never
overcome, as in Boswell, erotic or political radicalism will
inevitably end in punishment and defeat. Byron, another
archetypal Scottish Calvinist writer, was, therefore, by inheritance flawed at the heart of his radical will. The melodramatic,
intense, brooding wilfulness of his heroes, as Jacobson has
suggested, masked an underlying chaos of conflicting emotions. Or, as Edwin Muir has remarked:

> He preferred to paint this weak good-nature as something
> very like villainy, and thought it better to be thought wicked
> than feeble-willed. The fatal complications in which he
> involved himself during the five years of public success in
> London might as readily have been brought about by
> weakness as by double-dyed infamy; and indeed, seen
> through Byron's predestinarian spectacles, the one became
> indistinguishable from the other.[28]

Though Hazlitt has many profound and painful things to say
about Scotsmen and what religious repression did to them, he
never seems to have penetrated beneath Byron's aristocratic
surface to the tumult within and out of which the late poetry
developed. This led him to misunderstand the quintessential-
ly Scottish achievement of *Don Juan*. Ironically, Hazlitt defined
arguably the central characteristic of authentic Scottish poetry
without being able to appreciate it.

> The *Don Juan* indeed has great power; but its power is owing
> to the force of the serious writing, and to the oddity of the
> contrast between that and the flashy passages with which it
> is interlarded. From the sublime to the ridiculous there is but
> one step. You laugh and are surprised that any one should
> turn around and *travestie* himself: the drollery is in the utter
> discontinuity of ideas and feelings. He makes virtue serve as
> a foil to vice; *dandyism* is (for want of any other) a variety of
> genius. A classical intoxication is followed by the splashing
> of soda-water, by frothy effusions of ordinary bile. After
> the lightning and the hurricane, we are introduced to the

interior of the cabin and the contents of wash-hand basins. The solemn hero of tragedy plays *Scrub* in the farce. This is 'very tolerable and not to be endured.' The Noble Lord is almost the only writer who has prostituted his talents in this way. He hallows in order to desecrate; takes a pleasure in defacing the images of beauty his hands have wrought; and raises our hopes and our belief in goodness to Heaven only to dash them to earth again, and break them in pieces the more effectually from the very height they have fallen. Our enthusiasm for genius or virtue is thus turned into a jest by the very person who has kindled it, and who thus fatally quenches the sparks of both.[29]

Adherents of MacDiarmid will have observed that what Hazlitt here negates, the Scottish poet sees as the quintessential, creative zig-zag of Scottish poetry with its dynamic interaction of disparate elements.[30] Eliot also recognised this in Byron and sensed a 'flyting' satirical tone in *Don Juan* that he traced back to Dunbar. Eliot also recognised in the achievement of Byron's last, great mock epic a 'left-field', rapscallion element that MacDiarmid would have applauded as the true voice of Scottish poetry. Hazlitt forgave Byron much for the manner of his death. The mass of bad poetry, inflaming rather than purging his audience's doubtful social passions, can be equally forgotten in the face of his last poem, whose achievement Eliot so cogently defines:

I do not pretend that Byron is Villon (nor, for other reasons, does Dunbar or Burns equal the French poet), but I have come to find in him certain qualities, besides his abundance, that are too uncommon in English poetry, as well as the absence of some vices that are too common. And his own vices seem to have twin virtues that closely resemble them. With his charlatanism, he has also an unusual frankness; with his pose, he is also a *poète contumace* in a solemn country; with his humbug and self-deception he has also a reckless raffish honesty; he is at once a vulgar patrician and a dignified toss-pot; with all his bogus diabolism and his vanity of pretending to disreputability, he is genuinely superstitious and disreputable. I am speaking of the qualities and defects visible in his work, and important in estimating his work: not of the private life, with which I am not concerned.[31]

NOTES

1. See that incomparable contemporary poet-critic's *Children of the Mire: Modern Poetry from Romanticism to the Avant-Garde* (1974).
2. *The Complete Poetical Works of Percy Bysshe Shelley*, ed. Hutchinson (1956), Part the Fifth, stanzas x–xiii, 355–6.
3. *Red Shelley* (1980).
4. *Byron's 'Don Juan'*, ed. Steffen and Pratt (1957), II, 3–4.
5. 'Lord Byron', *Collected Writings*, ed. P. P. Howe (1932), vol. 11, 70–1.
6. 'Byron and Wordsworth', *Collected Writings*, vol. 20, 155–6.
7. 'Lord Byron', 70.
8. 'On the Aristocracy of Letters', *Collected Writings*, vol. 8, 209–10.
9. 'Lord Byron', 70.
10. 'Best Sellers of Yesterday: VI William Black', *Edwin Muir: Uncollected Scottish Criticism*, ed. Noble (1982), 222–7.
11. 'Byron' in *English Romantic Poets*, ed. M. H. Abrams (1960), 197.
12. Ibid., 203–4.
13. *Complete Poetical Works*, ed. J. J. McGann, vol. II, Canto the Third, Stanza 26.
14. 'On the Pleasure of Hating', *Collected Writings*, vol. 12, 129.
15. *Byron's Politics* (1987), 77.
16. *Miscellaneous Prose Works*, vol. XVII (1843), 359–60.
17. *Romantics, Rebels and Reactionaries* (1981), 118–19. We should probably read David for Géricault here.
18. *Adult Pleasures* (1988), 35.
19. Ibid., 37.
20. Quoted by Kelsall, 83.
21. Ibid., 86.
22. 'Lord Byron's *Tragedy of Marino Faliero*', *Collected Writings*, vol. 19, 44–51.
23. 'Address Intended to be Recited at the Caledonian Meeting', *Complete Poetical Works* (1981), vol. III, 270–1.
24. 'A Romantic Poet', *Edwin Muir: Uncollected Scottish Criticism*, 206.
25. 'Pulpit Oratory – Dr Chalmers and Mr Irving', *Collected Writings*, vol. 20, 113–22.
26. Both Eliot and Muir have accurate analyses of this ambivalence in Byron.
27. 'Burns, Blake and Romantic Revolt' in *The Art of Robert Burns*, eds. Jack and Noble (1982), 191–214.
28. 'A Romantic Poet', 207.
29. 'Lord Byron', *Collected Writings*, vol. 11, 75.

30. 'This Scottish strain is tremendously idiosyncratic, full of a wild humour which blends the actual and the apocalyptic in an incalculable fashion.' *Albyn or Scotland and the Future* (1927), 22.
31. 'Byron', 209.

Byron and Catholicism

WILLIAM J. DONNELLY

⚹ Byron, in his letters, said a considerable amount about Catholicism, particularly during his years of exile in southern Europe. I would admit that we cannot make too much of this material, which is qualified by the occasional flippancy of tone. Moreover, elsewhere in Byron's writings there are many more negative references to religion in general. Nevertheless, I have sought to place his remarks on Catholicism in the context both of Byron's Romanticism and his Scottish background.

First of all then, in speaking about religion, it can be argued that we are concerned with what is, in the widest sense, the central feature of Romantic poetry, reacting as it does against the emphasis on science and rationalism which had begun with the Renaissance and reached its zenith during the Age of Enlightenment in the eighteenth century. Collectively, Byron's fellow Romantic poets reject as inadequate a view of life based solely on science and reason and seek a more satisfactory definition of existence with reference beyond such terms. In so doing, they might all be deemed religious, in the broadest sense of the term. Thus, when Blake seeks 'To see a World in a Grain of Sand/And Heaven in a Wild Flower' he is essentially religious, in that reason sees a grain of sand in a grain of sand, while to science, a wild flower is a botanical specimen. In the same way, the epigraph at the beginning of Coleridge's *Ancient Mariner* declares the need to:

> . . . contemplate in my mind the idea of a greater and better world, lest the mind, grown used to dealing with small matters of everyday life should dwindle, and become wholly submerged in petty thoughts.[1]

In the poem itself, as in most of Coleridge's poetry, he seeks to put this into practice. That is, he seeks to express a consciousness of a world beyond that of material reality. The title of

44

Wordsworth's famous ode 'Intimations of Immortality from Recollections of Early Childhood' speaks for itself in this regard, and again it could be said to indicate his central concern in much of his major work. Even Keats, who rejects Christianity, is nevertheless obsessed in his poetry by the sad inadequacy of this mortal world, and tantalised by the vision, or the illusion, of an immortal solution. The odes to Psyche, to the Nightingale, to the Grecian Urn, for example, are all images of this. Finally Shelley, though militantly atheist, is concerned to condemn as inadequate a purely rationalist view of existence, and to assert instead the primacy of that which is beyond reason, that is the imagination, and the vision of human perfectibility which the imagination can envisage.

What then of Byron? In the light of all this, Byron is distinct and unique among Romantic poets in having little or no time for religious or visionary solutions to the human predicament. While he deeply abhors the sad reality of the human lot, he proposes no transcendental compensations. On the other hand, while with Shelley he feels the need to defy and resist tyranny at all cost, he does not really have any faith in Shelley's view of a perfectible world. His view of the human condition is summed up in the figure of the notorious Byronic hero – of Cain, of Manfred, of Childe Harold, among others – a hero defiant but damned, an outcast in this world, and in any other.

Now on several occasions Byron's distinct and unique viewpoint has been attributed to the influence of his Scottish Calvinist background. Angus Calder, for example, has recently written of the poet's 'secular Calivinism',[2] secular, in that he is no transcendentalist, yet Calvinist, in that he sees no salvation in another world, or in this one. I should like to suggest my understanding of the source of this condition, in that it may shed some light on our later discussion of Byron's response to Catholicism.

What does Byron himself say about his Calvinist back-ground? While for the most part he gloried in his Scottish upbringing, his references to its Calvinist aspect are all negative – I should say that we are speaking here of the dogma that Burns satirised so brilliantly, which has nothing to do

with Presbyterianism as we know it, which in many respects presents quite the opposite religious view. Thus he speaks of '. . . being early disgusted with a Calvinistic Scotch school where I was cudgelled to Church for the first ten years of my life.'[3] And again, 'I was bred in Scotland among Calvinists in the first part of my life – which gave me a dislike of that persuasion . . .'[4] However, perhaps more important than Byron's reaction to Calvinist religion was the effect upon him of the psychological ambience that Calvinism created, by which he would inevitably have been influenced.

Essentially, the pre-Reformation Church was, rightly or wrongly, a vehicle whereby the inescapable imperfection of the human being could be catered for. Paintings, statues, rituals, sacraments, and most obviously the sacrament of Penance, were all sources of intercession, mediation, reconciliation. By contrast, and again rightly or wrongly, post-Reformation Calvinism was an absolute denial of this function, with profound psychological implications. Such a denial leaves the imperfect individual, and imperfect humanity at large, uncatered for, unreconciled, irredeemable – but for those arbitrarily elected to salvation through the doctrine of predestination.

To such a terrible psychological burden there seem to be two notable responses. The one is honest while the other is dishonest, but perhaps understandable in the circumstances. The second of these is to deem oneself elected to salvation in a world in which most of one's fellows are hopelessly damned. The double standards which this mental shuffle necessitates has of course provided a rich seam for the Scottish writer, being the basis of Burns's kirk satires, Hogg's *Justified Sinner* and several works by Stevenson as well as the most obvious, *Jekyll and Hyde*, to go no further. It may well be the source of Byron's oft-repeated detestation of cant.

Alternatively, the honest response to the situation would be to conclude that our unreconciled imperfections must indeed render us irredeemable, that there is no solution to our predicament in this world or in any other: we can either despair, or defy. It might be argued then that the defiant, damned Byronic hero becomes more fully comprehensible in

the context of this psychological ambience created by Calvinism. While rejecting the Calvinist religion, he remains a 'secular Calvinist', and therefore his is a vision of unaccommodated man, in every sense of the term. Perhaps in the light of this peculiarly Scottish Romanticism, we can give some shape to Byron's responses to Catholicism.

It should be noted that, in a Britain remaining fairly hostile to Catholicism, Byron's attitude towards that religion was unusually sympathetic from the outset. In his poetry, in his letters, and in an address to Parliament, he supported Catholic emancipation. In this instance, however, we are more interested in his personal response, having taken up exile in Catholic Europe, and in particular Italy.

To begin with, it is notable that in contrast to his comments on his Calvinist past, his references to Catholicism are fairly expansive, largely positive, and even, on occasion, enthusiastic. A few examples will suffice to strike the tone. Speaking in 1821 of his daughter Allegra, he writes: 'It is besides my wish that she should be a Roman Catholic which I look upon as the best religion as it is assuredly the oldest of the various branches of Christianity.'[5] Again, he writes to Thomas Moore in 1882: 'I am no enemy of religion. As proof I'm educating my natural daughter a strict Catholic in a convent in Romagna: for I think people can never have *enough* of religion, if they are to have any. I incline, myself, very much to the Catholic doctrines.'[7] How genuine such statements are we can never be sure. A final example is perhaps particularly telling. Again writing to Moore, later in the same year, he states:

> As I said before, I am really a great admirer of tangible religion; and am breeding one of my daughters a Catholic, that she may have her hands full. It is by far the most elegant worship, hardly excepting the Greek mythology. What with incense, pictures, statues, altars, shrines, relics, and the real presence, confession, absolution – there is something sensible to grasp at.[7]

This, together with the attributed statement, 'I have often wished I had been born a Catholic. That Purgatory of theirs is a comfortable doctrine',[8] would seem to suggest the essence of Byron's attraction to Catholicism, in the light of the

psychological implications of Calvinist dogma. He is drawn precisely to those reconciliatory functions that his own religious background had denied him.

I think that there is a real attraction here to some degree, despite the occasional flippancy which it would be dishonest not to acknowledge. Thus, in the letter to Moore he continues: 'Besides, it leaves no possibility of doubt; for those who swallow their Deity, really and truly, in transubstantiation, can hardly find anything else otherwise than easy of digestion.' However, he goes on:

> I am afraid that this sounds flippant, but I do not mean it to be so; only my turn of mind is so given to taking things in the absurd point of view, that it breaks out in spite of me every now and then.

I think this letter strikes the balance. A degree of flippancy, of scepticism, is surely to be expected. A statement of blind faith would be infinitely more surprising. Quite apart from those aspects of his character already discussed, Byron inhabits a world that is post-Hume, post-Voltaire, and any discussion of belief must take place in the presence of alternatives. In such a world, the most sophisticated contemplation of the religious view is open to the charge of naïvety, even where it is the non-believer's understanding of belief that is the more naïve. The flippancy is perhaps simply an acknowledgement of this, and does not seem to contradict a degree of genuine attraction. Conversely, this is equally compatible with the conclusion that, all told, Byron probably remains a sceptic, when he is not a cynic to the last.

This is about as far as we can get in the matter of Byron's attitude to the Catholic religion. Insofar as religion appealed, Catholicism appealed, for reasons that may relate to his own Calvinist background. Just as we looked beyond Calvinism to the ambience it created, however, so beyond Catholicism we must consider the effect of the ambience created by Catholic Europe, and by Italy in particular. This, I should stress, is a matter distinct from religion, as we are speaking for the most part of influences of which no clergymen of whatever denomination would approve. As we are told in *Don Juan*:

'Tis a sad thing, I cannot choose but say,
  And all the fault of that indecent sun,
Who cannot leave alone our helpless clay,
  But will keep baking, broiling, burning on,
That howsoever people fast and pray
  The flesh is frail, and so the soul undone:
What men call gallantry, and gods adultery.
Is much more common where the climate's sultry.

Happy the nations of the moral North!
  Where all is virtue, and the winter season
Sends sin, without a rag on, shivering forth . . .[9]

The reasons for this contrast have to do with more than the
weather. For Byron, northern morality is something held
together by hypocrisy and cant, and it is their relative absence
that he responds to in his Italian years.

It is interesting to speculate how far this in turn contributed
to the full emergence of another poetic voice. It is worth
considering how far the changing tone of the poetry itself,
from the damned hero of *Childe Harold* to the urbane and
cynical comedy of *Don Juan*, reflects the two contrasting
psychological ambiences, and how far the emergence of the
latter voice, reflects Italy's facilitation of Byron, a relatively
accommodated man in what we might call a condition of
secular Catholicism.

I have said at the outset that I have no particular axe to grind
in this matter, and that I am conscious that the evidence is such
as to remain fairly open ended. Nevertheless, it is an interesting area for speculation, not only in itself, but in relation to
other aspects of Byron's background, as I hope this paper has
suggested.

NOTES

1. Epigraph to *The Rime of the Ancient Mariner*, from *Archaeolo-
   giae Philosophicae*, Thomas Burnet, 1692.
2. Angus Calder, *Byron* (1984), 52.

3. Letter to William Gifford, 18 June 1813. *Byron's Letters and Journals*, ed. Leslie A. Marchand (1973–82), vol. 3, 64.
4. Letter to Annabella Milbanke, 26 September 1813, *BLJ*, vol. 3, 11.
5. Letter to Richard Belgrave Hoppner, 3 April 1821. *BLJ*, vol. 8, 98.
6. Letter to Thomas Moore, 4 March 1822. *BLJ*, vol. 9, 119.
7. Letter to Thomas Moore, 8 March 1822. *BLJ*, vol. 9, 123.
8. *Medwin's Conversations with Lord Byron*, Ernest J. Lovell Jnr (ed.), 1966.
9. *Don Juan*, Canto I, stanza 63.

Byron and Scott

P. H. SCOTT

At first glance, you might suppose that it would be hard to find two men more different and incompatible in character than Sir Walter Scott and Lord Byron. Scott, outwardly at least, was a prudent, douce, respectable Edinburgh lawyer, devoted to the Roman, and Scottish, virtues of effort, stoical endurance and self-control. *Agere atque pati Romanum est*, as he remarked in his *Journal*.[1] What would such a man have in common with the reckless, outrageous, passionate, indulgent, convention-defying Byron? Of course, the truth is the opposite from what one might expect. Each liked and admired the other to a degree which must be rare between two writers who were clear rivals for public favour.

Each expressed his feelings about the other as a man and a writer frequently and consistently. Scott of Byron, in a letter of 10 January 1817 to John Murray: 'No one can honour Lord Byron's poems more than I do and no one has so great a wish to love him personally'.[2] Byron of Scott in his *Journal* on 17 November 1813: 'I like the man – and admire his works to what Mr Braham calls *Entusymusy*'[3] In his *Journal* a few days later, he said that Scott was 'undoubtedly the Monarch of Parnassus' and placed him on the top of his 'Gradus ad Parnassum'.[4] Again in 1821:

Scott is certainly the most wonderful writer of the day. His novels are a new literature in themselves, and his poetry is as good as any – if not better . . . I like him, too, for his manliness of character, for the extreme pleasantness of his conversation, and his good nature towards myself, person-ally. May he prosper for he deserves it. I know no reading to which I fall with such alacrity as a work of W. Scott's.[5]

Equally, Scott gives pride of place to Byron. He spoke of Byron's 'boundless Genius',[6] and said that he gave up writing

poetry because Byron beat him.[7] 'It is well for us', he remarked in a letter to his publisher Cadell, 'that he has not turned himself to tale telling, for he would have endangered our supremacy in that department.'[8]

This relationship, 'warm and cordial on both sides', as Leslie Marchand describes it,[9] began rather inauspiciously. Byron published his *English Bards and Scotch Reviewers* in March 1809 anonymously, but it was no secret that he had written it. The poem makes some mocking references to Scott's *Lay of the Last Minstrel* and *Marmion* and accuses him of writing for money:

> And thinkst thou, Scott! by vain conceit perchance,
> On public taste to foist thy stale romance,
> Though Murray with his Miller may combine
> To yield thy muse just half-a-crown per line?
> No when the sons of song descend to trade,
> Their bays are sear, their former laurels fade.

Scott seems to have come across the poem in August when he mentions it in the course of a letter to Robert Southey:

> In the meantime, it is funny enough to see a whelp of a young Lord Byron abusing me, of whose circumstances he knows nothing, for endeavouring to scratch out a living with my pen. God help the bear, if, having little else to eat, he must not even suck his own paws. I can assure the noble imp of fame it is not my fault that I was not born to a park and £5.000 a-year, as it is not his lordship's merit, although it may be his great good fortune, that he was not born to live by his literary talents or success.[10]

Scott evidently did not nurse a grievance, as many people attacked like this would have done. Little more than a month after the publication of Cantos I and II of Byron's *Childe Harold's Pilgrimage* in February 1812, he was praising the poem in letters to Joanna Baillie and J. B. S. Morritt. 'It is, I think, a very clever poem', he wrote to the former, 'but gives no good symptom of the writers heart or morals . . . I wish you would read it.'[11] To Morritt his praise was even stronger, although he still had reservations about the morality of the piece: 'Though there is something provoking and insulting both to morality and to feeling in his misanthropical ennui it gives nevertheless

an odd poignancy to his descriptions and reflections and upon the whole it is a poem of most extraordinary power and may rank its author with our first poets.'[12]

So far Scott and Byron had not met or even exchanged letters. They knew each other only by reputation and from their published works. As it happens, they were brought together very shortly after Scott had been writing these letters to his friends. Byron took the initiative, perhaps regretting his attack in *English Bards and Scotch Reviewers* or because of a generous impulse to tell Scott something which he know would please him. In June 1812, Byron had met the Prince Regent at an evening party where the Prince had talked about poetry at some length. He had displayed 'an intimacy and critical taste which at once surprised and delighted Lord Byron.' In particular, he spoke about Scott, whom he preferred 'far beyond every other poet of his time.' These phrases come from a letter from the publisher, John Murray, whom Byron had asked to pass on the report to Scott.[13] Scott responded by asking Murray to convey his thanks for this 'very handsome and gratifying communication', and took the opportunity to clear his name 'from any tinge of mercenary or sordid feeling in the eyes of a contemporary of genius.' He had *not* written *Marmion* under contract for a sum of money.[14] Thus encouraged, Byron replied by apologising for the satire, 'written when I was very young & very angry, & fully bent on displaying my wrath and my wit.'[15] In his reply, Scott invited Byron to visit Abbotsford, then a new and far from complete house.[16] From its beginning, their correspondence was friendly, warm and relaxed. Byron sent Scott a copy of his poem, *The Giaour* enscribed 'To the Monarch of Parnassus from one of his subjects'.[17]

In the event, Byron never visited Abbotsford, although Lady Byron did so in 1817. Eight years and much turmoil after the first invitation, Byron wrote from Ravenna to John Murray on 23 April 1820: 'My love to Scott . . . I hope to see him at Abbotsford before very long, and I will sweat his Claret for him'.[18] It was not to be. Their only meetings were in London in 1815 when Scott was passing through to and from Paris. Scott described them in a letter which he wrote in response to a

request from Tom Moore for facts and recollections which
would help him with his *Life* of Byron. It is worth quoting at
length because it says so much about Scott and Byron:

It was in the spring of 1815, that, chancing to be in
London, I had the advantage of a personal introduction to
Lord Byron. Report had prepared me to meet a man of
peculiar habits and a quick temper, and I had some doubts
whether we were likely to suit each other in society. I was
most agreeably disappointed in this respect. I found Lord
Byron in the highest degree courteous, and even kind. We
met for an hour or two almost daily, in Mr. Murray's
drawing-room, and found a great deal to say to each other.
We also met frequently in parties and evening society, so
that for about two months I had the advantage of a
considerable intimacy with this distinguished individual.
Our sentiments agreed a good deal, except upon the
subjects of religion and politics, upon neither of which I was
inclined to believe that Lord Byron entertained very fixed
opinions . . .

On politics, he used sometimes to express a high strain of
what is now called Liberalism; but it appeared to me that the
pleasure it afforded him, as a vehicle for displaying his wit
and satire against individuals in office, was at the bottom of
this habit of thinking, rather than any real conviction of the
political principles on which he talked. He was certainly
proud of his rank and ancient family, and, in that respect, as
much an aristocrat as was consistent with good sense and
good breeding. Some disgusts, how adopted I know not,
seemed to me to have given this peculiar and (as it appeared
to me) contradictory cast of mind; but at heart, I would have
termed Byron a patrician on principle.

Lord Byron's reading did not seem to have been very
extensive, either in poetry or history. Having the advantage
of him in that respect, and possessing a good competent
share of such reading as is little read, I was sometimes able to
put under his eye objects which had for him the interest of
novelty. I remember particularly repeating to him the fine
poem of Hardyknute, an imitation of the old Scottish ballad,
with which he was so much affected, that some one who

was in the same apartment asked me what I could possibly have been telling Byron by which he was so much agitated.

I saw Byron for the last time in 1815, after I returned from France. He dined, or lunched, with me at Long's, in Bond Street. I never saw him so full of gayety and good-humor . . .

Several letters passed between us – one perhaps every half-year. Like the old heroes in Homer, we exchanged gifts. I gave Byron a beautiful dagger mounted with gold, which had been the property of the redoubted Elfi Bey. But I was to play the part of Diomed in the Iliad, for Byron sent me, some time after, a large sepulchral vase of silver. It was full of dead men's bones . . .

I think I also remarked in Byron's temper starts of suspicion, when he seemed to pause and consider whether there had not been a secret, and perhaps offensive, meaning in something casually said to him. In this case, I also judged it best to let his mind, like a troubled spring, work itself clear, which it did in a minute or two. I was considerably older, you will recollect, than my noble friend, and had no reason to fear his misconstruing my sentiments towards him, nor had I ever the slightest reason to doubt that they were kindly returned on his part. If I had occasion to be mortified by the display of genius which threw into the shade such pretensions as I was then supposed to possess, I might console myself that, in my own case, the materials of mental happiness had been mingled in a greater proportion.[19]

Scott retained warm memories of his last meeting with Byron, which was on 14 September 1815.[20] Ten years later he referred to it in his *Journal*: 'I never saw Byron so full of fun, frolic, wit and whim: he was as playful as a kitten'.[21] But playful or not, Byron was then in the middle of his disastrous marriage. During the scandal of the separation which followed, almost all of Scott's correspondents took the side of Lady Byron.[22] Joanna Baillie even tried to persuade him to intervene over the financial settlement, since there was nobody, she said, whose good opinion Byron was more anxious to preserve.[23] Scott regretted the breakdown of the marriage, but he refused to be stampeded into the general

mood of condemnation. 'I was in great hopes', he told Morritt in February 1816, 'that the comfort of domestic security might tame the wayward irregularity of mind which is unfortunately for its owner connected with such splendid talent. I have known Lord Byron to do very great and generous things and I would have been most happy to find that he had adopted other and more settled habits'.[24] In another letter to Morritt in May: 'Lord Byron with high genius and many points of a noble and generous feeling has Child Harolded himself and Outlawd himself into too great a resemblance with the picture of his imagination'.[25]

Byron left Britain in April 1816 to escape both the bailiffs and the weight of public disapproval. By the end of June he had written the third canto of *Childe Harold* and it was published by John Murray in November. Scott responded by writing a review of some thirty pages for the *Quarterly* (published in February 1817). He knew that he was taking a risk. The poem was so autobiographical in content that it was impossible to comment on it without saying something about Byron's personal situation, and, as he told Murray, he did not want to cause him pain.[26] The review praised Byron for his 'wide, powerful and original view of poetry' and suggested that his 'family misfortunes' were the consequences of 'a powerful and unbridled imagination, . . .the author and architect of its own disappointments'. Scott concluded with the 'anxious wish and eager hope' that Byron would recover peace of mind for the exercise of his 'splendid talents'.[27] At any time this would have been a kind and understanding review. In the circumstances, the writing of it was not only delicate and tactful, but courageous.

Byron was deeply touched. 'He must be a gallant as well as a good man,' he wrote to Murray in March 'who has ventured in that place – and at this time – to write such an article anonymously . . . It is not the mere praise, but there is a *tact* & a *delicacy* throughout not only with regard to me – but to *others*'.[28] A week later he had heard that Scott was the author and wrote to Thomas Moore, 'you will agree with me that such an article is still more honourable to him than to myself'.[29] Byron never forgot this example of Scott's courage and

generosity. He mentioned it in a letter to Scott for the first time in January 1822: 'You went out of your way in 1817 – to do me a service when it required not merely kindness – but courage to do so . . . The very *tardiness* of this acknowledgement will at least show that I have not forgotten the obligation'.[30] On 29 May 1823, when Byron was preparing to leave Italy for his final journey to Greece, he took time to write to Henri Beyle (Stendhal). Beyle had praised Scott's writing in a pamphlet but said that 'his character is little worthy of enthusiasm'. Byron rose to his defence:

I have known Walter Scott long and well, and in occasional situations which call forth the *real* character – and I can assure you that his character *is* worthy of admiration – that of all men he is the most *open*, the most *honourable*, the most *amiable*. With his politics I have nothing to do: they differ from mine, which renders it difficult for me to speak of them. But he is *perfectly sincere* in them . . . *Believe* the *truth*. I say that Walter Scott is as nearly a thorough good man as man can be, because I *know* it by experience to be the case.[31]

There cannot be much doubt that when he wrote these words the *Quarterly* article was in his mind.

Scott proved his robust support for Byron once again by accepting in December 1821 the dedication of one of his boldest works, *Cain*. In his letter of acceptance to Murray, Scott said that 'Byron matched Milton on his own ground. Some part of the language is bold, and may shock one class of readers . . . But then they must condemn the Paradise Lost, if they have a mind to be consistent.'[32]

Byron's admiration for Scott as a writer was based at first on his poetry, but he became a great enthusiast for the novels as soon as he discovered them. The first of them, *Waverley*, was published on 7 July 1814. As early as the 24th of the same month Byron wrote to Murray: 'Waverley is the best and most interesting novel I have read since – I don't know when . . . besides – it is all easy to me – because I have been in Scotland so much – (though then young enough too) and feel at home with the people lowland & Gael.'[33] From then onwards Byron read all of Scott's novels as they appeared and, by his own repeated account, many times over. He said that he liked no reading so

well.[34] Although they were still anonymous, 'by the author of Waverley', Byron seems never to have had any doubts that Scott was the 'great unknown'. In a letter of February 1820, for instance: 'I have more of Scott's novels (for surely they are Scott's) since we met, and am more and more delighted. I think I even prefer them to his poetry.'[35] Or in his *Journal* on 5 January 1821: 'Read the conclusion, for the fiftieth time (I have read all W. Scott's novels at least fifty times) of the third series of "Tales of my Landlord", – grand work – Scotch Fielding, as well as great English poet – wonderful man I long to get drunk with him.'[36]

From 1814 to the end of his life Byron's letters are full of allusions to the novels, praise of them, and requests to Murray to be sure to send new ones as they were published. There are also frequent quotations from them, usually phrases of the dialogue in Scots, which show how familiar he was with them and how much at home with Scots vocabulary. He wrote to Scott himself about them from Pisa on 12 January 1822:

> I don't like to bore you about the Scotch novels (as they call them though two of them are wholly English – and the rest half so) but nothing can or could ever persuade me since I was the first ten minutes in your company that you are *not* the Man . . . To me those novels have so much of "Auld lang syne". (I was bred a canny Scot till ten years old) that I never move without them – and when I removed from Ravenna to Pisa the other day – and sent on my library before – they were the only books that I kept by me – although I already knew them by heart . . . I need not add that I would be delighted to see you again – which is far more than I shall ever feel or say for England or (with a few exceptions of "kith – kin – and allies") any thing that it contains . . . But my "heart warms to the Tartan" or to any thing of Scotland.[37]

Byron's response to the Waverley novels helps to explain the evident affinity between him and Scott. In his *Journal*, Scott describes himself as 'tolerably national'.[38] This was a calculated understatement, because Scott's passion for Scotland was a dominant force in his composition.[39] I think that there is strong evidence that Byron too was 'tolerably national', if not

to the same overwhelming degree as Scott. In his essay on Byron, T. S. Eliot said that Byron was a Scottish poet, and in comparing him with Scott remarked: 'Possibly Byron, who must have thought of himself as an English poet, was the more Scotch of the two because of being unconscious of his true nationality.'[40] On the contrary, I think that Byron makes it very clear, as in the letters just quoted, that he was perfectly conscious of his Scottishness.

Indeed when Byron said in Canto X of *Don Juan* that he was 'half a Scot by birth, and bred a whole one',[41] he was speaking the literal truth. He was 'half a Scot' because his mother was Scottish; he was 'bred a whole one' because he spent the first ten years of his life in Aberdeen and was educated at the Grammar School. I think it is generally agreed that the first few years of life are the most influential in determining character and attitudes. Many people, including Eliot and Grierson, have noted the influence on him of Scottish Calvinism.[42] It is also quite certain that he heard all around him in Aberdeen, and learned to speak, good Buchan Scots. His frequent quotation in his letters of phrases in Scots shows how he responded to it. No doubt he acquired an English veneer at Harrow and Cambridge; but, as Eliot says, he 'remained oddly alien' in English society.[43] I think that is because he was fundamentally Scottish.

There is a significant episode in Byron's correspondence with his half-sister, Augusta Leigh. She wrote in February 1817 to congratulate him on the assumption that he was 'P.P.' By this she meant Peter Pattieson, the fictitious character to whom Scott attributed *Tales of My Landlord*, the four volumes published in December, 1816, which contained *The Black Dwarf* and *Old Mortality*. The books had not yet reached Byron in Venice and he had no idea what Augusta was talking about. 'I am not P.P.', he replied, 'and I assure you upon my honour – & do not understand to what book you allude – so that all your compliments are quite thrown away.'[44] However after he had read the *Tales*, he wrote to John Murray in May: 'The *Tales of My Landlord* I have read with great pleasure – & perfectly understand why my sister & aunt are so very positive in the very erroneous persuasion that they must have been written

by me – if you knew me as well as they do – you would perhaps have fallen into the same mistake.'[45]

In his footnote to the first of these two letters, Leslie Marchand suggests that Augusta was misled by the character of David Ritchie in *The Black Dwarf* who was 'haunted by a consciousness of his own deformity'. Certainly, it is quite possible that this resemblance to Byron triggered off her conclusion. (Incidentally, another bond between Scott and Byron was, of course, that they were both lame and both compensated for their handicap by vigorous physical activity, Byron by his swimming and Scott on horseback.) But this point alone could hardly have led anyone to imagine that Byron was the author, unless they could also suppose he was capable of writing anything so essentially Scottish as these two novels and of writing superb dialogue in Scots. Although the *Tales* were not attributed on the title page to the 'author of Waverley', it was very clear to everyone that they were by the same writer. The implication of Augusta's conclusion, and of Byron's understanding of it, was that he was Scottish enough to be conceivable as the author of the first six, and some of the most Scottish, of the Waverley novels.

Yet another quality which Byron and Scott shared, and which they recognised and appreciated in each other, was a general approach to life which might seem inconsistent with their reputation as leaders of European romanticism. In a letter to Murray in March 1817, Byron said of Scott: 'He & Gifford & Moore are the only *regulars* I ever knew who had nothing of the *Garrison* about their manner – no nonsense – nor affectations look you.'[46] Elsewhere, he refers to Scott as a man of the world.[47] Scott said very much the same of Byron: 'What I liked about Byron, besides his boundless genius, was his generosity of spirit as well as purse, and his utter contempt of all affectations of literature.'[48] Scott was delighted by a story which Moore told him about Byron: 'While they stood at the window of Byron's Palazzo in Venice looking at a beautiful sunset, Moore was naturally led to say something of its beauty, when Byron answered in a tone that I can easily conceive, 'Ah come, d-n me, Tom, don't be poetical'.[49]

This feet-on-the-ground, no-nonsense attitude is remote from the Byronic hero, but it is entirely consistent with the

realistic, and sceptical attitude, the eighteenth-century rationalism, which formed one aspect of both Byron and Scott. Both
had been educated in cities, Aberdeen and Edinburgh, where
the atmosphere of the Scottish Enlightenment prevailed and it
left its influence on both of them. Its co-existence with
romanticism of diverse kinds might perhaps be an example of
what Gregory Smith called 'the Caledonian antisyzygy' and
identified as the 'combination of opposites', or 'the contrasts
which the Scot shows at every turn.'[50]

Another obvious link between the two men is, of course,
that they both wrote long, narrative poems. There are many
similarities in metre, style and manner, even if the setting and
content is very different and if Byron eventually added a new
dimension to this kind of writing. Byron probably got the idea
of writing poems of this kind because of Scott's example.
Lockhart had no doubt that this was so: 'indeed in all his early
serious narratives, Byron owed at least half his success to
clever though perhaps unconscious imitation of Scott.'[51]
Lockhart has another anecdote which shows how closely the
minds of the two poets sometimes resembled one another. In
London in 1828, he was with Scott listening to Mrs Arkwright
singing some of her own settings. After one of them, Scott
whispered to him 'Capital words – whose are they? – Byron's I
suppose, but I don't remember them.' Lockhart continues, 'He
was astonished when I told him that they were his own in *The
Pirate*. He seemed pleased at the moment, but said next minute
– "You have distressed me – if memory goes, all is up with me,
for that was always my strong point".'[51]

Before Byron decided in 1823 to go to Greece to help in their
revolution, he had considered other possibilities. In 1819 he
had thought in particular of South America, where states were
struggling for their independence.[53] In May 1822 he asked
John Murray to send him the *Lockhart Papers*, 'a publication
upon Scotch affairs of some time since'.[54] The papers were the
Memoirs of John Lockhart of Carnwath, a member of the
Scottish Parliament from 1703 to 1707 and a strong opponent of
the Union with England. Byron did not say why he wanted the
books, but since they contain the best contemporary exposure
of the sordid transaction of the Union, it might be that he

thought also of Scotland as a country where an independence movement might be stimulated or encouraged.

But, of course, it was Greece that he chose, and there he died on 19 April 1823. When the news reached Edinburgh, Scott wrote an obituary for the *Edinburgh Weekly Journal*, calling Byron, 'That mighty genius, which walked among men as something superior to ordinary mortality, and whose powers we beheld with wonder and something approaching to terror.'[55] So ended a relationship which did credit to both men and gave much pleasure and satisfaction to both of them. During the rest of his life, Scott's thoughts often turned to Byron: 'I very often think of him almost with tears', he told Moore in August 1825.[56] Once, shortly after Byron's death, he imagined for a moment that he saw him 'with wonderful accuracy' standing in the hall at Abbotsford.[57] He summed up his view of Byron in his Journal: 'This was the man – quaint, capricious, and playful, with all his immense genius. He wrote from impulse never from effort and therefore I have always reckond Burns and Byron the most genuine poetical genuises of my time and half a century before me.'[58] In 1832, as Scott was himself close to death and was hurrying home to Abbotsford from Italy, he visited Venice and went to see the balcony on which Byron had said to Moore, 'Damn me, Tom, don't be poetical.'[59]

ABBREVIATIONS

J.	Sir Walter Scott, *Journal* (Edinburgh, 1950)
G.	*The Letters of Sir Walter Scott*, ed. H. J. C. Grierson, 12 vols (London, 1932–37)
M.	*Byron's Letters and Journals*, ed. Leslie A. Marchand, 12 vols (London, 1973–82)
L.	J. G. Lockhart, *Memoirs of Sir Walter Scott*, 5 vols (London, 1900)

NOTES

1. J.   22 June 1826
2. G.   IV, 365
3. M.   III, 209
4. M.   III, 219–20

5. M. VIII, 23
6. J. 23 November 1825
7. L. V, 391
8. Edgar Johnson, *Sir Walter Scott, The Great Unknown* (New York, 1970), II, 1219 (quoting a letter from Scott to Cadell)
9. M. IV, 358
10. G. II, 214
11. G. III, 98–9
12. G. III, 114–15
13. G. III, 135 f.n. 1
14. G. III, 135–9
15. M. II, 182
16. G. III, 140–1
17. L. II, 508
18. M. VII, 83
19. L. II, 514–17
20. L. III, 24
21. J. 21 December 1825
22. Wilfred Partington, *Sir Walter's Post-Bag* (London, 1932), 114
23. Ibid., 116
24. G. IV, 184
25. G. IV, 234
26. G. IV, 364–5
27. *Quarterly Review*, XVI, dated October 1816 (but published in February 1817), 189, 207, 208
28. M. V, 178
29. M. V, 185
30. M. IX, 85–6
31. M. X, 189–90
32. G. VII, 37
33. M. IV, 146
34. M. VII, 48
35. M. VII, 45
36. M. VIII, 13
37. M. IX, 86–7
38. J. 22 November 1825
39. See Chapter 7 of my *Walter Scott and Scotland* (Edinburgh, 1981)
40. T. S. Eliot, 'Byron' in *From Anne to Victoria; Essays by Various Hands* (1937), 602
41. Stanza XVII
42. T. S. Eliot, op. cit., 604; Herbert Grierson, *The Background of English Literature* (1962), 152.
43. T. S. Eliot, op. cit., 617.
44. M. V, 171
45. M. V, 220

46. M.   V, 192
47. M.   IX, 30
48. J.   23 November 1825
49. J.   9 February 1826
50. G. Gregory Smith, *Scottish Literature: Character and Influence* (1919), 4
51. L.   II, 509
52. L.   V, 195–6 f.n.
53. M.   VI, 123 f.n. 1; 212
54. M.   IX, 156.
55. Sir Walter Scott, *Miscellaneous Prose* (Edinburgh, 1834), II, 343
56. G.   IX, 199
57. L.   V, 131–2 and f.n. 57;   J.   9 February 1826
58. J.   9 February 1826
59. Edgar Johnson, op. cit., II, 1248.

The Provost and his Lord: John Galt and Lord Byron

MARGERY MCCULLOCH

John Galt first met Lord Byron in Gibraltar in 1809. Both were attempting to escape the frustrations of London. Galt was a Scot, born and brought up in his early years in the Ayrshire town of Irvine – the 'Gudetown' of his novel *The Provost*. His family later moved to Greenock to take advantage of the prosperity of that increasingly thriving port, and Galt's first employment was in the world of commerce. He was an unusual business man, however, in that from his beginnings literary activity went hand in hand with business activity. Like many ambitious Scots in the nineteenth century, he soon found his way to London, where at first he was conspicuously successful neither in business nor in literature. His travels in Europe between 1809 and 1811, which provided a significant part of the material for his biography of Lord Byron, were in part an attempt to recover his nervous health after the failure of his business ventures and literary aspirations.

Byron – 'half a Scot by birth'[1] through his mother's family – had also left England in a state of disaffection. He had recently taken his seat in the House of Lords unaccompanied by supporters and after some initial difficulty in establishing his credentials. His collection of poems *Hours of Idleness* had been devastatingly dismissed by the *Edinburgh Review* in what Galt called 'bleak and blighting criticism'.[2] Unlike Galt, however, he had at least the satisfaction of leaving England in some notoriety, his satirical reply to the *Review* – *English Bards and Scotch Reviewers* – having attracted much attention and a rapid sale.

The impulse behind this essay, then, is John Galt's *The Life of Lord Byron*, written and published in 1830 and based largely on his meeting and travels with Byron in the Eastern Mediterranean between 1809 and 1811; and the associations and

65

reflections it has aroused in my own mind with regard to both author and subject, especially with regard to Byron's personality and poetry.

Galt's first sighting of Byron – whom he did not at that point recognise – was in the garrison library in Gibraltar:

> . . . a young man came in and seated himself opposite to me at the table where I was reading. Something in his appearance attracted my attention. His dress indicated a Londoner of some fashion, partly by its neatness and simplicity, with just so much of a peculiarity of style as served to show, that although he belonged to the order of metropolitan beaux, he was not altogether a common one.
>
> I thought his face not unknown to me; I began to conjecture where I could have seen him; and, after an unobserved scrutiny, to speculate both as to his character and vocation. His physiognomy was pre-possessing and intelligent, but ever and anon his brows lowered and gathered; a habit, as I then thought, with a degree of affectation in it, probably first assumed for picturesque effect and energetic expression; but which I afterwards discovered was undoubtedly the occasional scowl of some unpleasant reminiscence: it was certainly disagreeable – forbidding – but still the general cast of his features was impressed with elegance and character.   (59-60)

Later, having discovered the identity of the young man and having embarked with him and his travelling companion, Hobhouse, for Sardinia, Galt observes Byron's adoption of a demeanour which anticipates that of his hero, Childe Harold:

> Hobhouse, with more of the commoner, made himself one of the passengers at once; but Byron held himself aloof, and sat on the rail, leaning on the mizzen shrouds, inhaling, as it were poetical sympathy, from the gloomy rock, then dark and stern in the twilight.   (61)

In these early observations of Byron one can recognise the eye and mind of the creator of Micah Balwhidder and Provost Pawkie of Gudetown. Galt makes it clear at the outset of his biography that there will be no sensational preoccupation with or prying into Byron's personal life. Instead, he will confine himself 'as much as practicable, consistent with the end in

view, to an outline of his Lordship's intellectual features' (iii), and will attempt to 'consider him . . . as his character will be estimated when contemporary surmises are forgotten, and when the monument he has raised to himself is contemplated for its beauty and magnificence, without suggesting recollections of the eccentricities of the builder' (4).

Galt is a discerning biographer and the course he adopts accords well with his own qualities as a writer and man. As with the relations between Provost Pawkie and *his* lord, Galt is courteous but never subservient towards Lord Byron. Indeed, there is more than a little of Pawkie's canniness in the following report of Byron's touchiness and Galt's humouring of him after a theatrical performance in Cagliari:

When the performance was over, Mr Hill came down with Lord Byron to the gate of the upper town, where his Lordship, as we were taking leave, thanked him with more elocution than was precisely requisite. The style and formality of the speech amused Mr Hobhouse, as well as others; and when the minister retired, he began to rally his Lordship on the subject. But Byron really fancied that he had acquitted himself with grace and dignity, and took the jocularity of his friend amiss – a little banter ensued – the poet became petulant, and Mr Hobhouse walked on; while Byron, on account of his lameness, and the roughness of the pavement, took hold of my arm, appealing to me, if he could have said less, after the kind and hospitable treatment we had all received. Of course, though I thought pretty much as Mr Hobhouse did, I could not do otherwise than civilly assent, especially as his Lordship's comfort, at the moment, seemed in some degree dependent on being confirmed in the good opinion he was desirous to entertain of his own courtesy. From that night I evidently rose in his good graces; and, as he was always most agreeable and interesting when familiar, it was worth my while to advance, but by cautious circumvallations, into his intimacy; for his uncertain temper made his favour precarious.   (66)

It should be said that Byron's sentiments, at that time, did not exactly accord with Galt's. He is reported by Lady Blessington as regretting that he had not truly appreciated Galt

during these early meetings: 'his manner had not deference enough for my then aristocratical taste, and finding I could not awe him into respect sufficiently profound for my sublimer self either as a peer or an author, I felt a little grudge towards him . . .'[3]

Galt, on the other hand, is objective and fair in his portrait of Byron. He provides the reader with the kind of precise, relevant detail of events which enables one to enter into the happenings and form one's own judgements. He is also, as one would expect from his fiction, psychologically perceptive and looks below the surface petulance of his subject to the insecurity and sensitivity within. Byron's behaviour might often display 'the pride of rank' (154), but Galt's portrait shows us a man whose insecurity made him 'so skinless in sensibility as respected himself, and so distrustful in his universal apprehensions of human nature, as respected others' (175).

There seems to have been an intuitive understanding of Byron on Galt's part, an unforced sympathetic feeling for the man. It is interesting to note that while he always seeks out a psychological explanation for Byron's capriciousness, Galt has little patience with the unconventional ideas and behaviour of Shelley, with whom Byron later became friendly. While lamenting that Byron had 'a wayward delight in magnifying his excesses, not in what was to his credit, like most men, but in what was calculated to do him no honour' (224), Galt dismisses Shelley as having 'a singular incapability of conceiving the existing state of things as it practically affects the nature and conditions of man' (255). Nor could Shelley's work provide, for Galt, a counter-balance to his behaviour, as did Byron's. For Galt, Shelley's 'works were sullied with the erroneous induction of an understanding which, in as much as he regarded all the existing world in the wrong, must be considered as having been either shattered or defective' (256-7). Shelley, in Galt's view, 'was more of a metaphysician than a poet' (257).

Despite the differences in the character and life-style of Galt and Byron, there are significant underlying similarities which would have worked towards Galt's understanding of his subject. Both were role-players in their work, Galt through the

adoption of such personae as Balwhidder and Pawkie, Byron through his diverse poetic voices. Both were enemies of hypocrisy and were to become masters of the ironic presentation of self-delusion. There is, however, less of what Galt called Byron's 'gratuitous spleen' (190) in his own work, while Byron is reported by Lady Blessington to have said of Galt's novels that what he admired particularly 'is that with a perfect knowledge of human nature and its frailties and legerdemain tricks, he shows a tenderness of heart which convinces me that *his* is in the right place, and he has a sly caustic humour that is very amusing.'[4] This is a fair comment on Galt's presentation of a character such as Provost Pawkie, for however devious and self-reflexive that politician's activities may be, they are also on the whole directed towards the improvement of the community and are conducted with a sound common sense which points to his author's Enlightenment affinities.

This affinity with the eighteenth century is another bond between the two writers. Byron's alienated, angst-ful protagonists have become the archetypes of what we call 'the Romantic hero', but Byron himself looked to another age for *his* poetic heroes. The narrator's credo in Canto I of *Don Juan* is 'Thou shalt believe in Milton, Dryden, Pope;/Thou shalt not set up Wordsworth, Coleridge, Southey' (stanza 205); and although the attacks on the Lakers in *Don Juan* have the verve and pointedness of a current preoccupation, Byron's published disdain for their work and his admiration for the earlier Milton, Dryden and Pope, goes back as far as the 1809 *English Bards and Scotch Reviewers*.

Galt criticised Byron for aiming his satire too generally in that early work, but he himself shows a similar antipathy to the kind of sensibility epitomised by the 'gelatinous' Lakers, as he calls them in the biography (41). Most readers of Byron are familiar with the passage in Canto I of *Don Juan* where Juan for the first time feels the stirrings of love – what Galt was wont to call 'the passion' – but because his mother has edited out natural history from his education, he is unable to understand what is wrong with him. In his account of Juan's sufferings, Byron wittily combines an attack on the hypocrisy of educational attitudes with allusions to Wordsworth's philosophy of

nature, and with an ironic backward glance at his own Byronic heroes. In his characteristic fashion he builds up the portrait of his man of unfocused feeling stanza by stanza, accumulating detail and just when it seems about to reach its sublime climax, he undercuts the effect with a bathetic final line:

> He pored upon the leaves, and on the flowers,
>   And heard a voice in all the winds; and then
> He thought of wood nymphs and immortal bowers,
>   And how the goddesses came down to men:
> He miss'd the pathway, he forgot the hours,
>   And when he look'd upon his watch again,
> He found how much old Time had been a winner –
> He also found that he had lost his dinner.   (stanza 94)

Significant in the context of the relationship between Galt and Byron is a similarly comic passage from Galt's *Annals of the Parish*, where the Revd Balwhidder attempts to mitigate his grief at the loss of his first wife. Initially he occupies his mind by the erecting of a monument and the composing of an epitaph, but being once again in danger of 'sinking into the hyperchonderies', Balwhidder follows Wordsworth and Juan and 'often walked in the fields, and held communion with nature and wondered at the mysteries thereof.'[5] This inspired communing in turn leads him to think of writing a book, but finally common sense and Providence intervene to persuade him to take another wife, as Galt – like Byron – deflects his romantic musings with a domestic crisis which demands a more down-to-earth, social resolution. In the absence of a mistress, Balwhidder's 'servant lasses . . . wastered everything at such a rate, and made such a galravitching in the house, that, long before the end of the year, the year's stipend was all spent.' Worse still, 'one of my lassies had got herself with bairn, which was an awful thing to think had happened in the house of her master, and that master a minister of the Gospel.'[6] In the face of these worldly disasters, Balwhidder gives up 'sauntering along the edge of Eglesham Wood, looking at the industrious bee going from flower to flower, and the idle butterfly, that layeth up no store, but perisheth ere it is winter',[7] and sets about the search for a prudent helpmate.

Thus reason and good sense, a little tinged with self-delusion, triumph as in Byron over contemplation and the mysteries of the natural world.

This preference for Enlightenment good sense as opposed to the sensibility associated with Wordsworth and the Lakers is in evidence throughout Galt's biography of Byron. Speaking of the poet's early childhood in Aberdeenshire, Galt rejects theories which ascribe his poetic genius to his early association with the natural world. Yet, paradoxically, his accompanying comment that 'deep feelings of dissatisfaction and disappoint-ment are . . . the very spirit of his works, and a spirit of such qualities is the least of all likely to have arisen from the contemplation of magnificent nature, or to have been inspired by studying her storms or serenity' (19) takes us into the world of Wordsworth's 'Michael' or 'Resolution and Independence', where nature provides consolation and strength for the disquieted spirit. To the twentieth-century reader, Galt's viewpoint that Byron's delight 'in contemplating the Malvern Hills, was not because they resembled the scenery of Lochyna-gar, but because they awoke trains of thought and fancy, associated with recollections of that scenery,' brings him close to a perception of Wordsworth's poetic method, as does his further comment that the 'poesy of the feeling lay not in the beauty of the objects, but in the moral effect of the traditions, to which these objects served as talisman of the memory' (28).

In effect, Galt, like Byron, has affinities with both the Enlightenment and the Romantic periods in his attitude to the nature of poetry. Perhaps because of the different cultural and social ambience from which he came, and the nature of a literary tradition which in the eighteenth century produced Fergusson and Burns as opposed to Pope, Galt, unlike the aristocratically-inclined and Harrow-educated Byron, could more readily accommodate Wordsworth's insistence that 'a poet is a man speaking to men' and using the language 'really used by men.'[8] With Galt, however, there is an emphasis on the 'worldly' and the social not to be found in Wordsworth, which relates to the eighteenth-century conception of the poet's social role. In Galt's view, 'the greatest poets have all been men – worldly men, different only from others in

reasoning, more by feeling than induction.' He stresses that no greater misconception has ever been obtruded upon the world as philosophic criticism, than the theory of poets being the offspring of 'cooing lambs and capering doves' . . . The most vigorous poets, those who have influenced longest and are most quoted, have indeed been all men of great shrewdness of remark, and any thing but your chin-on-hand contemplators . . . Are there any symptoms of the gelatinous character of the effusions of the Lakers in the compositions of Homer?

He has no doubt as to where the poet should find the material for his work:

> Compare, the poets that babble of green fields with those who deal in the actions of men, such as Shakespeare, and it must be confessed that it is not those who have looked at external nature who are the true poets, but those who have seen and considered most about the business and bosom of man. It may be an advantage that a poet should have the benefit of landscapes and storms, as children are the better for country air and cow's milk; but the true scene of their manly work and business is in the populous city.   (40-2)

Such a comment is reminiscent of Byron's attacks on Wordsworth, Coleridge and Southey. Although Byron condemned them as 'epic renegades', ostensibly for turning Tory (*Don Juan*, Dedication, I), one senses that, like Galt, he had an innate dislike of the formal and philosophical nature of the attempt to 'make it new' by these older Romantics. Again like Galt, Byron is inevitably a man of his time, and, despite his dislike of Wordsworth's poetic method, in *Don Juan* and *The Vision of Judgment* especially he employs the everyday language and everyday subjects (although from a higher social stratum) which Wordsworth advocated. Yet, as with his biographer, he is strongly affiliated to the eighteenth-century concept of the social role of the artist and the artist as craftsman. Pope's 'Know then thyself, presume not God to scan,/The proper study of mankind is man'[9] has much relevance to the work and attitudes of both Galt and Byron. Childe Harold may seem to be at one with Wordsworth's solitary protagonists, shunning society and communing with

the natural world, nevertheless society is an ever-present touchstone in Byron's poetry as it is not in Wordsworth's. The society Harold rejects is always with us, as a result of his constantly expressed alienation from it. Similarly, the natural world serves not as a substitute for society, but as a reminder of Harold's parting from that society. In his observations on Byron in Athens and Ephesus, Galt comments that 'Lord Byron made almost daily excursions on horseback chiefly for exercise and to see the localities of celebrated spots. He affected to have no taste for the arts, and he certainly took but little pleasure in the examination of the ruins' (120). In Galt's view, Byron's imagination was one which 'required the action of living characters to awaken its dormant sympathies' (137).

Yet, although he looked to Pope for many of his poetic values, Byron's social satires reveal the dissimilarity in their worlds. However critical Pope may be of his society's morality and attitudes in action, his criticism comes from a stable base within that society. At root he and his society have the same moral values, and he can expect to find an audience. The situation is otherwise with Byron. Superficially his verse satires may remind one of Pope, but his is the comedy of manners of a subversive outsider, who knows that his views will outrage even while they entertain. Nor is there any sense in Byron's poetry of a reforming purpose, of a shock administered in order to return the audience to the ideal of shared values. Like his alienated, wandering heroes, Byron himself, even at his most entertaining and uncensorious, seems set apart from the society he observes and caricatures. This sense of apartness is something we do not find in Galt. Despite the long periods he spent outside Scotland – in London, Europe and later in Canada – and despite the problems of Scottish national and cultural identity brought about by the two Unions with England, Galt's ironic, dramatic enactments of Scottish small town and parish life are the work of an insider, who appears to be proceeding from a stable base within his society.

In addition to his observations on Byron the man, Galt makes acute comments throughout his biography on the nature of Byron's poetry. One of these is his recurring

insistence that Byron's best poetry is based on factual experience and not on imagination. For Galt, unlike Coleridge,
memory is not allied with fancy as a lower form of creative
activity, but is the very corner-stone of such activity. Galt
parted from Byron and Hobhouse at Malta in the autumn of
1809, and did not meet up with them again until the following
spring in Athens. He continues his account of Byron's travels
during this period by references to the detail of *Childe Harold's
Pilgrimage*, supported by his reading of Byron's and Hobhouse's accounts of their journeyings and his own memories
of their conversations. By using *Childe Harold* in this way, he
gives substance to his contention that Byron's is a poetry based
on fact and experience which demonstrates 'how little, after
all, of great invention is requisite to make interesting and
magnificent poetry' (102).

It would be wrong, however, to see Galt as the kind of
commentator who insists on a direct correspondence between
Byron's poetry and his life: he is well aware that the use to
which experience is put is part of the transforming imaginative
process. Byron's own journey from the town of Joannina to
Zitza in search of the vizier Ali Pashaw is seen to parallel that of
his hero Childe Harold, as does his kind reception by the prior
of the monastery where the travellers seek shelter for the
night. On the other hand, Galt comments that while 'many
traits and lineaments of Lord Byron's own character may be
traced in the portraits of his heroes, I have yet often thought
that Ali Pashaw was the model from which he drew several of
their most remarkable features' (88). In a piece of Pawkie-like
self-advertisement, Galt describes Byron's encounter with Ali
Pashaw – 'the Rob Roy of Albania' (89) – not only through the
detail of *Childe Harold* but also through his memories of his
own reception by Ali Pashaw's son at Tripolizza, justifying the
inclusion of his experience with the comment that 'the
ceremonies on such visits are similar all over Turkey, among
personages of the same rank' (84).

One passage from Byron's poetry which exudes the exotic
flamboyance of Hollywood invention, but which Galt maintains has its origins in a witnessed event, is the description of
the dogs devouring the dead in *The Siege of Corinth*:

From a Tartar's skull they had stripped the flesh,
As ye peel the fig when its fruit is fresh;
And their white tusks crunched o'er the whiter skull,
As it slipped through the jaws, when their edge grew dull,
As they lazily mumbled the bones of the dead,
When they scarce could rise from the spot where they fed;
So well had they broken a lingering fast
With those who had fallen for that night's repast.
And Alp knew, by the turbans that rolled on the sand,
The foremost of these were the best of his band:
Crimson and green were the shawls of their wear,
And each scalp had a single long tuft of hair,
All the rest was shaven and bare.
The scalps were in the wild dog's maw,
The hair was tangled round his jaw. . . .   (stanza 16)

This scene apparently originated in the sight of two dogs
gnawing a dead body under the walls of the Palace of the
Sultans in Constantinople, when Byron and Hobhouse had
left the comfort of their ship's cabin for an outing in the city.
Galt comments:

> This hideous picture is a striking instance of the uses to
> which imaginative power may turn the slightest hint, and of
> horror augmented till it reach that extreme point at which
> the ridiculous commences. The whole compass of English
> poetry affords no parallel to this passage. It even exceeds the
> celebrated catalogue of dreadful things on the sacramental
> table in Tam O'Shanter . . . The whole passage is fearfully
> distinct, and though its circumstances, as the poet himself
> says, 'sickening', is yet an amazing display of poetical power
> and high invention.'   (148)

On the other hand, Galt finds that although the descriptions in
the seraglio section of *Don Juan* 'abound in picturesque beauty,
they have not that air of truth and fact about them, which
render the pictures of Byron so generally valuable, independent of their poetical excellence.' The reason for this, in Galt's
view, is that these passages had to come entirely from
invention, the seraglio not being accessible beyond the courts
and official apartments. In consequence he finds 'a vagueness'

in these descriptions 'which has the effect of making them obscure, and even fantastical . . . his gorgeous description of the sultan's seraglio is like a versified passage of an Arabian Tale, while the imagery of Childe Harold's visit to Ali Pashaw, has all the freshness and life of an actual scene' (150-51).

One could argue that the impression of the 'obscure' and 'fantastical' in the seraglio section of *Don Juan* is in keeping with Juan's own sense of mystery with regard to his destination and its purpose, and with the expectations which his author wishes to arouse in his readers. It is interesting, however, that Galt's views are very much in line with his own practice as novelist. He insisted that *Annals of the Parish* and *The Provost* were 'theoretical histories'[10] not novels, and although they have the attributes of imaginative fiction they are in fact closely based on small town and parish life and records of their time. Like Byron's transformations of personal experience and observation in his poetry, they are artistic recreations which display a high order of imaginative activity.

Despite his view of the poet as a man of worldly affairs, Galt has little to say about *Don Juan*, which he calls 'a poetical novel' (263). He identifies the comedy of manners nature of the work, but seems strangely oblivious to Byron's serious attack on social hypocrisy: 'nor can it be said to inculcate any particular moral, or to do more than unmantle the decorum of society . . . affording ample opportunities to unveil the foibles and follies of all sorts of men – and women too' (263-4). He absolves it, however, from charges of immorality, remarking shrewdly that 'perhaps . . . there was more of prudery than of equity' in such criticism (266). He himself found it 'deficient as a true limning of the world, by showing man as if he were always ruled by one predominant appetite' (264). He appears to have been impervious to its sexual comedy.

In discussing Byron's love poetry, Galt draws attention to what he sees as the dichotomy between Byron's reputation as a womaniser, a reputation encouraged by Byron himself, and the lack of 'sensual images' in his love poetry. Instead Galt finds 'the icy metaphysical glitter of Byron's amorous allusions' (15). In his view, the poem 'She walks in beauty like the night' is 'a perfect example . . . of his bodiless admiration of

beauty, and objectless enthusiasm of love,' and a poem where the 'simile' has become the 'the principal'. He continues: 'There is upon the subject of love, no doubt, much beautiful composition throughout his works; but not one line in all the thousands which shows a sexual feeling of female attraction – all is vague and passionless, save in the delicious rhythm of the verse' (16). This analysis applies to *Don Juan* also, where Galt notes that 'not even in the freest passages' has Byron associated beauty or love 'with sensual images' (15).

'She walks in beauty' is a fine lyric poem, but it is, as Galt's comments suggest, a poem of idealised, unattainable love, as elusive as Shelley's search for intellectual beauty. As such it is the antithesis of the kind of sexual relationships we find most often depicted in Byron's verse: worldly, uncommitted, often cynical, surprisingly deficient in sensual detail; the reader is led to supply this from his/her imagination or experience and in doing so, is drawn into the sexual conspiracy.

There is nothing in Byron's love scenes to compare with the eroticism and sensuousness of Porphyro's stolen meeting with Madeline in Keats's 'The Eve of St Agnes', which succeeds in being both idealised and of the tangible, everyday world. Byron's nearest approach to the depiction of an idealised yet fulfilled love is perhaps the Haidée episode in *Don Juan*, where the innocent love of the young lovers is at one with their natural, island environment, set apart from the corrupting influence of the 'civilised' world. Yet even here Byron's register is ambivalent and unstable, as if he fears to commit himself to his idyll. The early descriptions of the lovers' delight in each other and in their surroundings are frequently deflected in Byron's characteristic ironical manner: 'Well – Juan after bathing in the sea,/Came always back to coffee and Haidee' (II, 171), while as the tale proceeds, a sense of imminent tragedy begins to prevail over the initial flippancy. I find it impossible to integrate Haidée's ending into the predominant comedy-of-manners context of *Don Juan*: 'She died, but not alone; she held within/A second principle of life, which might/Have dawn'd a fair and sinless child of sin' (IV, 70) seems to introduce into the poem a note of genuine human tragedy and unfulfilled potentiality which is foreign to the way in which Byron more

typically chooses to put forward his view of the human condition, and in consequence one cannot be drawn into the comedy of Juan's activities in quite the same way as before the Haidée interlude. A willing suspension of judgement has been disturbed and cannot easily be restored.

If Galt has little to say about *Don Juan*, then he says even less about *The Vision of Judgment*. In his criticism of the poem he shows himself to be a political conservative, and, like his own Provost Pawkie, stout in the defence of monarchy and government. He refers to 'the disgust which The Vision of Judgment had produced' and gives his opinion that 'much good could not be anticipated from a work which outraged the loyal and decorous sentiments of the nation towards the memory of George III' (270). As Galt seems to have missed the social criticism in *Don Juan*, so he seems to have overlooked the even stronger satirical element in *The Vision*, and is more than a little guilty of hypocrisy in his comments on the poem. George III had not been admired in life and Byron is certainly not over-generous towards him in death. Yet his attack is directed as much against the hypocrisy of the trappings of political mourning by an ungrieving nation – 'Who cared about the corpse? The funeral/Made the attraction, and the black the woe' (stanza 10) – as it is against the inadequacies of the king. Similarly his religious satire in the poem has as its target human beings who remake God in their limited, worldly image and make of His heaven an Establishment Club to which only those who follow the Establishment way in politics and religion will be admitted.

Galt did not consider Byron to be an atheist like Shelley. His view was that Byron

> had but loose feelings in religion . . . with him religion was a sentiment, and the convictions of the understanding had nothing whatever to do with his creed . . . He reasoned on every topic by instinct, rather than by induction or any process of logic; and could never be so convinced of the truth or falsehood of an abstract proposition, as to feel it affect the current of his actions.   (281)

He reports discussions between Byron and Dr Kennedy, a religious man who sought to convert the poet to orthodox

Christianity when he was in Cephalonia. While Byron may have to some extent been playing with the doctor, these reported conversations do reveal to us the Byron of the poetry and the Count Maddalo of Shelley's 'Julian and Maddalo', who rejected Julian's and his author's evolutionary optimism for the deterministic creed that things are as they are. According to Dr Kennedy, Byron put forward his belief in 'predestination . . . and in the depravity of the human heart in general, and of my own in particular.' He continued: 'Pre destination appears to me just; from my own reflection and experience, I am influenced in a way which is incompre hensible, and am led to do things which I never intended, (289-90). He departs from orthodox predestination, however, in his insistence that 'I cannot yield to your doctrine of the eternal duration of punishment' (292). This too is the attitude of the narrator of *The Vision of Judgment* who knows 'one may be damn'd/For hoping no one else may e'er be so' (stanza 14), and who in the end lets George III slip into Heaven during the tumult aroused by the 'grand heroics' of Southey's poem of praise (stanza 103).

Byron's Calvinistic leanings in religion bring us to an aspect of his life and work which Galt surprisingly does not develop in his biography: Byron's dual Scottish and English ancestry, and the effect on his personality and poetry of his spending the formative years of childhood in Aberdeenshire. In the essay 'Byron', T. S. Eliot draws attention to what he considers Byron's un-Englishness, finding in 'his delight in posing as a damned creature', his satirical 'flyting' – a mode foreign to English literature but an attribute of the Scottish tradition – and his particular way with the English language an apartness from the English tradition and its context: 'for what Byron understands and dislikes about English society,' comments Eliot, 'is very much what an intelligent foreigner . . . would understand and dislike.'[11]

Galt pre-dated Eliot in drawing attention to Byron's apart-ness from English traditions. Assessing Byron's achievement at the end of his biography, he comments on 'a great similarity' which 'runs through every thing that has come from the poet's pen; but it is a family resemblance, the progeny are all like one

another; but where', asks Galt, 'are those who are like them? I know of no author in prose or rhyme, in the English language, with whom Byron can be compared' (329). He finds that 'his verse is often so harsh and his language so obscure . . . he possessed not the instinct requisite to guide him in the selection of the things necessary to the inspiration of delight.' And finally Galt thought that Byron lacked 'a tuneful voice' (328).

All this suggests a writer who has anticipated Edwin Muir's twentieth-century injunction to Scottish writers to *adopt* the English language and tradition, rather than a poet who is at home in English. Although Galt ascribes the 'conception of the dark and guilty beings which he delighted to describe' to Byron's 'mother's traditions of her ancestors' (21), unfortunately he does not develop the comment in the context of the nature of these traditions and their distinctiveness from those of England. He was aware of the inhibiting effect of the 'translation' procedures forced by public taste upon Scots-speaking authors,[12] yet Galt's own work shows little of the stress arising from the sense of a divided culture which we find in Burns and Scott, for example. He seems on the whole to have accepted his role as a North British writer, and to have exploited successfully in his best work the linguistic potentialities of the two traditions. In turn, he ignores the possible influence of the cultural divide in his discussions of Byron's personality and poetic achievement.

Although Byron himself affects a longing for the 'Highland cave' and 'dusky wild' of his childhood in the early poem 'I would I were a careless child', and rejects 'the cumbrous pomp of Saxon pride', this poem is something of an exercise in rhetoric and role-playing which may well have taken its starting- point from Coleridge's 'To the River Otter', with its similar theme and final line: 'Ah! that once more I were a careless child!' (Indeed, Galt – who himself had a brush with Byron over possible plagiarism - tells us in the biography that it was 'an early trick of his Lordship to filch good things' (183-4).) On the whole, Byron appears to have accepted willingly his transformation from penniless, if honourably descended, Scotsman to English lord. It is noticeable that his satiric reply to the criticism of the *Edinburgh Review* was entitled '*English* Bards

and Scotch Reviewers' (my emphasis). He does not draw on the Scots language which must have been part of his childhood background. Nor does he draw on Scottish themes, nor, consciously, on the Scottish literary tradition in his work. While one recognises the Calvinist in Byron's deterministic philosophy and also, perhaps, in his early alienated heroes, on the whole Byron's Calvinism does not lead to the censoriousness associated with Scotland's reformed religion, but rather to a rejection of idealistic attempts to change human nature, and a comedy based on the acceptance of things as they are.

In addition, one does not find in Byron that ability to move easily from the everyday, domestic detail to the metaphysical or the conceptual which is a recurring feature in Scottish literature. Byron's apostrophe to freedom in Canto IV of *Childe Harold* – 'Yet Freedom! yet thy banner, torn, but flying,/ Streams like the thunder-storm *against* the wind' (stanza 98) – is a splendid rhetorical piece, but for me it remains mere rhetoric when constrasted with the freedom passage from John Barbour's *The Brus* which translates the idealistic concept into the conditions of everyday living.

On the other hand, as Galt and Eliot have pointed out, there are elements in Byron's poetry which do not fit easily into English literature. His particular form of satire with its extravagant moods, its wild lurches from urbane wit and irony to grotesque and often spiteful comic characterisation, brings to mind the Scottish tradition, as do his guilt-ridden heroes and his deterministic philosophy of existence. The wonder of the Alps and the mountains of Switzerland conjures up for Byron not a confirmation of the existence and power of God, as for Coleridge, nor the human mind's potential, as for Shelley. Standing at the foot of the Jungfrau, he watches the mountain torrent 'curving over the rock, like the tail of the white horse streaming in the wind just as might be conceived would be that of the pale horse in which Death is mounted in the Apocalypse' (quoted by Galt, 213). This seems a very Scottish response and image. He saw himself as Scott's heir: 'Sir Walter reign'd before me' (*Don Juan*, XI, 57). The Spenserian stanza he borrowed for *Childe Harold* came not directly from his knowledge of *The Faerie Queene*, but through the mediation of

the Scottish writers James Thomson and James Beattie, while
the *ottava rima* form first used in *Beppo* and which he made so
much his own in *Don Juan* and *The Vision of Judgment* came from
the Italian Pulci.

Although he claimed Pope as one of his masters, a comparison of Byron's couplets and satiric verse forms with the poetry
of Pope shows a formal unsettledness which patterns the lack
of a stable centre for his philosophical and moral perspectives.
Pope's couplets are formally and conceptually balanced. Even
when, like Byron, he is building up to a climax by an
accumulation of detailed effects, each stage has its own witty
or ironic point to make in addition to its apposite contribution
to the metaphor of the whole, as in this brief excerpt from the
description of the life cycle of Dullness in *The Dunciad:*

> Here she beholds the chaos dark and deep
> Where nameless Somethings in their causes sleep,
> Till Genial Jacob, or a warm Third day,
> Call forth each mass, a Poem, or a Play:
> How hints, like spawn, scarce quick in embryo lie,
> How new-born nonsense first is caught to cry,
> Maggots half-form'd in rhyme exactly meet,
> And learn to crawl upon poetic feet . . . (I, 55-62)

Byron, on the other hand, in *his* attack on what seems to him to
be literary dullness, does not succeed in achieving such
consistent conceptual, rhythmic and rhyming patterning:

> Bob Southey! You're a poet – Poet-laureate,
>   And representative of all the race;
> Although 'tis true that you turn'd out a Tory at
>   Last, – yours has lately been a common case;
> And now my Epic Renegade! what are ye at?
>   With all the Lakers, in and out of place?
> A nest of tuneful persons, to my eye
>   Like 'four and twenty Blackbirds in a pye . . .'
>   (*Don Juan*, Dedication)

Yet it would be unsafe to diagnose this more awkward verse as
stemming from a Scotsman's lack of identity with the English
language and tradition. It could as readily relate to the conflict

between the more formal verse conventions of an earlier period and the new, freer movement in ideas and syntax which one associaties with Romantic poetry. One of the hallmarks of Wordsworth's poetry is his ability to use blank verse as a conversational, reflective, meditative medium which enables his thoughts to flow forwards and backwards freely from line to line, turning, continuing, stopping where he will. Despite Byron's dislike of Wordsworth's poetic methods, his best verse also has this flexible immediacy, but it can often seem in conflict with the restrictions imposed by a tighter verse and rhyming pattern.

In the essay 'Byron as a Scottish Poet', the Scottish poet and critic Tom Scott puts forward a seductive case for Byron's membership of the Scottish tradition by translating the opening of *Beppo* into the kind of light Scots often used by Burns and MacDiarmid:

> It's kent, at least it sould be, that thurchoot
> Aa countries o the Catholic persuasion,
> Some weeks afore Shrove Tuesday comes aboot,
> The folk aa tak their fill o recreation,
> An buy repentance, ere they grouw devoot,
> Houevir hiech their rank, or laich their station,
> Wi fiddlin, feastin, dancin, drinkin, maskin,
> An ither things that may be had for askin.[13]

Verse form apart, this does seem to sit happily alongside Burns's 'Holy Fair', while that 'an the mair murkily the better' in the stanza which follows the above reminds one of MacDiarmid's similar ironic use of parenthetical comments as, for example, in his 'an no' for guid' in the poem 'Lourd on my hert.'[14] On the other hand, Byron's *Beppo* seems to me to read equally well in its original English form – surely an almost unique achievement for any Scottish or English poet. The difference seems to be one of tone. Perhaps because of the characteristics of the Scots language, the narrator of Tom Scott's *Beppo* appears to be speaking from the *inside*. One expects him to go on to be part of the reveller's tale to be told and to draw the reader into the revelry with him. The English narrator, on the other hand, while uncensorious, even

approving of the marital goings-on, remains an observer, keeping an aristocratic distance from the happenings, even while being entertained by them. In general, Byron's links with the Scottish tradition seem to be more implicit, related to sensibility and cast of mind, rather than overtly literary in nature.

The latter part of Galt's biography of Byron is inevitably based not on first-hand experience but on the reminiscences of others and on Byron's own journals. It appears that the two men fell out in 1813, the year of Galt's marriage and of Byron's continuing triumphs after the publication of the first two cantos of *Childe Harold*. Galt says little about the quarrel, leaving it to be assumed that marriage had put an end to his bachelor travels and acquaintanceship with Byron. Byron is reported by Lady Blessington as speaking admiringly of Galt in later years and regretting that he had taken 'no pains to cultivate his acquaintance further than I should with any common-place person, which he was not.'[15]

Galt's biography of Byron is certainly not commonplace. To the end he maintains the objectivity, percipience and unmalicious irony which is characteristic of his portrait. He views Byron's transformation from portrayer of heroes to man of action with some foreboding, a foreboding justified by the tragic events of Missolonghi. Galt's is necessarily a fragmentary biography, but it remains one of the fairest and most perceptive accounts of Byron and his work. As with Provost Pawkie, Galt has done well by his lord – and himself.

NOTES

1. Lord Byron, *Don Juan* Canto X, stanza 17.
2. John Galt, *The Life of Lord Byron* (London, 1830), 43. Page numbers for subsequent quotations from Galt's *Life* will be given in parentheses in the text.
3. *Lady Blessington's Conversations of Lord Byron*, ed. with introduction and notes by Ernest J. Lovell (1969), 146.
4. Ibid.
5. John Galt, *Annals of the Parish* (1821; Edinburgh: Mercat Press, 1980), 27.

6. Ibid., 27–8.
7. Ibid., 27.
8. William Wordsworth, Preface to *Lyrical Ballads*.
9. Alexander Pope, *Essay on Man*, Epistle II, 1–2.
10. John Galt, *Autobiography* (1833), II, 155–6.
11. T. S. Eliot, 'Byron', *English Romantic Poets: Modern Essays in Criticism*, ed. M. H. Abrams (1975), 262, 273.
12. See John Galt, 'Biographical Sketch of John Wilson', *Scottish Descriptive Poems*, ed. J. Leyden (1803), 14.
13. Tom Scott, 'Byron as a Scottish Poet', *Byron: Wrath and Rhyme*, ed. Alan Bold (1983), 29.
14. Hugh MacDiarmid, *To Circumjack Cencrastus, Complete Poems 1920–1976* (1978), 204.
15. *Blessington*, 146.

Lord Byron and Lord Elgin

DOUGLAS DUNN

Byron's encounter with the activities of Lord Elgin witnesses an episode in Byron's European and revolutionary mentality. It was political as well as poetic. Extreme, unorthodox and original, it shows Byron out of step with his times and far in advance of conventional opinion. It is an incident in Byron's work that called forth both satire and hortatory lyricism. Satire is more than poetry in its bad mood. Byron's satire can, at times, look like vindictive posturing, waves of sheer tantrum whose distempers preclude the sublime. Satire is poetry when it has decided on action and on the practical, and in the work of a poet as restless, itinerant and topical as Byron, it reaches beyond routine discourtesy and insult as if intent on an effective intervention. Although perhaps thinned out by the action of time, the values of Augustan satire survived in Byron's instinctive repertoire; they co-existed with the new melancholy his poetry invented, and with his libertarian hope and determination. Together these three sources of feeling and style characterized the poetry that was to have such a devasting effect on the literature of Europe, reflected in its music, opera, drama, painting and sculpture in the reactionary aftermath of Napoleon's defeat. As poetry, it is a fiery and elusive phenomenon: sometimes agonized, sometimes elegiac, sometimes languid, radical, republican or loud in its revolutionary encouragement, or Napoleonic, combative, soaring and inspired.

Thomas Bruce, seventh Earl of Elgin and eleventh Earl of Kincardine, was born in July 1766. He was therefore twenty-one years older than Byron. His father died in 1771 and Thomas Bruce's elder brother William became the sixth Earl. He died only two months later so that Thomas Bruce became

seventh Earl rather sooner than anyone had bargained for. Like Byron, he was half-a-Scot: his mother was the daughter of a London banker. Money and heiresses figure in Elgin's story almost as much as in Byron's. While Byron could find excitement in the Gordon history of his mother's side, Elgin could point with satsifaction to a Bruce ancestry which had provided a Scottish dynasty. Like Byron, he was at Harrow school, though not for long, being transferred quickly to Westminster. Subsequently, Elgin studied at St Andrews, and then in Paris. Whatever else he might have been, he was far from the half-witted archaeological impresario and shipping-agent made out by his legion of detractors.

He opted for a military career in 1785. For a time he served as an ensign in the Foot Guards, then after what looks like rapid promotion, he commanded a regiment that he raised himself, the Elgin Highland Fencibles. He was lieutenant-colonel by the age of twenty-nine, and, by the time of his death, a major-general. We cannot be sure of it, but senior military rank before the age of thirty suggests a certain degree of competence.

Opportunities for a military career in Elgin's lifetime could hardly be said to have been lacking. Yet he never saw active service. His health was uncertain. He was a martyr to rheumatism. A political and diplomatic career must therefore have looked like an answer to a young man as ambitious as Elgin appears to have been, or who at least considered himself serviceable. He could have chosen the life of many Scottish aristocrats, and done nothing.

At the age of twenty-four, in 1790, he approached Henry Dundas whose manipulative and not entirely benign political genius dominated Scottish public life. In those days Scotland was served in the House of Lords by sixteen representative peers. To get there, a Scottish aristocrat had to be elected, which meant substantially that he required Dundas to appoint him. Elgin achieved this, perhaps as a stepping stone to more active work: at any rate, during the seventeen years that Elgin held his seat he was abroad for most of them. It suggests, too, that Elgin was ambitious and that his political views were predictably those of his class. A year later he was appointed to

the Embassy to the Emperor Leopold II. From Envoy Extraordinary in Vienna, he went in the same capacity to Brussels, and after that to Berlin, at times rushing back to London to occupy his seat in the House of Lords where his contributions continued to recommend him to the government party. Serious European business was afoot in the 1790s, and Elgin must have been considered capable as well as politically and personally suitable. Nothing, so far, suggests that he was a fool.

In the 1790s Elgin also instigated costly improvements to his house, Broomhall in Fife near Dunfermline. His mother pointed out the extravagance of her son's plans. 'Considering your Taste and style of living a prodigious House will be a monstrous burden,' she wrote. Not long after, Elgin married Mary Nisbet of Dirleton. Her family was worth something in the region of £18,000 a year, a lot more than Elgin's estates produced – £2000 a year seems the likeliest assessment of Elgin's independent wealth. No doubt it mattered that the future Lady Elgin was lively and good-looking. Elgin himself seems to have been dull, conscientious and prone to pomposity and stiffness. What also mattered were the future Lady Elgin's expectations. Alterations to Broomhall had made it necessary for Elgin to borrow heavily, and as well as an heiress he needed for the shorter term a prestigious and well-remunerated appointment. His selection as Ambassador Extraordinary and Minister Plenipotentiary to the Sublime Porte of Selim III, at a salary pf £6000 a year plus expenses, must therefore have looked almost as handsome as Mary Nisbet and perhaps as sublime as the Porte. It also represented promotion, and yet another sign that Elgin was considered a useful man to have in a diplomatically tricky and potentially significant corner of the world, already penetrated by French influence.

Before the recently married couple set off for Turkey, it was suggested to Elgin by Thomas Harrison, Elgin's dilatory architect at Broomhall, who was soon to be sacked, that his ambassadorship presented a golden opportunity for introducing to Britain authentic materials for the study of classical architecture and art. Harrison appears to have suggested the

making of accurate drawings, plaster casts, mouldings, mea-
surements and other reproductive and survey work, some of it
of an artistic nature, carried out on the spot, and with
systematic expertise. To some extent it would duplicate work
already published, and this was the reason given by Greville,
the Foreign Secretary, for refusing Elgin government funds for
the purpose.

Elgin committed himself to the idea of improving public
taste and appreciation of the arts, but it seems proper to
suggest that public taste might have held less appeal for Elgin
than the means through which he could make a name for
himself in a field of endeavour that could earn him the British
peerage he desired. No one, least of all Elgin himself, had in
mind the physical dismemberment of architectural sculptures
from the Parthenon and their shipment to Britain.

To fulfil this self-chosen, unofficial but enthusiastic part of
his mission, for which the government refused to pay, Elgin
interviewed a number of British artists. J. M. W. Turner was
one, but he asked for too much, and insisted also on keeping
his drawings and paintings, which failed to suit Elgin's pocket
and possessiveness. Thomas Girtin and William Daniell also
applied. It was *en route* to Turkey, however, at Sicily, that Elgin
met his artist, Lusieri – later a friend of Byron's when he visited
Greece a decade later. Lusieri was recommended by Sir
William Hamilton, husband of Emma, the Lady Hamilton of
Nelsonian fame. He was a connoisseur and antiquarian whose
collections had inspired Josiah Wedgewood and English
pottery. Also employed was Theodor or Fydor Ivanovich, a
colourful drunkard of obscure, exotic origins, probably a
Kalmuck or Tartar. Given that Lusieri was talented but
incorrigibly lazy (as well as loyal), and that the Kalmuk Fydor
was seldom sober enough to do what was required at the
requisite tempo, Elgin looks to have been courted by bad luck
from the start. Other employees turned out to be unreliable
and even treacherous. Drawings, Lusieri's responsibility, and
engravings, the Kalmuk's function, never materialized in a
state other than promised or unfinished.

As part of the Ottoman Empire in Europe, Greece had
stagnated into a condition of wearied resignation and cynical

survival. Athens was described by John Galt in *Letters from the Levant* (1813) as like several villages pressed into one with a population of around 10,000 people. It was a backwater, scraping into the top fifty of the towns of European Turkey. There was a Turkish garrison as well as a mosque on the Acropolis. A little over a hundred years before, Venetian artillery had devastated the Parthenon, setting off the Turkish arsenals and magazines. Ruthless as the Turks could be, the Greeks preferred them to the Venetians and their mercenary hirelings. In the past the Turks had taken boys for the corps of Janissaries, normally one per family, subjecting them to a thorough military and religious education. As in Bosnia and elsewhere in Turkish Europe, these professional soldiers of Christian origin rose to positions of high rank in the Empire. Young women were enlisted into the harems of officers and administrators. Able bodied men could end up pulling an oar in Turkish galleys; they could find themselves similarly hard-worked in Venetian vessels. Turkish capitation tax meant itself to be taken literally: pay up and you kept a head on your shoulders. But by the early nineteenth century tyranny was at the whim of local disdars and governors, while religious oppression had always been the exception rather than the rule. Most towns were sleepy and depressed; now and again life was animated by the visitation of brutality. Some districts were ungovernable, autonomous regions controlled by *klephts* or brigands whose armed bands were subsequently to form important units of the Greek revolutionary armies.

Elgin's employees were in Athens in August 1800, Lusieri arriving a little later, but Elgin was in Constantinople where he was engaged in countering the deviousness of the Levant Company and contending with its incumbents for whom Elgin's authority weakened their own. He was kept busy negotiating with the lethargic, ritualistic Turkish government, purchasing supplies for the British forces engaged in the war in Egypt. It was a responsibility which he seems to have discharged with fastidiousness, some of it at his own expense, always a risk, and perhaps even an expectation, borne by Britannic ambassadors of the time.

Reports from Athens also engrossed Elgin's attention. For

years the Turks had been casual pilferers of the Parthenon, usually for building materials, sometimes recycling statuary into mortar for new constructions. Whole temples had been blown up and cleared away. General Morosini's bombardiers of the 1680s had left their marks too, and some might even have been Scots, for the Venetians recruited Scottish troops for the wars of Candia (1649-1669). Also, the French were there, and their Consul had it in mind to remove choice examples of the classical artistry in stone, which was a potential threat to Elgin's quieter intentions as well as his career prospects. Parts had already been looted by sundry travellers and antiquarians. As work started on scale drawings and surveys, the local disdar laid a ban on any activity of the sort until a *firman* or note of authority could be produced from Constantinople.

Elgin pressed for this document with great vigour. Egypt, nominally a Turkish province, had been retaken; British favour ran high in Constantinople. There wasn't much that a Turkish government would deny a British ambassador. Bearing in mind the reports he had received from Athens – the French were after them; the Turks were disrespectful; the statuary seemed in danger – Elgin looked for, and got, the widest terms he could, or else Turkish vagueness conspired to provide the document he wanted. For an official instrument, the *firman* is remarkably unclear, leaving room for Elgin's interpretation – that he was at liberty to remove from the Parthenon what were decided to be of value to his philanthropic or self-regarding purposes. It seems never to have crossed his mind that a *firman* in strict terms preventing dismantlement and removal by the French or defacement by anyone else would have been more appropriate. That the French were at the time interned by the Turks might have led Elgin to believe that he had the chance to escalate their own greedy intentions, a patriotic opportunity for British gain which would enhance his prestige at home. It might look like the archaeological version of a victory. Other evidence suggests that if these thoughts ran through his mind, then so too did a more private avariciousness succeed them.

It was some time before Elgin set eyes on the objects of his goodwill towards art and Elgin, the statuary that was costing him a fortune in wages, tackle and gear. He visited Athens in

the Spring of 1802 but only for as long as his duties permitted. His interest was intense, but distant; he seems to have acted in the matter like a grandee, behaving, perhaps, a little above his grandeur. On their return to England, the Elgins decided to travel much of the way through France. During their journey the war reopened, having been temporarily armisticed by the Treaty of Amiens in March 1802. They were interned. Napoleon, for whom the acquisition of European plunder was a serious business, held Elgin personally responsible for the fact that the Marbles were Elgin's and not his. The unfortunate Elgin was marooned in France for three years. He claimed that the French offered to release him in return for the Marbles. Elgin refused, and it forms a nice scene to imagine him weighing up his patriotic duty, his possible éclat in London, and his personal magnification, against an offer of being set at liberty. The invitation is credible when you consider the Napoleonic appetite for imperial aggrandisement.

It was a stroke of bad luck worsened by the terms of Elgin's subsequent parole, which tied him to an obligation to return to France whenever or if that might be demanded: it ruled out a continued diplomatic career for as long as the war continued. Lady Elgin added to his misfortunes. While in France she fell in love with another internee, Robert Ferguson of Raith, perhaps the original Raith'Rover, a Fife landowner of great wealth, a Fellow of the Royal Society and a Whig. There was a messy divorce – first a trial in London, and then another in Edinburgh in March 1808. Elgin came out of it with £10,000 in damages, useful money considering the almighty amounts disbursed for the sake of the Marbles, but it might have looked a pittance when contrasted with Mary Nisbet's inheritance, to say nothing of his growing debts and persistent creditors. He also had four children to look after.

Reports of Elgin's Parthenon scoop, added to the first sensational disclosures of his divorce in December 1807, which was widely publicized, made Elgin well known. It was a negative celebrity; scandal and an almost comically severe run of bad luck subtracted from the grateful acclaim that he might have believed he deserved. Among artists, what he had brought to Britain was welcomed, often ecstatically; its sculp-

tural genius was upheld as exemplary in exactly the manner that Elgin must have hoped would be the case. Turner approved, as did Flaxman, Chantry, Benjamin West and many others, including Haydon, later the friend and encourager of Keats on whom the impact of the Parthenon sculptures was to play a significant part in the hastening and strengthening of his poetry. Fuseli, Nollekens, and Canova, the greatest sculptor of his day, were also more than impressed. Most of these artists had benfited from the Italian archaeological and restoration industry of the eighteenth century in which the Scottish artist and excavator Gavin Hamilton played a large part. What Elgin imported from Greece looked like the real thing to many in the best position to judge – heroic, inspiring, inestimable. The vigour and genius of Phidias and Praxiteles in three dimensions left British artists gaping in awe-struck wonder.

Dissenting voices contradicted this display of appreciation. Significantly, they came from connoisseurs and collectors, not artists. Richard Payne Knight was their leader, an arbiter of taste and a man whose envy and malice led him easily into doltishness. 'You have lost your labour, my lord Elgin,' he is reported as having said. 'Your marbles are overrated: they are not Greek. They are Roman of the time of Hadrian.' Perhaps only an English critic can be so wrong and still retain the power of spokesmanship. As a leading voice in the Society of Dilettanti, Knight's opinion carried weight with those in no position to judge for themselves or predisposed to welcome yet another stroke of bad luck in Elgin's mounting catalogue of calamities.

With all this going on it was inevitable that Byron should have become familiar with Lord Elgin and his deeds, his misfortune and financial embarrassments, his cuckolding, and the suspicion, now growing stronger, that Elgin had misused his ambassadorial trust in obtaining ambiguous rights from the Turkish government preparatory to his acquistion of the Marbles for himself, not for the nation. Byron's first satirical punch was struck in *English Bards and Scotch Reviewers*, which appeared in two editions in March and October 1809 before he departed on the tour that would take him to Greece. His

farewell peroration tells us what he will not do – he will not try
to make a name for himself by collecting lumps of antique
stone:

> Let ABERDEEN and ELGIN still pursue
> The shade of fame through regions of Virtu;
> Waste useless thousands on their Phidian freaks,
> Mis-shapen monuments and maimed antiques;
> And make their grand saloons a general mart
> For all the mutilated blocks of art . . .

Lord Aberdeen had been in Greece and returned with a few
stones – 'Athenian' Aberdeen he was called: he was not
entirely on Elgin's side in the controversy that had already
begun and was to continue to oppose him for several years
more, as Elgin attempted to off-load his burdensome collection
on the British government. Detectable in Byron's lines is a low
opinion of antiquities in a damaged condition and never mind
if they were sculpted by Phidias. His view was to be changed
when he experienced Greece and learned of the indigenous
value of antiquity to a suppressed nationality. 'The shade of
fame', too, hints at a topical hunch that Elgin's intentions had
less to do with art and were more concerned with limelight and
applause.

True to the sentiments of *English Bards and Scotch Reviewers*,
Byron was an unconventional voyager. With the rest of
Europe closed by warfare to young men on hedonistic or
educational journeys, Greece had become a favourite destination. High-born, classically educated young men were there
in numbers. Many held opinions on Elgin's removal of the
Marbles, rarely in his favour. Most, though, were happy to
thieve portable souvenirs of their own. Their estimates of
the contemporary Greeks were low. One traveller observed
that Arcadian shepherds had 'almost reverted to the
balenephageous state of their primitive ancestors.' That is,
they ate acorns. They were unworthy of the civilisation in
which these young visitors were schooled and of whose
language they knew more than the vast majority of Greeks
themselves. Contemptible, shabby and impoverished, the
inhabitants of a town like Athens offered a dispiriting contrast

to the glory represented by visible ruins. Where was their art, poetry, philosophy and power? Romaic, or *Demotiki*, modern Greek, was hardly in accord with the tongue of Homer and Euripides. Even more amusing or disgusting was that the natives called themselves *Romaioi*, or Romans, as they had done since the time of the eastern Roman empire and its Byzantine successor. Not even proper Christians, they were inferior, clearly, mere docile provincials of a backward despotism in which the only right they enjoyed was the right to be taxed. Lazy, servile, superstitious, greedy, untrustworthy, unwashed, degenerate and debased – to the comfortable nomads who descended on Greece, its people had nothing to recommend them.

'Plausible rascals', Byron called the Greeks. 'They are perhaps the most depraved and degraded people under the sun.' He liked them. A bi-sexual amorist, Byron's depravity was itself astounding, making up for what it lacked in cynicism with a voluptuary performance largely free of hypocrisy and erotic cant. He was aware of the political imperfections of the Greeks, but had the capacity of mind to notice that where the Greeks lived was Greece and that the long withering of Ottoman hegemony was worth a more active response than languid regret. He perceived how liberty and the dignity it bestows might transform native indolence. Whatever was contemptible about the inhabitants had been encouraged by generations of fatalism.

'Hobhouse rhymes and journalises. I stare and do nothing,' Byron reported. 'Do I look like one of these emasculated fogies?' he expostulated when invited to join a trip to nearby ruins. 'Let's have a swim.' But he was not idle. He began writing *Childe Harold's Pilgrimage* at Jannina in Epirus on 31 October 1809, finishing the second canto at Smyrna on the Turkish mainland on 28 March 1810. He began *The Curse of Minerva* in Athens; there is a likelihood that it was suggested by a poem of John Galt's called *The Atheniad* which Byron read in manuscript.

He wrote some shorter poems too, for example 'Maid of Athens, Ere We Part' or the translation of a 'Greek War Song by Rhigas Pheraios', in which he writes of 'The Turkish

tyrant's yoke' and appeals to Sparta and Athens to rise from their slumber and defeatism. No other British traveller identified with a Greek cause; no other British traveller noticed that there was one.

Lusieri, Elgin's artist-in-residence in Athens, was a frequent companion, guide and friend. Byron developed a sexual friendship with Nicolo Giraud, Lusieri's half-Greek brother-in-law. He saw at first hand Lusieri's dutiful defence of a second shipment of Marbles, holding off the covetous designs of Fauvel, who for years had been the French Consul in Athens and under instructions to lay his hands on as many treasures as possible with which to enrich the Louvre, the Musée Napoleon. With Lusieri and Nicolo Giraud, Byron sailed for Malta on board the ship *Hydra* which transported the second haul of Elgin's Marbles. Doubtless too, Byron picked up from Lusieri something close to a realistic picture of Lord Elgin's intentions.

At this point in the story we encounter an opportunistic near miss by John Galt, who was then in Athens. Aware of Elgin's shaky finances and that his bankers in Malta might easily refuse the necessary transaction, Galt instructed his agent in Malta to purchase the Marbles should Elgin's credit be refused. 'Here was a chance of the most exquisite relics of art in the world becoming mine,' Galt wrote in his *Autobiography* in 1833, 'and a speculation by the sale of them in London would realise a fortune,' he added, with a retrospective commercial sorrow not often associated with novelists. Had Elgin's credit not still looked good, the 'Galt Marbles' might now be on display beside Elgin's in the British Museum.

On his return to Britain, Byron offered *Childe Harold's Pilgrimage* to the publisher Miller, who turned it down on the grounds of its attack on Elgin. Miller was Elgin's publisher: he had issued Elgin's *Memorandum on the Subject of the Earl of Elgin's Pursuits in Greece* late in 1810. Byron was therefore unfamiliar with Elgin's version of events as well as his claim for reimbursement of expenses before the Marbles could be presented in the national collections.

Byron carried letters to Elgin from Lusieri, which he saw to it were delivered. In a letter dated 31 July 1810, to his former travelling companion, Hobhouse, Byron said,

Lord Elgin has been teazing to see me these last four days. I wrote to him, at his own request, all I knew about his robberies, and at last have written to say that it is my intention to publish (in Childe Harold) on that topic, I thought proper, since he insisted on seeing me, to give him notice that he might not have an opportunity of accusing me of double dealing afterwards.

As yet the hapless Elgin could have had no very sure idea of how much damage a Byronic intervention might wreak on his chequered fortunes. He was probably encouraged by the opinion of his former secretary, William Richard Hamilton, who believed that Byron's negative opinion would do Elgin's cause the world of good. Hamilton had prised the Rosetta Stone from French hands after their defeat in Egypt. That, too, is in the British Museum, where its presence seems strangely uncontroversial.

That was the closest Elgin and Byron came to meeting. Byron's mother died on 1 August, and his friend Matthews soon after. By 10 August he was writing again to Hobhouse: 'I am very lonely, and should think myself miserable were it not for a kind of hysterical merriment, which I can neither account for nor conquer.' Mourning, laughing and wining, Elgin and the Marbles were far from Byron's mind.

Canto II, the Greek Canto of *Childe Harold's Pilgrimage*, contains only one passage of satire and it deals with the pillage of the Parthenon. Understanding the Canto demands that we attend to its view of history. It is elegiac as well as lively in its incitements: it reads like a reprimand delivered against time. 'Ancient of days! august Athena! where,/Where are thy men of might? thy grand in soul?' Childe Harolde ruminates on the ruins of Greece. 'Is this the whole?' he asks. 'A school-boy's tale, the wonder of an hour!' Structural and rhetorical cunning hold Byron's stanzas to an insistent narrative travelogue in which the past is visited as much as the present, and the spirit as much as the fact:

> Look on its broken arch, its ruin'd wall,
> Its chambers desolate, and portals foul:
> Yes, this was once Ambition's airy hall,

> The dome of Thought, the palace of the Soul:
> Behold through each lack-lustre, eyeless hole,
> The gay recess of Wisdom and of Wit
> And Passion's host, that never brook'd control:
> Can all, saint, sage, or sophist ever writ,
> People this lonely tower, this tenement refit?

Trembling, questioning and elegiac, the mood continues, half lament and half accusation, until by Stanza XI he courts the topical and contemporary.

> But who, of all the plunderers of yon fane
> On high, where Pallas linger'd, loth to flee
> The latest relic of her ancient reign;
> The last, the worst, dull spoiler, who was he?
> Blush, Caledonia! such thy son could be!
> England! I joy no child he was of thine:
> Thy free-born men should spare what once was free;
> Yet they could violate each saddening shrine,
> And bear these altars o'er the long-reluctant brine.

QUOD NON FECERUNT GOTI, HOC FECERUNT SCOTI, a Latinate wag had chipped into a column on the Parthenon. But Byron, like Elgin, was half-Scots. Why this gratuitous sally against Elgin as a Scot when Englishmen had already proved themselves adept at cultural burglary? It was convenient; it spiced his detestation of Elgin with mischief directed against Byron's maternal country. Conceivably, it was an act of instinctive revenge for humiliations experienced in his childhood.

> But most the modern Pict's ignoble boast,
> To rive what Goth, and Turk, and Time hath spar'd:
> Cold as the crags upon his native coast,
> His mind as barren and his heart as hard,
> Is he whose head conceiv'd, whose hand prepar'd,
> Aught to displace Athena's poor remains:
> Her sons too weak the sacred shrine to guard,
> Yet felt some portion of their mother's pains,
> And never knew, till then, the weight of Despot's chains.

Elgin's argument was that the Turks were careless, casual wreckers of the Parthenon, while the Greeks couldn't care less. Other Greeks, however, looked on the visible residue of previous civilisation as supports to revolutionary action, as declarations of independence reared in stone. Byron ignored French plans for the removal of the Marbles. 'Tell not the deed to blushing Europe's ears,' he wrote. Europe was not blushing; it was bleeding. French embarrassment stemmed from their failure to take the Marbles for Napoleon's gratification. Satire breezed over these complicating factors and introduced a vituperative note of opinion to the otherwise persuasive movement of poetry.

Before the notion of 'imperialism' was coined (it enters the language in the 1850s), Byron had perceived its meaning.

> The ocean queen, the free Britannia bears
> The last poor plunder from a bleeding land . . .

'I have some early prepossession in favour of Greece,' Byron wrote, 'and do not think the honour of England advanced by plunder, whether of India or Attica.' Elgin, as a diplomat, a soldier, and a man of high orthodox beliefs and ambitions, probably failed to understand the gist of Byron's complaint.

> Cold is the heart, fair Greece! that looks on thee,
> Nor feels as lovers o'er the dust they lov'd . . .

Byron chastises British scholars, travellers and gentlemen, satisfying their classical educations with insensitive sight-seeing. In the lyrical embrace of these lines there is a sense, too, of European culture regretted as disrespectful of a civilisation that contributed largely to its identity. It is part of Byron's lament for the cruelty of time and the ironic displacements of history that relegate whole civilisations to the status of memory, fragments and plundered monuments. In that context, his abusive treatment of Elgin acts as a trigger to the topical. 'Fair Greece! sad relic of departed worth!' he exclaims by Stanza LXXIII.

> Immortal, though no more! though fallen, great!
> Who now shall lead thy scatter'd children forth,
> And long accustom'd bondage uncreate?

'Spirit of Freedom!' he shouts, 'lost Liberty!' His verse is not only sung at a Romantic pitch, but it is practical; there is nothing inactive or beaten about Byron's history.

> For foreign arms and aid they fondly sigh,
> Nor solely dare encounter hostile rage,
> Or tear their name defil'd from Slavery's mournful page.

Ten years before the Greek wars of independence, Byron wrote like an astute, lyric propagandist, editorialising in song, the breathless volume of his incitements directed as much to the Greeks as anyone else:

> Hereditary bondsmen! know ye not
> Who would be free themselves must strike the blow?
> By their right arms the conquest must be wrought?
> Will Gaul or Muscovite redress ye? no!
> True, they may lay your proud despoilers low,
> But not for you will Freedom's altars flame.
> Shades of the Helots! triumph o'er your foe!
> Greece! change thy lords, thy state is still the same;
> Thy glorious day is o'er, but not thine years of shame.

When Byron woke up to find himself eating the breakfast of a celebrity, Elgin's rise from bed that same morning was to discover his notoriety even more widely distributed. As the years passed before June 1816 when the House of Commons decided to pay Elgin – by a margin of two votes out of one hundred and sixty-two votes cast – Byron's own notoriety had been disclosed in the scandal of his failed marriage. Rumours of sodomy and incest made it impossible for Byron to remain in England and after April 1816 he was never to return. It is just possible that Byron's scandal saved the day for Elgin when his case was debated by truculent parliamentarians. Elgin was voted £35,000. By then his debts incurred in bringing the Marbles to Britain amounted, plus interest, to £90,000 according to Elgin's own figures. A few pieces remain at Broomhall and at one time, in a moment of desperation, Elgin considered installing the entire collection there, although, earlier in Turkey, at the height of his enthusiasm, he seems to have intended his Parthenon swag solely to adorn his house.

Statements testifying to their beauty and importance cut very little ice with the government. In those days artists enjoyed as much prestige as tradesmen and their opinions were easily discounted. It did not matter that Goethe, for instance, decorated his house in Weimar with Haydon's drawings of the statuary, believing that the works carried out under the direction of Phidias proved what Goethe's aesthetics had come to be. Later, it would count for little that Keats had been inspired by the controversial statuary, not only in two sonnets, but as his gifts grew and his imagination lived more with what he had experienced of Greek art, in *Endymion*, 'Ode on a Grecian Urn, and *Hyperion*.

Classicism in British art and architecture was already running out of impetus. By 1830 a Gothic design was chosen for the House of Parliament, much to Elgin's disgust. It was left to the Scottish architect James 'Greek' Thomson to bring the style to its most original achievement in this country. Ironically, one of his finest buildings was Elgin Place Congregational Church at 240 Bath Street in Glasgow. Calton Hill Monument in Edinburgh, an unfinished attempt to reconstruct the Parthenon, was undertaken at the suggestion of Elgin. It was also a dream of the Scottish artist H.W. Williams, who published two volumes of his Greek travels in 1820.

Byron's angriest attack on Elgin appeared in *The Curse of Minerva*. A privately printed version was circulated in 1812 but the complete poem remained unpublished until after Byron's death. Augustan measures guide Byron's satire from one immoderate sentiment to another: the impression is of rancour and vindictiveness disciplined by the equity that lurks within the balance of heroic couplets and antitheses. Iambic justice, the drum roll of English tradition in its encounter with the literature of antiquity, rallies to its assault, which will be personal, and all the while the reasonableness of the verse carries with it enough propriety and a sense of redress to clothe Byron's bad temper with satirical privilege.

> "Mortal!" ('twas thus she spake) "that blush of shame
> Proclaims thee Briton, once a noble name;

First of the mighty, foremost of the free,
Now honoured *less* by all, and *least* by me:
Chief of thy foes shall Pallas still be found.
Seek'st thou the cause of loathing? – look around.
Lo! here, despite of war and wasting fire,
I saw successive tyrannies expire;
'Scap'd from the ravage of the Turk and Goth,
Thy country sends a spoiler worse than both.
Survey this vacant, violated fane;
Recount the relics torn that yet remain:
These Cecrops placed, *this* Pericles adorned,
That Adrian rear'd when drooping Science mourn'd.
What more I owe let Gratitude attest –
Know, Alaric and Elgin did the rest.
That all may learn from whence the plunderer came,
The insulted wall sustains his hated name:
For Elgin's fame thus grateful Pallas pleads,
Below, his name – above, behold his deeds!
Be ever hailed with equal honour here
The Gothic monarch and the Pictish peer:
Arms gave the first his right, the last had none,
But basely stole what less barbarians won.
So when the Lion quits his full repast,
Next prowls the Wolf, the filthy Jackall last:
Flesh, limbs, and blood the former make their own,
The last poor brute securely gnaws the bone.
Yet still the Gods are just, and crimes are crost:
See here what Elgin won, and what he lost!
   Another name with *his* pollutes my shrine:
   Behold where Dian's beams disdain to shine!
   Some retribution still might Pallas claim,
When Venus half-avenged Minerva's shame."

Elgin's 'crime' is roundly cursed in the passage: Minerva claims the intervention of Venus on her behalf, her half-vengeance referring to Mary, Lady Elgin, by then Elgin's ex-wife. Supernatural properties, Britain's alleged shame, or a pairing like 'Alaric and Elgin' contrasted with Pericles and Hadrian, or the wolf and the jackal, conspire to demean Elgin

as a barbarian worse than a Goth, and to dignify Byron's
animosity as the justice of the Gods. Byron's satirical tactics are
faultless: high-flown historical purpose transacts with squalid
bile, and neither seems to mock the other.

> She ceas'd awhile, and thus I dared reply,
> To soothe the vengeance kindling in her eye:
> 'Daughter of Jove! In Britain's injur'd name,
> A true-born Briton may the deed disclaim.
> Frown not on England; England owns him not:
> Athena! no; thy plunderer was a Scot.
> Ask'st thou the difference? From fair Phyle's towers
> Survey Boeotia; – Caledonia's ours.
> And well I know within that bastard land
> Hath Wisdom's goddess never held command:
> A barren soil, where Nature's germs, confin'd
> To stern sterility can stint the mind;
> Whose thistle well betrays the niggard earth,
> Emblem of all to whom the land gives birth;
> Each genial influence nurtur'd to resist,
> A land of meanness, sophistry and mist:
> Each breeze from foggy mount and marshy plain
> Dilutes with drivel every drizzly brain,
> Till burst at length each wat'ry head o'erflows,
> Foul as their soil, and frigid as their snows:
> Then thousand schemes of petulance and pride
> Despatch her scheming children far and wide;
> Some East, some West, some – every where but North,
> In quest of lawless gain, they issue forth
> And thus – accursed be the day and year!
> She sent a Pict to play the felon here.'

Elgin's felonious connoisseurship touched Byron on a raw
nerve. Before he went to Greece he was already contemptuous
of 'emasculated fogies', the geriatric castrati of country-house
antiquarianism. Once there, he noticed that the removal of relics
robbed a place of its poetry. In his 'Letter on the Rev. W. L.
Bowles's Strictures on the Life and Writings of Pope', he wrote:

> ... to whatever spot of earth these ruins were transported, if
> they were *capable* of transportation, like the obelisk, and the

sphinx, and the Memnon's head, *there* they would still exist in the perfection of their beauty, and in the pride of their poetry. I opposed, and ever will oppose, the robbery of ruins from Athens, to instruct the English in sculpture; but why did I do so? The *ruins* are as poetical in Piccadilly as they were in the Parthenon; but the Parthenon and its rock are less so without them. Such is the poetry of art.

It was a theft, too, that by Byron's lights insulted the cause of Greek independence and illustrated the careless manner of British superiority and self-interest. Elgin's nationality – as a nobleman whose ambitions were directed chiefly at British preferment, it is tempting to see his Scottishness as border-line – was a convenient catalyst to Byron's own ambiguous, marginal Scottishness, which he disliked and distrusted except when it suited him to appear otherwise. Yet in such lines as,

> Each genial influence nurtured to resist;
> A land of meanness, sophistry and mist . . .

Byron was perhaps exercising more Scottishness than he knew. Their exaggeration is undeniable, but they summarize what many Scottish writers have perceived and contested, as do these lines:

> Yet Caledonia claims same native worth,
> And dull Boeotia gave a Pindar birth;
> So may her few, the letter'd and the brave,
> Bound to no clime and victors of the grave,
> Shake off the sordid dust of such a land,
> And shine like children of a happier strand;
> As once, of yore, in some obnoxious place,
> Ten names (if found) have saved a wretched race.

Minerva's curse then falls on the head of the already ruined Elgin, referring, in all likelihood and with considerable cruelty, to Elgin's epileptic son:

> "First on the head of him who did this deed
> My curse shall light – on him and all his seed:
> Without one spark of intellectual fire,
> Be all the sons as senseless as the sire."

Not even Byron and Minerva, however, could prevent Elgin's son, the eighth Earl of Elgin, from a career of great integrity, his diplomacy taking him to Jamaica, Canada, China and Japan, and finally as Viceroy to India. In case such an event might happen, Byron covered himself:

> If one with wit the parent brood disgrace,
> Believe him bastard of a brighter race:
> Still with his hireling artists let him prate,
> And Folly's praise repay for Wisdom's hate;
> Long of their Patron's gusto let them tell,
> Whose noblest, *native* gusto is – to sell:
> To sell, and make – may Shame record the day,
> The State receiver of his pilfer'd prey . . .

True-born Briton or not, not even Byron could get out of that one – although by a mere two votes, Westminster appropriated the Marbles, and, at least officially if not historically, Elgin was cleared of wrong-doing and of crossing the line between ambassadorship and entrepreneurial interests in art and architecture.

Elgin remarried. His second wife, Elizabeth Oswald of Dunnikeir in Fife, bore eight children to add to the four presented by Mary Nisbet. £18,000 of the expenses coughed up by the government were immediately transferred to creditors who had craftily assigned their requirements in advance to the government itself. Elgin's request to the government was limited to the recovery of his costs; he did not add to them an estimate of their value. 'I have been activated by no motives of private emolument,' he declared, 'nor deterred from doing what I felt to be a substantial good, by considerations of personal risk, or the fear of calumnious misrepresentations.'

Elgin's statement is tinged with retrospective self-justification. Fear of French depredations, an ambiguous *firman* produced at a moment when British favour ran high with the Turkish government, opened up to Elgin an opportunity in which his original, benign intentions took second place to personal aggrandizement. Enough of this was known at the time for Byron to feel sure of his opinions, and for

parliament to come to such a close decision. It looks possible that governmental propriety and face-saving came to Elgin's partial rescue, and that parliamentary criticism gave it a close run.

Elgin's petitions to the government and public opinion must have appeared tedious and incessant, the desperation of a debtor, the turn-around of a man who thought he had trumped his fortunes with a stunt of artistic, patriotic daring, selflessness disguising private, acquisitive pride. Onlookers would have been divided between the comedy of the episode as it afflicted Elgin, and its offensiveness as it afflicted the public purse and Greek integrity. So many enemies in his life suggests that Elgin might have been an unpleasant man. After Constantinople he was not even much to look at: he lost most of his nose to an unidentified disease. His picture in the National Portrait Gallery shows him leaning self-confidently on his sword: his expression conveys pride and impatience, the hand on his hip suggests more than a hint of swagger and aristocratic disdain.

John Galt suggested to Byron that he might have overstepped the boundaries of satire. Byron replied, '& if you will prove to me that Ld. Elgin's *is* "the error of a liberal mind" the "Muse" shall forthwith eat her own words although they choak her – & me into the bargain.'

Elgin signed an early petition in favour of the Greek nationalists, as did Byron. Unlike Byron he did not join the Greek Committee, an organization of men and women later to be known as the Philhellenes. British policy was discouraging, and Elgin, still hopeful of a British peerage – he had been returned as a representative Scottish peer in 1820 – would in all likelihood have imitated it. Castlereagh and Wellington, for instance, were anti-Greek.

Among the volunteers who served with the Greek land and naval forces, the proportion of Irish and Scots was noticeably high. Colonel C. M. Woodhouse, an authority on this subject, has described it as a surrogate of Scottish and Irish nationalism. Whatever Byron might have thought of Caledonia, its libertarians, adventurers and freebooters were not slow to enlist in a fight for freedom, as many saw it, or an opportunity

for military hedonism, as it was seen by others. Admiral Lord Cochrane harnessed Greek seamanship to brilliant effect. Thomas Gordon of Cairness rose to the rank of General in the Greek armies. One or two even changed sides, sickened by Greek atrocities, only to be sickened yet again by Turkish brutality.

Unwittingly, Elgin, at Byron's tender mercy, contributed towards convictions and events distant from his own mind and intentions. History is never simple. Kapodistrias, for example, a leader of the Greek movement, had helped the Tsar draw up the terms of the Holy Alliance. Elgin's activities roused Byron's animosity and helped Byron to identify his sympathies, lending a topical note to *Childe Harold's Pilgrimage* (Elgin is the only living person mentioned) that braced his verse with an incident a British public could hang on to. The poem contributed to the beliefs of a generation, disposing many towards a favourable view of the struggle that led to the first new nation state of nineteenth-century Europe.

The controversy for which Elgin must be held responsible is still unresolved. In his book of Greek travels, Hobhouse reported:

> I cannot forbear mentioning a singular speech of a learned Greek of Ionnina who said to me, "You English are carrying off the works of the Greeks, our forefathers – preserve them well – we Greeks will come and redemand them."

They have been asking for their return ever since the 1830s. Successive British administrations have refused to countenance the idea. As Scots we have an interest in encouraging the restitution of the Parthenon sculptures, and in disproving the Byronic assessment of a nationality that contributed to Greek nationhood. It would be a decision that expressed coincidentally that we cared for our own culturally depleted country.

Byron: An Edinburgh Re-Review

JON CURT

In the early years of the nineteenth century, the recently founded *Edinburgh Review* was not primarily a literary journal. Instead, the majority of the contributors concerned themselves with the political and economic developments of the day – both words having much broader implications then than they do now. Nevertheless, in the interests of being a well-rounded periodical, some attention had to be given to advances in the literary field, although this was usually regarded as an opportunity to deride such respected and established poets as Wordsworth and Coleridge. Only Scott could reasonably expect to treated with the sort of courtesy and consideration most writers feel they deserve. In this way, *The Edinburgh Review* had already gained the reputation of savaging most of the literature that came its way, prompting Shelley to say some years later 'In my opinion, the *Edinburgh Review* is as well qualified to judge the merits of a poet, as Homer would have been to write a commentary on the Newtonian System.'[1] Of course, Shelley had an axe to grind with most reviewers, but in the case of this particular journal there may be more than a germ of truth in his opinion. Given this intolerance of even the highly regarded authors of the day, how much more damning could it be when confronted with a small collection of verse, unpromisingly entitled *Hours of Idleness* and written, according to the title page, by 'George Gordon, Lord Byron, a Minor'? This, surely, was a lamb to the slaughter.

The actual review appeared in *The Edinburgh Review*, number XI, in January 1808, and was written, anonymously, by Henry Brougham, whose cultural preferences were well demonstrated two years later when he became an English Member of Parliament. As a piece of criticism, it is essentially a highly

sarcastic attack on Byron's false modesty about publishing the poems at all, as expressed in the 'Preface'. Indeed, Brougham consistently ignores the verse and quotes so extensively from what is, after all, a short preface, that Byron can be forgiven for feeling that the review was more of an attack on his person than his poetry. The last paragraph is typical:

But whatever judgement may be passed on the poems of this noble minor, it seems we must take them as we find them, and be content, for they are the last we shall ever have from him. He is at best, he says, but an intruder in the groves of Parnassus; he never lived in a garret like thorough-bred poets; and 'though he once roved a careless mountaineer in the Highlands of Scotland', he has not of late enjoyed this advantage. Moreover, he expects no profit from his publication; and whether it succeeds or not, 'it is highly improbable from his situation and pursuits hereafter' that he should again condescend to become an author. Therefore, let us take what we get and be thankful. What right have we poor devils to be nice? We are well off to have got so much from a man of this lord's station, who does not live in a garret, but 'has the sway' of Newstead Abbey. Again, we say, let us be thankful; and with honest Sancho, bid God bless the giver nor look the gift horse in the mouth.[2]

It is hardly surprising that Byron felt this review so strongly, although to his credit, he hid his anger from all but his closest friends until such time as it could be profitably vented in the poem that was to become *English Bards and Scotch Reviewers*. With admirable resignation, he wrote to a friend, John Becher, on 26 February 1808:

. . . You know the System of the Edinburgh Gentlemen is universal attack, they praise none, and neither the public nor the author expects praise from them, it is however something to be noticed, as they profess to pass judgement only on works requiring public attention.

The next month, again to Becher, Byron is still presenting the front of a man determined to make the most of his bad luck: 'For my own part, these "paper bullets of the brain" have only taught me to stand fire; and as I have been lucky enough upon the whole, my repose and appetite are not discomposed'[3]

(28 March 1808). But on 27 February he had written to his lifelong friend, John Cam Hobhouse, in what we feel sure was a more honest tone: "As an author, I am cut to atoms by the E Review, it is just out and has completely demolished my little fabric of fame.'

Both sides of Byron's attitude to bad reviews – his own sensitivity and the swaggering bravado he assumed – can be seen in his reaction to the news of Keats's death. Not knowing the facts, he genuinely believed, for some time at any rate, that he had been slain by an unfavourable review in the *Quarterly*. Writing to John Murray, he says:

> You know very well that I did not approve of Keats's poetry, or principles of poetry . . . but as he is dead, – omit *all* that is said *about him* in any MSS of mine . . . I do not envy the man – who wrote the article – your review people have no more right to kill than any other foot pads. However – he who would die of an article in a review – would probably have died of something else equally trivial . . .[4]

Recollecting his first review in his journal five years later, Byron confesses to a feeling of anger and hints at the intriguing possibility that *English Bards and Scotch Reviewers* was actually started on the day the review was published:

> I remember the effect of the *first* Edinburgh Review on me. I heard of it six weeks before, – read it the day of its denunciation – dined and drank three bottles of claret, . . . neither ate nor slept the less, but, nevertheless, was not easy till I had vented my wrath and my rhyme, in the same pages against everything and every body.[5]

That 'wrath and rhyme' eventually found its way into the poem published in 1809, *English Bards and Scotch Reviewers*, in which Byron first found his satiric tone which, as he continued to hone and refine it, became the essence of his greatest poetry. In the 'Preface' to the poem, Byron makes explicit the targets of the satire:

> As to the *Edinburgh Reviewers*; it would, indeed, require a Hercules to crush the Hydra; but if the Author succeeds in merely 'bruising one of the heads of the serpent', though his own hand should suffer in the encounter, he will be amply satisfied.[6]

Ironically, the 'one head' he thought he was bruising – that of Francis Jeffrey, who is addressed by name in the 'Postscript' to the second edition, and the attack on whom occupies 101 lines of the poem[7] – was not, as we have seen, the reviewer of *Hours of Idleness*.

Exactly ten years after that first review, it was indeed Jeffrey who reviewed *Beppo* in *The Edinburgh Review*, number XXIX, February 1818. Of course, by then Byron was far and away the most popular poet of the day, and was not to be mocked as some 'noble minor'. Quite apart from the praise and admiration for the style of the poem, the review is characterised by the uncertainty that Byron is actually the poet in question, since *Beppo* was published anonymously and was Byron's first attempt at a poem in the *ottava rima* stanza which he was to make his own in the following years. Jeffrey, obviously with a clear recollection of what happened the last time Byron thought he had reviewed his poetry, squirms uncomfortably through several pages before a closing paragraph in which he fawns:

> We are not in the secret of this learned author's incognito; and at our distance from the metropolis, shall not expose ourselves by guessing. We cannot help thinking, however, that we have seen him before, and that 'We do not know that fine Roman hand'. At all events, we hope we shall see him again: and if he is not one of our old favourites, we are afraid we may be tempted to commit an infidelity on his account, – and let him supplant some of the less assiduous of the number.[8]

To point out the open contradiction between Brougham's and Jeffrey's reviews is obvious: Brougham sighing his relief at 'the high improbability from his sitation and position hereafter, that he should again condescend to become an author', compared to the rather sycophantic hope expressed by Jeffrey, that 'we shall see him again . . . one of our old favourites', but this is to do both reviewers an injustice. The fact remains that *Beppo* is a much better poem than anything in *Hours of Idleness*, and perhaps, when faced with yet another collection of adolescent verses, penned by a self-confessed 'minor', Brougham was right to concentrate his attack on the slightly

absurd posture of the young lord who so obviously wanted to be taken seriously.

Before leaving *The Edinburgh Review* behind, and moving on to the end of the nineteenth century, it is worthwhile pointing out that at no time in the years of its composition and publication did the *Review* attempt an assessment of *Don Juan*. Byron's plays were all considered – many at greater length than they deserved – but his *magnum opus* never figured in the pages of the illustrious journal. This may be insignificant, but one might speculate as to possible reasons for this apparent oversight: indifference to the work seems unlikely; fear of further reprisals for a hostile review, especially now that Byron and Jeffrey were the best of friends, seems unnecessarily cautious; which leaves the possibility that in this most decorous of European capitals, it might be best to pretend that evil Lord Byron's profligate poem did not exist at all.

Edinburgh figured once again in the forum of Byron criticism and commentary in 1896, when Professor George Saintsbury published his *History of Nineteenth Century Literature*, which, as the title suggests, purports to cover one hundred particularly productive years of writing in fewer than five hundred pages. Even the logistic constraints of such a project fail to explain why only five pages are devoted to Byron – and why those few comments should constitute the fullest consideration of the poet by the man who was certainly the greatest academic critic of his generation in Britain. This ignoring, or even ignorance, of Byron is not untypical of the late Victorians, who felt that he had too vulgar a style for his work to be what was then considered poetry of the first order.

Saintsbury begins with the standard comparison between Sir Walter Scott and Byron – Scott having been usurped as the most popular poet of the day by the scuccess of *Childe Harold's Pilgrimage* and Byron's subsequent Eastern Tales. Not surprisingly, Scott comes off best: 'Indeed, Scott, with all his indifference to a strictly academic correctness, never permitted himself the bad rhymes, the bad grammar, the slipshod phrase in which Byron unblushingly indulges.'[9] It was to be sixty years until another great Edinburgh professor, W. W. Robson, was to point out that Byron was the exponent of a technique,

or lack of technique, which permitted him not only to tolerate second-rateness, but to elaborate it with gusto. As for Saintsbury's claim of bad grammar, it is interesting to place it alongside Byron's own defence of *Don Juan* in a letter to Douglas Kinnaird, where he writes: 'As to *Don Juan*, confess, confess – you dog, and be candid, – that it is the sublime of *that there* sort of writing – it may be bawdy, but is not good English? It may be profligate, but is it not the life, is it not the thing?' (26 October 1819).

Saintsbury then dwells on Byron's enormous popularity, and makes of this another strand of his argument – namely, that popularity on that sort of scale must equal vulgarity. Using this as his main premise, and with a gigantic leap, the logic of which hardly conceals its absurdity, he then dismisses the whole of continental Europe with a single sweep:

> Is he a poetic star of the first magnitude, a poetic force of the first power at all? There may seem to be rashness, there may even seem to be puerile insolence and absurdity in denying or even doubting this in the face of such a European concert as has been described and admitted above. Yet the critical conscience admits of no transaction; and after all, as it was doubted by a great thinker whether nations might not go mad like individuals, I do not know why it should be regarded as impossible that continents should go mad like nations.[10]

Having accounted for those injudicious enough to actually *like* Byron's poetry, or those, like Goethe and others, who allowed themselves to be influenced by it, Saintsbury turns his attention, albeit very briefly, to the verse itself: 'His verse is to the greatest poetry what melodrama is to tragedy, what plaster is to marble, what pinchbeck is to gold.'[11]

It is tempting to treat Saintsbury as dismissively as he himself treats Byron, but that ignores the important possibility that he started reading Byron from the beginning, and worked his way through. To do this would mean starting with *Hours of Idleness*, which, as Brougham observed, is not high art in any shape or form. After that, the methodical reader is faced with the inconsistencies of *Childe Harold's Pilgrimage* and the relentless gothic egoism of the Eastern Tales. By the time one reaches the refreshing linguistic adventures of the *ottava rima* poems,

one's palate may well have become so jaded as to miss their worth altogether. To be fair to Saintsbury, he has, like so many others, found himself with the wrong yardstick by which to measure Byron's later poems and has lapsed into some kind of moralistic disapproval. As we have seen, Byron maintained that *Don Juan* was 'the sublime of *that there* sort of writing', the '*that there*' being underlined to emphasise the point that this was something contemptuous of tradition and convention, at the same time as being linguistically legitimised by its use of colloquial language and expression. Saintsbury's problem with *Don Juan* was something to which Byron was subjected throughout his life as well as after it. He was only too aware of this block on a proper appreciation of his work, and wrote emphatically to his publisher, John Murray, to justify Cantos I and II of *Don Juan* after an uncertain reaction by the British public: 'You have been careless of this poem because some of your Synod don't approve of it – but I tell you – it will be long before you see anything half so good . . . I have read over the poem carefully – and I tell you *it is poetry*.'[12] This is the same 'carelessness' exhibited by Saintsbury. Because he disapproved of both the style and content of *Don Juan*, he dismissed it in a matter of paragraphs rather than stop and actually look at the verse itself, and in this respect, he was failing in his duty as a critic. The general feeling in his short comment on Byron is that here was something he knew was, in a sense, genuinely 'beyond' him, and the brevity of his remarks is the manifestation of that awareness. He completes his discussion with this candid admission:

> I have read Byron again and again; I have sometimes, by reading Byron only and putting a strong constraint upon myself, got nearly into the mood to enjoy him. But let eye or ear once catch sight or sound of real poetry, and the enchantment vanishes.[13]

As Bernard Beatty observes in his book *Byron's Don Juan*, ' "Thinking men" on the whole have not made conspicuously good readers of Byron's poetry.'[14]

However, one 'thinking man' who managed to read and appreciate Byron without 'putting a strong constraint' on himself, was Saintsbury's successor as Professor of English

Literature at Edinburgh, H. J. C. Grierson. His Warton Lecture on English Poetry, delivered on 24 November 1920, helped to haul Byron criticism out of the trough of Victorianism where it had languished for so long.

Grierson's lecture, which was published in the *Proceedings of the British Academy*, is a model of common sense and historical context. He confesses a slight bewilderment at the way his predecessor 'pursued Byron's reputation with a curious rancour which has not coloured his often equally severe criticism of Wordsworth and his heresies,' but then goes on to show that if Byron is put in a tradition, *not* of the 'greatest poetry' as Saintsbury would have it, but of Donne, Swift and especially Burns, then we can come to a much fairer assessment of his work.

Even on the well-trodden ground of the Scott/Byron comparison, Grierson sees that Byron's value has been misunderstood for too long. In Scott, scenery and setting are everything, but for Byron:

> His central theme is the infinite worth of love and courage and endurance . . . Byron did in those poems, as Tennyson said, 'give the world another heart and new pulses, and so we are kept going' . . . he delineated, as Wordsworth and none of his contemporaries did, passion and energy.[15]

It is this 'passion and energy' that brings him into line with that alternative tradition in poetry, the tradition of Donne, Rochester, Swift and Burns: poets who appear to be almost physically present in their best work. As Grierson says:

> Byron was a lover, masculine and passionate, as Donne and Burns had been before him . . . Just as Burns was a great poet who was also a peasant, a peasant who really lived the life and shared the joys of peasants . . . so Byron was a poet who was also a man among men, and a man of the world, seeing the world with which he was at war, through the eyes of the world.[16]

By virtue of his Scottish upbringing, Byron always had a great affection for the poetry of Burns, but his most ecstatic comments on the man were reserved for his letters, particularly the 'unpublished and never-to-be-published' ones. As he wrote in his *Journal* on 13 December 1813: 'They are full of oaths and obscene songs. What an antithetical mind! –

tenderness, roughness – delicacy, coarseness – sentiment, sensuality – soaring and grovelling, dirt and deity – all mixed up in that one compound of inspired clay!' It is significant that what appealed most to Byron was the curious eclecticism in Burns which so many readers, both then and now, have found in his own poetry and personality.

Grierson is less explicit about the nature of a comparison with Swift, and concentrates on their shared political interests and involvements rather than stressing any obvious point of poetical similarity. So, in the final paragraph of his lecture, Grierson says:

> But pure humour and sincere generous passion are (*Don Juan's*) finest qualities, the humour, the love of banter, the passionate hatred of cruelty and injustice which were as characteristic of the real Byron as of the real Swift, with whom he had so much in common.[17]

This is to emphasise their common humanity, but there is also a poetic point to be made between the two. Byron's best tales – most notably *Mazeppa* – are written in the octosyllabic couplets that Swift used almost exclusively, which bowl along with such an impression of speed and informality that it is a short step from there to the *ottava rima* stanzas of *Beppo* and *Don Juan*. This is almost contrary to Byron's stand on the heroic couplet, which he championed in his early output and continued to defend throughout his life, in the works of such second-rate poets as Campbell and Rogers, although it had very obviously had its day. Whilst it is true that some of Scott's verse tales are also written in the octosyllabic couplet, and therefore must be considered as a possible influence on Byron, there is a significant anecdote in Leslie Marchand's biography which adds weight to the Swift side of the balance. One of Byron's shipmates on his last voyage to Greece, James Hamilton Browne, recorded in voyeuristic detail the movements and activities of the poet, and at one point observed that he was reading Swift and 'thus supposed he was proposing to write another canto of *Don Juan*.'[18] Surely, as Grierson came close to saying, it was the rapid, incisive and colloquial couplets of Swift, rather than the decorative, evenly stressed narratives of Scott, that influenced the later works of Byron.

Grierson had edited Donne's poems in 1912, and it was that edition which helped re-establish not just Donne himself, but the whole 'Metaphysical' period into the accepted area of study and commentary. With our hindsight, the comparison between Donne and Byron seems obvious, but Grierson's penetrating analysis in 1920 was something quite new. Of Donne, he says: '(the) songs and sonnets have reasserted their worth after a long interval, because passion made Donne a subtle and at times even a profound thinker, and because his style, if harsh and careless, is never banal, and often splendidly felicitous.'[19] All these comments on Donne's style, especially, as we have seen, the one concerning his carelessness, have been made at some time or another about Byron – although usually with a less complimentary intent.

Later, talking about the Haidée and Juan interlude in Canto II of *Don Juan*, he claims: 'There is nothing like it in English poetry except some of the songs of Burns and the complex, vibrant passion, sensual and spiritual, of Donne's songs and elegies.'[20] Again, it is that 'complex, vibrant passion', which is the driving force behind an established but often unrecognised tradition in English and Scottish poetry, that Grierson is right to identify in Byron. He, more than anyone else before or since, combined genuine talent with personal charisma to create his own potent and enormously appealing poetic tone.

Byron was a true internationalist: born in England, raised in Scotland, finding his cultural home in Italy and his destiny in Greece – there is no one country that can claim a greater right to him than any other. Nevertheless, there has always been a Scottish view of him and his work: from Brougham and Jeffrey in *The Edinburgh Review* itself, through the impatience and disapproval of Saintsbury, and the wisdom and perspective of Grierson, right up to the clarity of W. W. Robson, whose 1957 article 'Byron as Improviser'[21] is still rightly considered to be one of the most influential in the field. Edinburgh is, in some sense, Byron's capital city – he was proud of his Scottish upbringing and spoke with a marked accent until the day he died – but in this most conservative of northern capitals, it is the advantage of *not* having the hot-blooded passion of the Mediterranean nations, nor the *laissez faire* of Regency Lon-

don, that has enabled Edinburgh to keep Byron in clear focus.
There has always been, and will always be, an Edinburgh view
of Byron.

NOTES

1. Shelley to Byron, 20 November, 1816. The review which
   had prompted this remark was of Coleridge's 'Christabel'.
2. *The Edinburgh Review*, number XI (January, 1808), 285–9.
3. It has been pointed out that whenever Byron felt strongly
   about a subject in his letters, he invariably quoted from
   Shakespeare. The quotation here comes from *Much Ado
   About Nothing* II iii, and suggests that despite his dismissive
   tone, Byron is feeling the bad review quite keenly.
4. Byron to John Murray, 30 July 1821.
5. *Journal*, 22 November 1813.
6. Preface to the second edition, *English Bards and Scotch
   Reviewers*, 1809.
7. From 438–539.
8. *The Edinburgh Review*, number XXIX (February, 1818),
   302–10.
9. *A History of Nineteenth Century Literature* (1896), 79.
10. Ibid., 80.
11. Ibid.
12. Byron to John Murray, 31 August 1821.
13. *A History of Nineteenth Century Literature*, 81.
14. *Byron's Don Juan* (1985), 222.
15. *Proceedings of the British Academy*, IX. The Warton Lecture on
    English Poetry, number XI, 'Lord Byron: Arnold and
    Swinburne', 10.
16. Ibid., 11–12.
17. Ibid., 31.
18. In Leslie A. Marchand, *Byron: A Biography* (1958), III, 1095.
19. Grierson, 10.
20. Ibid., 23.
21. W. W. Robson, 'Byron as Improviser' in *Proceedings of The
    British Academy*, 1957

Byron, Scott, and Nostalgia

J. DRUMMOND BONE

The cliché is that in every Conradian bar in the Far East some Scotsman drips tears into his drink, bewailing the good old Broomielaw – or that in quiet tea-parties in the Cotswolds soft accents sigh for the Moray Firth. And of course it goes without saying that *that* Broomielaw and *that* Moray Firth are nowhere to be found, and *were* nowhere to be found. Nostalgia is a matter not only of a place and a time, but of the passing of time in general, and of a feeling of dislocation in general. It localises that feeling of every day being cast from the garden, of existential loss. Or indeed it particularises the sense of loss as the sense of existence itself, for it could be said of our caricature Scots that their only true loss would be the loss of loss, for their identity is bound up with this emotion. But then of how many other peoples could this, in different ways, be said? Nostalgia is certainly a feeling which is positively indulged; though a feeling of loss, it is also a recovery of the past, and an odd affirmation of continuity in which we do not exist in isolated 'presents', but in the stream of time. I shall look at an extract from Scott and a few from Byron, examining their treatment of time and place lost, and noticing a gradual shift from nostalgia presented as the artistic recuperation of some quite definite 'real' loss – in which the recovery becomes a positively pleasant or even useful emotion – to nostalgia presented as the loss, not of something definite, but of the loss of the ability to *prevent the emotion of loss*, a nostalgia which does not have the perhaps sentimental ease of loss turned into positive feeling, but an absence indeed.

The following is taken from the Introduction to Canto II of *Marmion*, and was written in 1808. It takes the form of a verse letter from the author to the Rev. John Marriott, but it is a very stylised author.

119

The scenes are desert now, and bare,
Where flourish'd once a forest fair,
When these waste glens with copse were lin'd,
And peopled with the hart and hind.
Yon Thorn – perchance whose prickly spears
Have fenc'd him for three hundred years,
While fell around his green compeers –
Yon lonely Thorn, would he could tell
The changes of his parent dell,
Since he, so grey and stubborn now,
Wav'd in each breeze a sapling bough;
Would he could tell how deep the shade
A thousand mingled branches made;
How broad the shadows of the oak,
How clung the rowan to the rock,
And through the foliage show'd his head,
With narrow leaves and berries red;
What pines on every mountain sprung,
O'er every dell what birches hung,
In every breeze what aspens shook,
What alders shaded every brook!

'Here, in my shade,' methinks he'd say,
'The mighty stag at noontide lay:
The wolf I've seen, a fiercer game,
(The neighbouring dingle bears his name,)
With lurching step around me prowl,
And stop, against the moon to howl;
The mountain-boar, on battle set,
His tusks upon my stem would whet;
While doe, and roe, and red-deer good,
Have bounded by, . . .

From Yair – which hills so closely bind,
Scarce can the Tweed his passage find,
Though much he fret and chafe and toil
Till all his eddying currents boil, –
Her long-descended lord has gone,
And left us by the stream alone.

And much I miss those sportive boys,
Companions of my mountain joys,
Just at the age 'twixt boy and youth,
When thought is speech, and speech is truth.
Close to my side, with what delight
They press'd to hear of Wallace wight,
When pointing to his airy mound,
I call'd his ramparts holy ground! . . .

When, musing on companions gone,
We doubly feel ourselves alone,
Something, my friend, we yet may gain;
There is a pleasure in this pain:
It soothes the love of lonely rest,
Deep in each gentler heart impress'd.
'Tis silent amid worldly toils,
And stifled soon by mental broils;
But, in a bosom thus prepar'd,
Its still small voice is often heard,
Whispering a mingled sentiment,
'Twixt resignation and content.[1]

The past before this forest clearance is clearly intended as a historical one. If the thorn could speak, that past would be recoverable. But it is also linked with images of holiness (Wallace's 'holy' ramparts) – and of innocent community (his young companions 'close to his side'), and the land has the plenty of a Paradise (in the listing of the trees and game).

Scott had clearly been reading his Wordsworth fairly closely – the 'sportive boys' may have something to do with the 'little lines of sportive wood' in 'Tintern Abbey', and there are other verbal echoes of the 'Intimations' Ode too, as well as the generally similar tone.[2] Quite explicitly, the sense of Paradise lost is seen not as a wholly unpleasant sensation, on the contrary, it informs the spiritual emptiness of the present ('There is a pleasure in this pain/It soothes the love of lonely rest'). The lost past is fuller than the alive now, and clearly with the reader's interest in that past, the sense of loss is filled by the plenitude of memory. Our own feelings as readers are

identified with the loss made positive, and we feel – arguably sentimentally – reassured.

Now this is a relatively easy process for literature, since it exists inevitably as a positive and not a negative quantity. As good Derrideans we know that if you seek presence in language you will find only absence – the odd corollary of which is that the absence of absence is presence, of a sort.[3] It is difficult to be negative, for the negative not to be turned into a positive just by virtue of its expression. A sense of meaning-lessness becomes a meaningful sense of meaninglessness if we read about it. Let us note again the line 'When thought is speech, and speech is truth'. Here is a picture of an ideal language, a guaranteed chain from truth to thought to speech, where speech *is* presencing. This for Scott's speaker is clearly a possibility, though now lost to him along with his idyllic childhood. There is scarcely even a hint that though it was possible, it is now and for ever impossible; for us, as for the narrator, this more general reading is suppressed in the (oddly) more hopeful 'such feelings pure,/They will not, cannot, long endure' (not quoted above).[4] The statement of loss here carries with it the emotion of its potential recovery. This sense of intermingled loss and the recovery of either presence in language or paradise in life – the two are, if not identical, very nearly so – is the stuff of nostalgia, at least if we accept that nostalgia is intrinsically un-self-critical. But the frequent admixture of irony – even in the case of the maudlin exile on his bar-stool – might suggest that this is not a total account of the nostalgic dynamic.

Byron takes up the same identification as Scott of the loss of the past with the loss of the ability of language to live. In the fragment 'Harmodia' we find:

'The things that were' – and what and whence were they,
Those clouds and rainbows of thy yesterday? . . .
Such is the past – the light of other days
That shines but warms not with its powerless rays
A moonbeam someone watches to behold
Distinct but distant – clear – but deathlike cold.
Oh, as full thought comes rushing o'er the mind,

Of all we saw before – to leave behind –
Of all – but words – what are they? can they give
A trace of breath to thoughts while yet they live?
No – Passion, Feeling speak not – or in vain –
The tear for Grief, the groan must speak for Pain . . .
The strife once o'er, then words may find their way,
Yet how enfeebled from the forced delay.[5]

Memory is not a means to recovery of the past here – on the contrary, memory reinforces the sense of alienation from the past ('Distinct, but distant'). Words reinforce the absence of the feelings of which they speak, and only inarticulate noise seems to carry the life of a thought. Thoughts are dead by the time they are expressed in language, as the past is realised as past in memory. This is not a matter of an incident of nostalgia, but of loss as the condition of life in time. There is no previous life of perfect communication which has been lost, no sense of a full but lost past, only the sense of passing. What memory laments here is memory itself. It is rather the inevitable inscription of loss than the preservation or recovery of some specific past and place. This is an early example of Byron's bleaker use of nostalgic emotion – bleaker that is than Scott's – but it is arguably simply a different thing from essential nostalgia – an interpretation of memory which denies the pleasantness which is perhaps central to the nostalgic moment. If certain uses of nostalgia uncritically recuperate the past in the guise of mourning its loss, so perhaps Byron's use of memory here is equally monovocal in its nihilism. It is not really a *critical* nostalgia.

'Harmodia' evolved into one of the best known of the *Hebrew Melodies*, 'Sun of the Sleepless'. The genre in which Byron found himself writing without question both produced and constrained the characteristic tone of the poems of exile and nostalgia in *Hebrew Melodies*. To a large extent his model is Moore, and this of course gives us a Celtic connection. Clearly the Jewish diaspora was a subject to which Byron was already susceptible.[6] But this heroic context enforced techniques to which one might say he was all too readily susceptible, having exercised them to the point of mannerism in *Childe Harold* I-II

and the already written members of the Turkish Tale sequence.[7] These all tend to produce a version of the past lost as absolute recovery. 'On Jordan's Banks' is a case in point:

> On Jordan's banks the Arab's camels stray,
> On Sion's hill the False One's votaries pray,
> The Baal-adorer bows on Sinai's steep –
> Yet there – even there – Oh God! thy thunders sleep:
>
> There – where thy finger scorch'd the tablet stone!
> There – where thy shadow to thy people shone!
> Thy glory shrouded in its garb of fire:
> Thyself – none living see and not expire!
>
> Oh! in the lightning let thy glance appear!
> Sweep from his shiver'd hand the oppressor's spear:
> How long by tyrants shall thy land be trod?
> How long thy temple worshipless, Oh God?[8]

The essence of the technique is to freeze time and concentrate space into a point. We note the repetitive 'There' – exactly 'there', and the climactic rise from 'thy glory' to the absolute revelation 'thyself', and the repeated negatives defining this moment – 'none living', 'not expire'. The aporia captures the moment of the lightning, out of grammatical time, and the insistent questions do not elicit a temporal answer, but on the contrary invoke an immediate presence at the climax and the end of the poem, on the dramatic direct appeal to the presence of God. For the purpose of future comparison we note too the firm consonant rhyme, with its closed ending and precise moment of enunciation – trod, God. Here, now. The ecstasy of loss has become an act of recovery. It might again be true that such ecstasy is only tangentially nostalgic however. More obviously to our point is the 'By the Rivers of Babylon'[9] in which the withholding of the song becomes the preservation of the past as still alive. That is, the loss of the place in which to sing the song becomes through the act of denial (through an affirmation of loss, in other words), a recovery of the meaning of itself. Again the rhetoric is full of evers and nevers, and precise moments. Still one does not want to say that even most

of the *Hebrew Melodies* works the same way. The last stanza of
'The Wild Gazelle', and curiously enough that fine variation
on a sonnet which begins with the definitive phrase 'It is the
hour'[10] are more genuinely negative, for example, and 'Sun of
the Sleepless' itself is one of Byron's finest evocations of loss,
using many of the techniques we shall look at below, and
which are perhaps more characteristic of the late poems.

Byron's experience of exile became real enough, of course.
His was not only a displacement in time, a sense of constantly
lost youth – Byron felt old from childhood as the merest glance
at his letters will show[11] – but an exile in physical fact. In
stanzas 8-10 of *Childe Harold* IV, relatively new to the exile's
state, he contemplates his relationship with Britain:

> I've taught me other tongues – and in strange eyes
> Have made me not a stranger; to the mind
> Which is itself, no changes bring surprise;
> Nor is it harsh to make, nor hard to find
> A country with – ay, or without mankind;
> Yet was I born where men are proud to be,
> Not without cause; and should I leave behind
> The inviolate island of the sage and free,
> And seek me out a home by a remoter sea,
>
> Perhaps I loved it well: and should I lay
> My ashes in a soil which is not mine,
> My spirit shall resume it – if we may
> Unbodied choose a sanctuary. I twine
> My hopes of being remembered in my line
> With my land's language: if too fond and far
> These aspirations in their scope incline, –
> If my fame should be, as my fortunes are,
> Of hasty growth and blight, and dull Oblivion bar
>
> My name from out the temple where the dead
> Are honoured by the nations – let it be –
> And light the laurels on a loftier head!
> And be the Spartan's epitaph on me –
> 'Sparta hath many a worthier son than he.'

> Meantime I seek no sympathies, nor need;
> The thorns which I have reaped are of the tree
> I planted, – they have torn me, – and I bleed:
> I should have known what fruit would spring from such
>   a seed.[12]

The 'quotation' from Milton, which he also used at the end of *Manfred*, somewhat undercuts the possibilities of belonging implied in 'not a stranger', already rather tenuously dependent on the still *'strange* eyes' (my emphasis), and this sense of isolation is increased in the qualification 'or without mankind'. The lost land can be reached in death, and through the poet's art becoming part of its tradition. The rhetoric which suggests the contrary – that he might be forgotten, condemned to an exile from not only his land but the tradition of his language for eternity – is duplicitous, for it makes heroic the imagined absence of heroic status. We note the grandiose repetition of the phrases 'And . . .', the quotation from the classical epitaph (precisely from the heroic age), and the really rather blasphemous possibilities in the image of the thorns, together with the blustering emphasis (particularly in the penultimate line) on the first person, with its emphatic assumption of responsibility for its own fate. The nostalgia here is still productive of an emotion which, at least in its force and assertion, can be characterised as positive. Its effect on the reader is to render isolation and loss as heroism and plenitude, all the more so since the imagined loss is two-fold – a real one in the past, and a possible further one in the future. Both, however, are accommodated into the grandeur of their present assumption. The pause at the end of the first stanza quoted, and the enjambement of the next, are ringingly resolved by the full close of the last, with the 'I''s dramatic assertion of its understanding, the climax of a movement which has really run through the three stanzas. But just as the *Hebrew Melodies* partly conditioned the presentation of nostalgia and exile by virtue of its 'national poem' genre, so here one suspects that Childe Harold the character is imposing on his author.

Even by 1816 – in the 'Epistle to Augusta' for example –

another tone is beginning to dominate.[13] We however shall pursue it in a later form, in an example from *Don Juan*:

> No more – no more – Oh! never more on me
>   The freshness of the heart can fall like dew,
> Which out of all the lovely things we see
>   Extracts emotions beautiful and new,
> Hived in our bosoms like the bag o' the bee:
>   Think'st thou the honey with those objects grew?
> Alas! 'twas not in them, but in thy power
> To double even the sweetness of a flower.[14]

The stanza begins with the characteristic absolute negation, but the repetition, with the rhetorical pause of the dash and the 'Oh!' which 'lapses' again from inarticulacy into words, spreads it into time – the negative becomes an ongoing state in this repetition, and this impression of time passing is helped by the comparative suggestion of 'more'. The climax of this negative journey is reached by way of the 'Oh', which might take us back to the sighs of 'Harmodia', or on to the syncopes and singulti of *Don Juan* XV, 'emblems of emotion' stacked against the great Ennui.[15] Here is the real loss, indicatable, but not expressible. But this inexpressible is not uncritically allowed to remain hanging outside of time, for the rest of the stanza becomes almost a gloss upon it. The two syllables of 'never' – as much as its more specific negative – make it stand as the climax of the articulate line, though it is the vowel assonance of the 'o' which has completely dominated the line until this point which really paces the rise of the line to the inarticulate climax of 'oh'. This climax is in fact dissolved by the change of vowel sound on 'never'. The next three lines take us into the world of nostalgia, and the tense is notably present in effect – can, see, extracts. The enjambement keeps the reader moving on apace in the search for the delayed verb, and it is partly the release of this control of breath, as it were, which produces the remarkable modulation on 'fall', a release with an ambiguous overtone played up by the two-edged possibility in its alliterative connection with 'freshness' – whether that is a link or a contrast – and a release saved from a trite stasis by the fact that one cannot savour it at the end of a line. Line three

highlights the simplicity of 'lovely' emerging from the context of the monosyllabic grammatical particles which precede it, and inviting us to be aware of its full force – things which invite love. Interestingly, too, Byron is royally unconcerned that what he says in lines 3-4 does not make clear sense, given what he is about to say. Here the heart extracts emotions 'out of all the lovely things we see', but in the very next line we learn that these emotions are 'hived in our bosoms like the bag o' the bee', and for a moment it is quite unclear whether they are, as the metaphor demands, stored after extraction, or whether as the grammar really suggests, they are somehow 'hived' at the moment of extraction. The past participle, hanging at a transition of tense, contributes to the awkwardness. The doubt of course is at the heart of the stanza – in youth we believe the emotions to spring out of the 'lovely things', but in age, the stanza tells us later, we learn that we have created them ourselves. This shift is packed into lines 3-4-5, and line 6 then draws the problem to the reader's attention – did you notice? did that surprise you? – 'Think'st thou the honey with those objects grew?' The tense at the end of the line slips now firmly back into the past, and we are returned to the world of loss, the negative. All the rhymes of this stanza, relatively close in relationship to play down the possible energy gained from emphasis at the line-ending, have been open, allowing the voice to diminuendo at its own speed, and that remains true even of the consonantal ending of the couplet, whether we pronounce it in Scots or English. But here the diminuendo effect is more strongly and unambiguously marked by the double rhyme or elision (I do not want to get involved in the argument about which it should be – the last line is almost impossible to decide on, because of the other possibility for elision in 'even') – the point is that the lines have an irresistible removal of emphasis at their end, and scarcely really 'end'. Here too the possible inversion of the effect of the answer 'No' is avoided in the introductory avoidance of a direct answer – alas. What is gone too has no universal pretension to it – it is 'only' the sweetness of a flower (the indefinite article here too is so right). In its very modesty, of course, lies the real sorrow.

Here what is lost is not, though it appears to be, a real Eden –

it is the ability to turn the world we know into an Eden which has gone. Loss as the human condition is, then, an experience of loss of real or hypothetical energy of the heart, spiritual energy, imagination – not the loss of a real or hypothetical country, or time, or object of love. Here loss is genuinely a state of mind, an inability to love. But even more importantly, this loss is carefully never allowed to monumentalise itself into the presence of loss, so to speak, it does not turn the experience of absence into the experience of presence, a lived experience of sadness into a literary experience of sentimental fullness. We are not allowed to dwell on the experience of loss for so long that it becomes an absolute and flips into a positive.

Of course in the next stanza Byron actually more or less makes this point explicit:

No more – no more – Oh! never more, my heart,
    Can thou be my sole world, my universe!
Once all in all, but now a thing apart,
    Thou canst not be my blessing or my curse:
The illusion's gone for ever, and thou art
    Insensible, I trust, but none the worse,
And in thy stead I've got a deal of judgment,
Though heaven knows how it ever found a lodgement.[16]

He returns to the vocabulary of single absolutes to describe how his emotions once seemed to him. But that absolute world of 'sole worlds', 'universes', 'all in alls', is now gone – not 'filled' however by absolute emptiness, not replaced by a chiasmic reversal of itself. No, the gap is somewhat covered by 'a deal of judgment' in exchange – and note too that half-way quantity, 'a deal'. Nor of course is it filled by the memory of a once complete, though now vanished, past, for the memory of completeness is qualified by the fact of memory. Taking the two stanzas together there is, however, a feeling for the joys of the past, but a feeling too that they are inevitably past. These two intertwine to give the sense that the nostalgic experience itself, if self-aware, affirms the true passage of time at the same time as it attempts to overcome it. And in tragic, ironic, sweetness it locates the passage of time in our human inability to possess our own experience.

NOTES

1. J. Logie Robertson, ed., *The Poetical Works of Sir Walter Scott*, Oxford, 1894, 100–2.
2. *Tintern Abbey*, ll. 15–16. The general sense of the transition from boyhood to youth is also paralleled in ll. 65–83.
3. 'History and knowledge . . . have always been determined . . . as detours *for the purpose* of the reappropriation of presence.' Or: 'All signifiers . . . are derivative with regard to what would wed the voice indissolubly to the mind or to the thought of the signified sense, indeed to the thing itself . . .'. Jacques Derrida, *Of Grammatology*, trans. G. C. Spivak, The Johns Hopkins University Press, 1976, 10 and 11. 'Absence' is as un-presence-able as any 'thing' else.
4. Scott, op. cit., 102.
5. All Byron quotations are from J. J. McGann, ed., *Lord Byron: The Complete Poetical Works*, Clarendon Press, Oxford. Hereafter McGann. Here vol. III, 1981, 275.
6. For discussion of the circumstances surrounding the composition of *Hebrew Melodies* see F. Burwick and P. Douglass, eds., *Byron and Nathan: A Selection of Hebrew Melodies*, North Carolina University Press, 1988; and of the national genre in particular, Thomas L. Ashton, ed., *Byron's Hebrew Melodies*, Routledge and Kegan Paul, 1972, 3–61.
7. *The Giaour, The Bride of Abydos, The Corsair*, and *Lara*. I have discussed these techniques in 'The Rhetoric of Freedom', in Alan Bold, ed., *Byron: Wrath and Rhyme*, Barnes and Noble and Vision, 1983, 166–85.
8. McGann, III. 293.
9. McGann, III. 308–9.
10. McGann, III. 297–8. The poem begins with a quatrain but proceeds in couplets. It is of 14 lines. The space of the quatrain rhymes is collapsed into the 'simultaneity' of the couplets, and the residual sonnet possibility is dissolved in this structural looseness — 'As twilight melts beneath the moon away' (l. 14).
11. For example Leslie Marchand's *Byron's Letters and Journals*, Murray, 1973–1982, letter of October 15, 1816, to Augusta, V, 119–20.
12. McGann, vol. II, p. 127.
13. McGann, vol. IV, 1986, pp. 35–40. Stanzas 8–11 are particularly to the point, in which Byron swings between the beauty of Lake Leman before him, and the memory of Augusta and the lake at Newstead in his memory. The

attempt to accommodate the memory to the present, and its failure, reaches towards an emptier nostalgia than the more public *Childe Harold* III.

14. McGann, vol. V, 1986, 77. This stanza and the one following it (i.e. 214 and 215) were added by Byron at proof stage. They were not however written *before* the subsequent stanzas, as is sometimes alleged, and were among the last stanzas added to the first Canto. Indeed the reprise of the train of thought from 213 is clear at 216. 214 and 215 essentially expand the thought of stanzas which though not in Byron's first draft predated them. See McGann, V. 665.

15. McGann, V. 589.

> Ah! – What should follow slips from my reflection:
>   Whatever follows ne'ertheless may be
> As àpropos of hope or retrospection,
>   As though the lurking thought had follow'd free.
> All present life is but an Interjection,
>   An 'Oh!' or 'Ah!' of joy or misery,
> Or a 'Ha! ha!' or 'Bah!' – a yawn, or 'Pooh!'
> Of which perhaps the latter is most true.
>
> But, more or less, the whole's a syncopé,
>   Or a singultus — emblems of Emotion,
> The grand Antithesis to great Ennui,
>   Wherewith we break our bubbles on the ocean,
> That Watery Outline of Eternity,
>   Or miniature at least, as is my notion,
> Which ministers unto the soul's delight,
> In seeing matters which are out of sight.

16. McGann, V. 77–8. Andrew Cooper puts Byron's refusal to polarise hope and despair nicely: 'Byron's skepticism is less a definite philosophic rationalism than a perpetual process of pragmatic adjustment.' *Doubt and Identity in Romantic Poetry*, Yale University Press, 1988, p. 145.

'The Island': Scotland, Greece and Romantic Savagery

ANGUS CALDER

The Island is Byron's last narrative poem, a tale set in the South Seas. It contains a passage very commonly cited as proof that Byron was still, or increasingly, attached to Scotland. It refers to the love between a Polynesian woman and a Scottish sailor:

> Both children of the isles, though distant far;
> Both born beneath a sea-presiding star;
> Both nourish'd amidst nature's native scenes,
> Loved to the last, whatever intervenes
> Between us and our childhood's sympathy,
> Which still reverts to what first caught the eye.
> He who first met the Highlands' swelling blue
> Will love each peak that shows a kindred hue,
> Hail in each crag a friend's familiar face,
> And clasp the mountain in his mind's embrace.
> Long have I roamed through lands which are not mine,
> Adored the Alp, and loved the Appennine,
> Revered Parnassus, and beheld the steep
> Jove's Ida and Olympus crown the deep;
> But 'twas not all long ages' love, nor all
> *Their* nature held me in their thrilling thrall;
> The infant rapture still survived the boy,
> And Loch-na-gar with Ida look'd o'er Troy,
> Mix'd Celtic memories with the Phrygian mount,
> And Highland linns with Castalie's clear fount.
>
> (Canto The Second, XII)

What is the real significance of this passage?

In April 1823 Byron wrote to the young Count D'Orsay, whose acquaintance he had just made, as a weary middle-aged man of the world:

132

. . . Though I love my country, I do not love my country-
men – at least, such as they now are . . . Live while you
can; and that you may have the full enjoyment of the many
advantages of youth, talent, and figure, which you possess,
is the wish of an – Englishman, – I suppose, but it is
no treason; for my mother was Scotch, and my name and
family are both Norman; and as for myself, I am of no
country.[1]

Byron was in what proved to be his last phase of intensive
creativity. He had not long finished a satire in couplets, *The
Age of Bronze* and was working still on the English cantos of
Don Juan. He had found time to compose *The Island* during five
weeks from mid-January to mid-February. Such variety of
subject matter and styles, and the pace at which he wrote
at this time, demonstrated his chameleon-like capacity. He
had not one literary personality, but several, and he developed them with reckless fluency. *The Island*, in parts, was as
sweetly lyrical as *Juan* was worldly-wise and acidulous.
Composing faces to meet the faces that he met, he presented
a very different Byron to D'Orsay from the one who had
written to Walter Scott just over a year before, exclaiming
that he carried the Waverley novels everywhere although
he already had them by heart, and who had insisted in the
Tenth Canto of *Juan* that he was 'half a Scot and bred a whole
one.'

It is not in Byron's letters, or in the few scattered references
to Scotland in his vast poetic output, that we can find evidence
to convince us that Byron was a Scottish poet, or, indeed, to
prove the contrary. He *said*, to D'Orsay at least, that he was not
a *traitor* to England when he attacked its prevailing values
because he was not *really* English anyway, but whether he *felt*
as identified with Scotland as he intimated to Scott is not
possible to estimate. This does not mean, though, that we
cannot follow T.S. Eliot, Leavis and other distinguished
literary scholars in arguing that the character of his verse is
largely, or wholly, explicable only in terms of his half-admitted
Scottishness.

There are three overlapping bases for such an argument.
One is provided by undisputed biographical facts. He was

brought up until the age of ten in Aberdeen by a mother proud of ancient Scottish lineage, who spoke with a strong Buchan accent. He was seduced by a Scottish serving girl when aged, probably, only nine. He attended schools in Aberdeen. In view of all this, his admiration for Burns and devotion to Scott involve more than his subscribing to current literary fashion: they would speak to and reach such a reader as they could not touch a purely English writer. Hence – our second basis – the arguments which account for features of his styles (I use the plural) by relating them to Scottish tradition are inherently plausible. I think that Tom Scott was unwise to accept T.S. Eliot's rather silly case that Byron writes English like an 'intelligent foreigner', and I am not sure that Eliot's more seductive notion that Byron's attacks on Southey are in the 'flyting' tradition of Dunbar and Montgomerie can be sustained as Tom Scott assumes – where is the evidence that Byron read Middle Scots verse? But Tom Scott's own point that Byron learnt how to use the Spenserian stanza not from *The Faerie Queene*, but from Thomson and Beattie, both eighteenth-century Scots writing in English, is firm enough, and his own experiments in transcribing passages from Byron into Scots are delightfully convincing.[2] The comparison of Byron with Burns was first made by Walter Scott; and when Byron wrote in enthusiasm about Burns's letters in his journal in 1813, he used terms which others have applied to him: 'What an antithetical mind! – tenderness, roughness, delicacy, coarseness – sentiment, sensuality – soaring and grovelling, dirt and deity – all mixed up in that one compound of inspired clay.'[3] Roderick S. Speer, who quotes this, has argued for Byron's Scottishness on a third basis, an ideological one. Irrespective of his subject matter, which is only very rarely Scottish, Byron displays throughout his work traits familiar in Scottish tradition. He believes in human fatedness. (We might call this belief Byron's 'secular Calvinism'.) The pressure exerted by this fatalism finds outlet in 'self assertive exuberance and extravagance'. (If we are doomed, we may as we have fun when we can: and why take life seriously at all?) Finally, Byron has an ambivalent appreciation of the 'antithetical' nature of human kind – 'soaring and grovelling, dirt and deity' – as Lyndsay and Burns

had had, as MacDiarmid and Grassic Gibbon would have after him.[4]

If this line of argument is right – and I think it is very plausible – Byron's specific references to Scotland in his last verse tale *The Island* are no more and no less 'Scottish' than his handling of 'The Prisoner of Chillon' or of English manners in *Don Juan*. His Scottishnness is assumed to be involuntary, the product of his earliest social conditioning in which balladry and the Bible intermingled. My own thrust here is to show that the high romantic sentiment attached to these Scottish references is interesting less because of its autobiographical slanting than because it shows Byron, like Scott before him, developing the discourse of poetic tartanry – but, being Byron, he gives this discourse an individual and rather extreme application.

The Island has not enjoyed much critical attention, still less esteem. Bernard Blackstone, who does consider it at some length, calls it 'the escapist poem *par excellence*.'[5] But Michael Cooke argues that the regression to infantile irresponsibility which Blackstone detected is in fact only on the poem's surface. The poet allows his young lovers to escape to an idyllic sanctuary from the sphere of civilised crime and retribution, but their absolution is 'not shared by Byron'.[6] The poem in fact presents two opposite versions of morality and does not attempt to reconcile them; like earlier verse tales by Byron (but unlike Scott's narrative poems), *The Island* creates moral suspense and refuses finally to resolve it. It offers a fictitious version of the fate of the *Bounty* mutineers, in which all perish at last, hunted down by British justice, except a young Scotsman who is rescued by a Polynesian Princess.

How seriously did Byron take this performance? He wrote *The Island* at a time when he was very serious indeed about producing more and more cantos of *Don Juan* in defiance of British reviewers, of the public's prudery, of his own friends' judgement, and that of his publisher John Murray. He had fallen out with Murray and meanwhile, with admirable self-sacrifice, shouldered as a kind of legacy from his beloved dead friend Shelley the burden of helping Leigh Hunt (who had come to Italy and was now dependent on him), and

Hunt's publisher brother John. He had contributed his great *Vision of Judgement* to the first number of *The Liberal*, a magazine edited by Leigh and brought out by John Hunt. At odds with Murray, he instructed his agent to give *The Island* to John, and also the last eleven finished cantos of *Don Juan*. One might infer from the circumstances that Byron reverted to the form of a short verse tale with a Byronic hero (Fletcher Christian) of his old type, mixing high romantic passages about love with exciting action and surprisingly firm patriotic, pro-British sentiment, in the hope that its sales might enable him to get the sponger Leigh Hunt, with his prudish wife and intolerable children, off his hands. He showed it to Leigh Hunt, who was critical, then coolly responded that he knew the poem was somewhat tame, but that he did not want to 'run counter to the reigning stupidity altogether – otherwise they will say that I am eulogising *Mutiny*.' It seems that he was courting sections of the public whose conventional moral views he despised. He presented a different, though equally cool face to his agent Kinnaird, sending the poem to him with the comment that it was too long for *The Liberal*, 'not good enough perhaps to publish alone,' but would 'make a respectable figure' in a future collection of his writings'.[7] (Nevertheless it was published alone a few months later.)

Byron had often belittled his own work, though, and it is easy to see what attracted him to the story of the *Bounty* Mutiny. He acknowledged two sources for his poem – Lieutenant William Bligh's own account of his voyage and the Mutiny (1790), and an interesting work on the Tonga Islands recently compiled by a London doctor named John Martin (1816), which would have given him copious information about Polynesian customs and language. A third element which he fused into the poem was his current obsession with the struggle of the Greek people for liberation. He left for Greece, abandoning even *Don Juan*, to serve that struggle in July 1823, just six months after writing *The Island*. The stanza which I quoted at the outset of this piece digresses, of course, from the South Seas to Greece. Let us see what he does with these three elements.

The case can quite easily be made that William Bligh, far

from being the brutal despot of Hollywood legend, was a humane commander by the standards of his time. His own account of his voyage to Tahiti to collect breadfruit plants for transmission to Jamaica, where plantation owners hoped to feed their slaves on the fruit, certainly exaggerates the good order prevailing before the sudden mutiny when the ship was homeward bound, on 28 April 1789. But Byron had every reason to hail him as a 'gallant chief' at the outset of his first Canto, which describes Bligh's abrupt arousal at dawn by the mutineers, and the grim scenes which preceded the division of the crew. Eighteen men, with Bligh, were forced into a small boat with scanty provisions; the courage and skill with which Bligh steered it 3900 miles to the Dutch East Indies are evoked in Byron's ninth stanza. His account follows Bligh's own closely. How could the creator of *The Corsair* have resisted the last exchange between Bligh and Fletcher Christian, the Mutiny's leader and (probably) sole instigator? Christian has 'seemed as if meditating destruction on himself and every one else.' Bligh upbraids him, reminding him of the kindnesses which he has shown him. To which Christian replies, ' "That, – Captain Bligh, – that is the thing; – I am in hell – I am in hell." '[8]

As for the twenty-five men left aboard, some were clearly reluctant to stay. Bligh is surprisingly generous when attributing motives to the mutineers proper: '. . . I can only conjecture, that the mutineers had flattered themselves with the hopes of a more happy life among the Otaheitans [Tahitians], than they could possibly enjoy in England.' The men had established 'female connections'; Polynesian women were 'handsome, mild and cheerful in their manners and conversation', and had 'sufficient delicacy to make them admired and beloved.' The local chiefs liked the sailors and pressed them to stay, even promising them 'large possessions'. Sailors who had no prospects of a better life on shipboard or at home in Britain 'imagined it in their power to fix themselves in the midst of plenty, on one of the finest islands in the world, where they need not labour.'[9] This seems excessively fair and essentially true. When he wrote his account, Bligh did not know the sequel.

Christian sailed the *Bounty* from the Tongan island off which

he had mutinied back to Tahiti. On the way, his ship called at Tubuai, one of a group known now as the Austral Islands (a French dependency). The natives were hostile, but the mutineers decided to settle there. They recruited twenty-four Tahitians, returned, and attempted to build a fort. Here Christian himself now faced mutiny, due to the shortage of women in his party, and his men voted two to one to return to Tahiti. After murdering scores of native islanders in a nasty punitive expedition, the *Bounty* men duly sailed back to the scene of their earlier fond 'connections'. Only nine of the twenty-five then then proceeded with Christian and some Tahitians to Pitcairn Island, where they were safe from British justice, but soon fell to murdering each other, till only one adult male, a white, was left alive – just sufficient to preserve a tiny patriarchal community. [10]

This was accidentally discovered by an American ship in 1808, and a British ship called at Pitcairn in 1814. So Christian's destination and death were known to many in Britain when Byron wrote his poem. However, he seems to have missed this new information. He believed that Tubuai, which he misspelt Toobonai, had been the last known port of call for Christian, and made that the scene of his poem. Furthermore, he took the liberty of assuming that the last four mutineers, including Christian, had been hunted down by an avenging British naval vessel. Something like this had indeed occurred on Tahiti, where Captain Edward Edwards had arrived in the *Pandora* nearly two years after the Mutiny, some of the *Bounty* men there had at once given themselves up, but a few had fled and had been pursued. Eventually ten men were brought home alive to stand court martial, six were sentenced to death, of whom two were pardoned. No gory last stand by four reckless men such as Byron recounts in his poem ever occurred.

There had been several Scots among the mutineers. McIntosh had been forced to stay on the *Bounty*, had begged Bligh to remember that he had no hand in the conspiracy, and was eventually pardoned. McCoy went with Christian to Pitcairn, where, as a former employee in a whisky distillery, he continued to manufacture some creature and was either pushed, or fell drunkenly, off a cliff. But the Norse name of

Byron's young hero, Torquil, and a reference to the Pentland Firth, suggest that Byron hazily supposed that Orkney was mountainous – for an Orcadian midshipman, George Stewart, was certainly Byron's model. He came from what Bligh called a 'creditable' parentage in Orkney, where Bligh himself had been warmly entertained by Stewart's people. He had a liaison with a high-born Tahitian, who died, of a broken heart it was said, after he had surrendered to Edwards. Their child was eventually brought up by missionaries. Stewart himself had perished in the shipwreck of Edwards' *Pandora* off Australia, with three other mutineers. Though he had shared whole-heartedly in the mutiny, and may even have put the idea into Christian's head, it suited a fellow midshipman, when he was eventually court martialled, to claim that the dead Stewart had been innocent. By 1831, when Sir John Barrow compiled what seemed to be a definitive account of the mutiny and its sequel, Stewart's purity of soul was taken for granted.[11] But Byron, very Byronically, makes Torquil defiant till almost the last, when just as pursuers close in for the kill, he is spirited away by Neuha, to a secret and very remarkable cave.

Such a cave as Byron describes existed, not off Tubuai, but in the Tongan group. He learnt about it in John Martin's *Account of the Natives of the Tonga Islands* (1816). Martin had never been to the Pacific, but constructed a readable and convincing book from what he was told by William Mariner who, as a young sailor aged fourteen on a privateer sailing in 1805, was spared when Tongan natives seized his ship and murdered most of the crew, was 'adopted' by an island chief, and lived for four years in the Polynesian fashion. His lord once went to the small island of Hoonga to shoot birds and rats. Thus Mariner came to enter a wonderful cave, invisible from sea or land because its only entrance was always below the surface of the ocean: a submarine passage eight or nine feet long led to a space some forty feet high by forty feet wide, with shelves on its sides where one could sit and lie. Here, as the hunting party drank *Kava*, Mariner heard the tale of the cavern's discoverer, a young chief diving for turtle, who kept its existence a secret. His island had a tyrant chief. When a lord revolted against him, all his family were condemned to death. The young

turtle-hunting chief loved one of the rebels' daughters and persuaded her to flee with him in a canoe to the secret cavern, where he sustained her with mats, dress and food until it was safe to carry her to Fiji.

Byron, despite transferring the cavern to another island group, sticks closely to Martin's working of Mariner's story which the learned doctor freely embellishes with European sentiment – 'How happy were they in this solitary retreat! Tyrannic power now longer reached them. Shut out from the world and all its cares and perplexities . . . themselves were the only powers they served . . . (etc.)' Byron adopts (Canto 4, Stanza 6) the comparison, which Martin attributes to Mariner, of the Cavern's interior to that of a Gothic cathedral.[12]

He also transferred to the Austral Islands some Tongan words from the lexicon provided by Martin in an appendix, and, in another act of 'poetical liberty', self-confessed in a footnote, one of the songs set down from Mariner's memory. Anyone wondering why Macpherson's 'Ossianic' writings had carried such conviction sixty years before will find a kind of explanation here. Contemporary conventions dominate 'translation' at any one time. 'Ossian' might well have looked similar had it been wholly genuine. Witness the process by which an orally transmitted song, given in Tongan and word-by-word translation in Martin's appendix, is mediated through Martin's age-of-sentiment prose into Byron's freely Romantic version, 'sung' by Neuha to Torquil in three stanzas at the start of his second canto. Byron claimed that he had 'altered and added but . . . retained as much as possible of the original'.

The would-be literal:

As our minds (are) reflecting the great wind whistles towards us from the great (lofty) Toa trees in the inland upon the plain.
Is (to) me (the) mind large, beholding the surf below, endeavouring in vain to tear away the rocks firm. . . .

becomes in Martin's version in the main text:

. . . The whistling of the wind among the branches of the lofty *toa* shall fill us with a pleasing melancholy: or our minds shall be seized with astonishment as we behold the

roar of the surf below, endeavouring but in vain to tear away
the firm rocks . . .
then Byron's:

> And we will sit in twilight's face, and see
> The sweet moon glancing through the tooa tree,
> The lofty accents of whose sighing bough
> Shall sadly please us as we lean below;
> Or climb the steep, and view the surf in vain
> Wrestle with rocky giants o'er the main.

Anyone who has heard Ravel's *Chansons madécasses* well sung
knows that in the rhetoric of Byron's period, serious pro-
'native', anti-colonial feeling may lie locked – Ravel in the
1920s set Evariste Parny's late eighteenth century 'Madagas-
can' prose poems and elicited from them keen lyricism,
shocking by fierce protest. Bernard Smith, in his authoritative
work on *European Vision and The South Pacific*, approves of
Byron's imaginative projection of Neuha, his Polynesian
Princess: she and Queequeg in *Moby Dick* are 'literary conventions, but they are conventions within which it was possible
for poet and novelist to go much farther in their endeavour to
understand the native and the native point of view'.

Farther, that is, than was permitted by the eighteenth-
century stereotype of the Noble Savage, present in Parny's
poems as in Diderot's *Supplement to Bougainville's Voyage*. By
the early nineteenth century, missionaries, professionally
committed to writing about and depicting pagans as ignoble,
had created a potent pejorative rhetoric. In the hands of the
evangelical poet James Montgomery, in his *Pelican Island*
(1827), the ignoble savage looks not altogether unlike Byron's
corsairs and Turks

> Their features terrible; when roused to wrath
> All evil passions lightened through their eyes . . .
> Their visages at rest were winter clouds
> Fix'd gloom, whence sun nor shower could be foretold:
> But, in high revelry, when full of prey,
> Cannibal prey, tremendous was the laughter.

Smith argues that the *romantic* savage, in poetry, fiction and

art, drew valuably on both the 'noble' and the 'ignoble' stereotypes. The romantic savage combines a love of freedom and a devoted patriotism with a temperament reacting 'violently and immediately to experience.' He is brave, emotionally profound, childlike, warmly generous. Smith shows how such a convention could be related to the cult of the Greeks which Byron promoted and exemplified. The artist Augustus Earle, travelling in the Antipodes in the late 1820s, despised Australian aborigines as 'the last link in the great chain of existence which unites man with the monkey', but thought the Polynesian Maoris perfectly beautiful; they reminded him of the Greeks of Homer's day, standing on the threshold of a glorious future. In the same year another traveller, D'Urville – the very Frenchman who had persuaded the French government to buy the Venus de Milo when stationed in the Aegean on survey work in 1820 – compared the Maori *pa* with the Greek *polis* and meditated on the succession of races which had emerged from obscurity to play brilliant roles on the world stage. Once Gauls and Britons – now Russians and North Americans – next, perhaps, Maoris united by a great lawgiver.[13]

The comparison of savage and Greek went back at least as far as J.J. Winckelmann's *Thoughts on the Imitation of Greek Works in Painting and Sculpture*, an influential text translated from German to English by Henry Fuseli in 1765 just as, not wholly by coincidence, 'Ossian' was beginning to make an impact. Winckelmann thought that the lithely athletic American Indian was a living equivalent of Homer's Achilles. The Greeks followed nature in their dress. Their arts imitated nature and were characterised by 'noble simplicity'.[14] Byron echoes Winckelmann when writing about the Apollo Belvedere in *Childe Harold* (Canto IV CLXI-CLXIII). He might seem in *The Island* to echo Robert Wood's *Essay on the Original Genius and Writings of Homer* (1769), which declared that 'while manners were rude, when arts were little cultivated and before science was reduced to general principles, poetry had acquired a greater degree of perfection than it has ever since obtained'.[15] For this is what Byron says in Canto the Second, Stanza V after Neuha has rendered her 'song of

Toobonai'. A 'simple ballad' outweighs in effect the monuments of empire:

> . . . The first, the freshest bud of Feeling's soil.
> Such was this rude rhyme – rhyme is of the rude –
> But such inspired the Norseman's solitude
> Who came and conquer'd: such, wherever rise
> Lands which no foes destroy or civilise
> Exist: and what can our accomplish'd art
> Of verse do more than reach the awaken'd heart?

C.M. Woodhouse has suggested that Byron's greatest gift to the pan-European philhellenic movement was to get people interested in *modern* Greeks, in people as well as antiquities. He studied their literature and refuted the views that it was non existent and that their speech was debased. He befriended them while remaining cheerfully ready to acknowledge their human failings.[16] What is involved in his treatment of Torquil and Neuha on the one hand, and Fletcher Christian on the other, is an ideological manoeuvre of great complexity. He invokes an ideal antique Greece and conveys it, via British and Polynesian figures, to the aid of the freedom struggle of living Greeks in 1822. Scotland, Gaeldom, Norse-ness, I submit, are there as a lubricant, rather than for their own sake, though they facilitate, besides Philhellenism, the exploration of other themes which preoccupied Byron.

In the first Canto, Byron's 'antithetical mind' sets up a complexly paradoxical relationship between the mutineers in general and the conception of a golden age. Amplifying cues from Bligh, Byron sees them as attracted back to Tahiti by 'the care of some soft savage', and by the liberty and equality possible in a society without money, a '. . . general garden, where all steps may roam/Where Nature owns a nation as her child.' We notice that, as in Byron's praise of primitive song, a notion of progress is subtly admitted – such a notion as D'Urville played with in New Zealand. The Norsemen, inspired by 'rude song', came and conquered, carrying world history forward. Tahiti is a child *nation*.

The British mutineers are conquerors, but from a corrupted

civilisation. The island is 'gentle', but these men are 'wild' and seek 'repose' through other's 'woes', when they expel Bligh and his companions naked on to the boundless ocean. Christian's dreadful reply to Bligh – 'I am in hell! In hell!' – shows that like Conrad the Corsair, or Alp the renegade besieging Corinth, he represents a quasi-heroic mentality torn by the conflict between conscience and will – 'volumes lurked below his fierce farewell.' However, the will to return to the island of kindly Nature, 'the goldless age where gold disturbs no dreams', as Tahiti was before Europeans 'bestow'd their customs' and 'left their vices', is presented at the Canto's end as paradoxically positive: 'And yet they seek to nestle with the dove/And tame their fiery spirits down to love.'

In the second Canto we find that Neuha and Torquil have achieved a love which represents an ideal fusion of European and Tahitian values, but on the basis of similar childhood conditioning by seas and mountains. Neuha is 'in years a child' though 'in growth a woman'; she is voluptuous yet faithful, energetic but selfless. Torquil has 'taught her passion's desolating joy' – the lovers are suspended until the end of the poem in the ecstasy of love's early stages, but this, as readers of *Don Juan* know, cannot last. Neuha is innocent as a lake before an earthquake arrives to

> . . . tear the naiad's cave,
> Root up the spring, and trample on the wave
> And crush the living waters to a mass,
> The amphibious desert of the dank morass.
> And must their fate be hers? The eternal change
> But grasps humanity with quicker range
> And they who fall but fall as worlds will fall,
> To rise, if just, a spirit o'er them all.

I am intrigued by the way in which Byron flirts with Christianity in that last line, and would be interested to see an interpretation of the poem in terms of the intimations of immortality which it proffers. [17] But here I am concerned with what he does with Scotland.

Torquil, being 'Hebridean', is attuned to mountains and wind. He is naturally bold – if born in Arabia he would have

been a rover, if fixed in Chile, a proud Indian chief – and 'On
Hellas' mountains, a rebellious Greek . . .'. But this very
energy, had he been 'bred to a throne', might have made him
unfit to reign; it is admirable in its place, even when seizing
thrones, but preys on itself if 'rear'd' to rule. Torquil represents the human capacity and appetite for power:

> A soaring spirit, ever in the van,
> A patriot hero or despotic chief,
> To form a nation's glory or its grief.

Here on Toobonai he has momentarily been *tamed* by Neuha,
into 'a blooming *boy*, a truant mutineer.'

We can now return to my opening quotation. Neuha and
Torquil are both '*children* of the isles'. Byron in youth had
written that curiously viable lyric which begins 'I would I were
a careless *child*/Still dwelling in my mountain *cave*.' Now he
associates his own early Highland memories on the one hand
with the fragile happiness of Torquil and Neuha – 'the half
savage and the whole'; on the other with the sublime scenes of
Greece which he can still behold with 'infant rapture' because
of his early love of mountains. So far Greece represents *nature*,
which taught Byron to 'adore' its landscapes. A few stanzas on
he plays with nature-worship – 'who thinks of self when
gazing on the sky?' – which would seem quasi-Wordsworthian,
did not Byron connect this with the carnal ecstasy of the
devoted lover.

Byron had been writing *Juan* for years and was wholly
practiced in his 'Juanist' poetic personality, which permitted him to move rapidly between the loftiest romantic
sentiment and the coarsest cynicism. *The Island* differs from
earlier exotic tales in that it contains 'Juanist' transitions. A
bluff English sailor's tobacco wafts into paradise, and an
English voice calls 'what cheer', Ben Bunting has come to warn
Torquil that a British ship has arrived to apprehend the
mutineers. And Torquil at once reacts with Gaelic-Norse
martial valour:

> '. . . We will die at our quarters like true men'.
> 'Ey, ey! for that 'tis all the same to Ben.'

Torquil appeals to Neuha:

> '. . . Unman me not; the hour will not allow
> A tear; I'm thine whatever intervenes!'
> 'Right' quoth Ben; 'that will do for the marines.'

Byron's note here reads: 'That will do for the marines but the sailors won't believe it is an old saying.' Ben's colloquial understatements, and his prosaic pipe, make Torquil's heroics seem ridiculous. A canto which starts in the Golden Age ends in bathos. What is Byron up to?

Well, if his practices work – and some might question whether they do – the effect is to modernise and humanise the fierce, 'half-savage' will for action and power which Torquil represents, and which is needed to free present-day Greece. The other mutineers have been 'crushed, dispers'd or ta'en' by the time the third Canto opens, but Torquil, Ben and Jack Skyscrape are still at large with Fletcher Christian. In an impressively sombre passage, intensely 'Byronic' as opposed to 'Juanist' we are told that the clubs and spears of their native allies have been useless against British guns; purely 'savage' courage, even that of Homer's heroes, cannot prevail in the present day.

> Even Greece can boast but one Thermopylae,
> Till now, when she has forged her broken chain
> Back to a sword, and dies and lives again!

Thermopylae will recur in the fourth Canto. When Torquil has safely escaped, thanks to to his Princess, to the secret cave, his three comrades make their last stand:

> '. . . and with that gloomy eye,
> Stern and sustain'd, of man's extremity,
> When hope is gone, nor glory's self remains
> To cheer resistance against death or chains, –
> They stood, the three, as the three hundred stood
> Who dyed Thermopylae with holy blood.
> But, ah! how different! 'tis the *cause* makes all,
> Degrades or hallows courage in its fall.

O'er them no fame, eternal and intense,
Blazed through the clouds of death and beckon'd hence;
No grateful country, smiling through her tears,
Begun the praises of a thousand years . . .

They will not be immortalised in Homeric epic, nor even perhaps in 'songs of Toobonai'. Christian, conscious of shame and guilt though he is, remains defiant till his last shot is expended, then plunges from a cliff and his body is 'crush'd into one gory mass', an inhuman shapelessness.

Torquil and Neuha live on to feast with the chiefs of 'the yet infant world'. We know from Byron that childlike love, like the primitive spear, is doomed to fail, the first for existential, the second for historical reasons. But Greece can be freed if virtues of the Romantic Savage – selflessness, patriotism, courage – can be wedded to animal tenacity like Christian's, and will to power like Torquil's. The guilt-making *violence* required for success will be sanctified by the primitive innocence displayed at an extreme in the 'childhood cave'. The freshness of natural man is endowed to the cause of Hellas by deft ideological elision. Torquil is a convenient agency for this gift, evoking Ossian's heroes and the courage of Highlanders at Culloden: the Gael transformed from doomed Noble Savage to potentially world-sweeping Romantic Savage (or Highland Soldier, helpful at Waterloo). Byron's ideological sleight could be summed up like this:

Ancient Greeks were romantic savages launching from Golden Age into triumphant nationhood with patriotic heroism, in the era of chariots and spears.

They dwelt with sea and mountains common to the formation of Polynesian and Gaelic-Norse Scot and their values: the narrator, claiming his own share of mountains, is the reader's window on to the landscapes of modern Greece, where mountain-sea, heroic-patriotic values are innate or latent.

Neuha's selfless (and triumphant) love represents innocence, will to sacrifice. This is a necessary ingredient in a heroic cause . . .

But so is Torquil's tenacious will to power, allied till almost

the last with Christian's untameable animal courage, not yet perverted, however, by Christian's anti-social self-consciousness and psychic torment.

Torquil connects in turn with the bluff martial competence of the modern British tar, Ben Bunting . . . Whose appreciation of tobacco is shared by the narrator . . . The latter's ardour for the Greek cause is therefore not that merely of a feckless romantic. But he sees that the Homeric valour and innate patriotism of the savage are necessary to ennoble and inspire the struggle.

I wrote of Byron's 'tartanry' for shorthand purposes. He takes Scott's successful (*Lady of the Lake*) romanticisation of Gaeldom off on his own, libertarian route. He may have despised tartan. *The Age of Bronze*, the topical satire which he had just finished when he began *The Island*, ends with raucous merriment over Sir William Curtis, London Alderman, donning the tartan on his trip to Edinburgh with George IV – that royal visit which Scott stage managed. MacDiarmid could hardly have been sharper:

> My muse 'gan weep, but, ere a tear was spilt,
> She caught Sir William Curtis in a kilt!
> While throng'd the chiefs of every Highland clan
> To hail their brother, Vich Ian Alderman!
> Guildhall grows Gael, and echoes with Erse roar
> While all the Common Council cry 'Claymore!'
> Too see proud Albyn's tartans as a belt
> Gird the gross sirloin of a city Celt,
> She burst into a laughter so extreme,
> That I awoke – and lo! it was *no* dream!

But a touch of Ossian made it possible for Byron to transcend the fatalistic *impasse* of the wonderful Haidée episode in *Don Juan*.

Haidée's rescue of Juan and their free, quasi-childlike love prefigure the Neuha-Torquil relationship. Her father's combination of piracy with patriotism anticipates the collusion of raw energy with a good cause which *The Island* will implicitly advocate. 'Valour was his and beauty dwelt with her' – and both, Byron tells us, are celebrated in 'rude' song. But Haidée's

death from horror and grief makes it hard to fuse her natural generosity with a successful, modern, heroic cause. The resourceful Neuha survives, having negotiated the first phase of coming to terms with modern reality in the era of triumphant commercial imperialism. Moral suspense is not resolved, unless by sleight of hand: Torquil's energy remains ambiguous, and Byron refuses to judge whether Christian has gone to hell. But neither libertarian belligerence nor natural heroism have been undermined. The option of idealistic struggle for nationhood remains open. Though we know by the end of the poem that 'civilisation' since 1790 has destroyed Polynesian Golden Age values, these are directed, along with those of Homer and 'Ossian', towards battle with the oppressive modern Turk.

NOTES

1. *Byron's Letters and Journals*, ed. Leslie A. Marchand, (1973–), vol. 10, 156.
2. Tom Scott, 'Byron as a Scottish Poet' in A. Bold (ed), *Byron: Wrath and Rhyme*, (1983), 17–37.
3. *BLJ*, vol. 2, 376–7.
4. Roderick S. Speer, *Studies in Scottish Literature XIV* (Columbia, South Carolina), 196–206.
5. Bernard Blackstone, *Byron III: Social Satire, Drama and Epic* (1971), 43.
6. M.G. Cooke, *The Blind Man Traces the Circle* (1969), 211.
7. *BLJ*, vol. 10, 89–90, 117–118.
8. William Bligh, *A Voyage to the South Sea . . .*, London: George Nicol, 1792 edn. 154ff.
9. Ibid., 162.
10. I follow the careful, demythologising account of the Mutiny in Gavin Kennedy's *Bligh*, 1978.
11. Sir John Barrow, *The Eventful History of the Mutiny and Piratical Seizure of HMS Bounty* (London, 1831), 88–9.
12. John Martin MD, *An Account of the Natives of the Tonga Islands . . .*, (Edinburgh, 1827, 3rd edition), I, 216–224.
13. Bernard Smith, *European Vision and the South Pacific* (1985, 2nd edition), 317–332.
14. Lorenz Eitner, ed. *Neoclassicism and Romanticism 1750–1830* (1971), I, 4–13.

15. Quoted in Hugh Honour, *Neo-Classicism* (1968), 64.
16. C.M. Woodhouse, *The Philhellenes* (1969), 10, 40–45.
17. There are others in 1:6, 2:16, 4:1, 4:3–4, 4:6; the association of the secret cave with the grave (its exterior) and a Gothic church (interior) reinforces this element, and could be seen as a key to the whole poem.

Byron landing from a Boat

MICHAEL REES

The only illustration to Moore's life of Byron in 1830 was an engraving of Sanders' oil painting, which shows Byron landing from a dinghy. This portrait, probably begun in 1807 when the poet was nineteen (as Moore's caption indicates), has contributed greatly to his image as a Romantic wanderer. Attended by Robert Rushton, the 'little page' of *Childe Harold's Pilgrimage*, Canto I, and son of a tenant at Newstead Abbey, Byron gazes defiantly and moodily into the distance, personifying the future Harold or Byronic hero, against a background of sea and cliffs which suggest the Highlands of Scotland.[1]

Artist and subject alike had Scottish origins. George Sanders (1774–1846) was born at Kinghorn, Fife, and educated in Edinburgh, but went to London in 1805 and became fashionable among the nobility: by 1806 he already had to refuse sitters. Byron, whom he painted several times, apparently commissioned the portrait of himself alone in 1807. A full-length engraving of this formed the frontispiece to the Moore-Wright collected edition of Byron in 1832, although W. Finden's *Illustrations of the Life and Works of Lord Byron* in 1834 included only a half-length engraving 'at the age of 17'. The poet might not have approved. In 1812 he made John Murray destroy a poor engraving from a Sanders miniature of him, intended for a new edition of *Childe Harold*, since 'the frontispiece of an author's visage is but a paltry exhibition'.[2]

In April 1809 Byron brought Rushton to London. It was doubtless then, at the studio in Vigo Lane[3], that Sanders completed the painting, perhaps copied from the earlier version, and added Rushton beside a dinghy with a larger boat behind and water, steep rocks, and clouds.

Byron had left Scotland in 1798 at the age of ten and never

returned, although *The Records of Invercauld* quote the gillie who claimed to have accompanied him to the summit of Lochnagar in 1803[4]. In August 1807 Byron revived his plan to revisit Scotland; he would travel to Edinburgh by carriage, tour the Highlands and take a boat to the Hebrides.[5] Nothing came of these plans, but a nostalgic allusion to them and to the days of 'Auld Lang Syne' may have lain behind his instructions to Sanders. Before leaving England for the Levant with John Cam Hobhouse in June 1809, Byron paid 250 guineas for the work, a canvas measuring 44½ by 35⅛ inches. By 1811 this was Sanders' standard fee for a full-length portrait.[6]

In letters to his mother from Falmouth, Constantinople and Greece, Byron referred five times to the painting, forever expected from Sanders, 'a noted limner'.[7] At last Catherine Gordon Byron announced its arrival 'after a *great* deal of trouble. Saunders [sic] said he kept it to show as an honor and credit to him, the countenance is *angelic* and the finest I ever saw and it is very like. Miss Rumbold (Sir Sidney Smith's Daughter in law) . . . fell quite in love with it . . .'[8] Byron's supposed letter to his mother of 1 July 1810, stating that the Sanders portrait 'does not *flatter* . . . but the subject is a bad one', is a forgery.[9] During Hobhouse's return journey from Greece, where he had left Byron in July 1810, he wrote to the latter from Cadiz: 'General Graham commander in chief here has seen your full length at Saunder's [sic] – he was praising it very much indeed – I could not help saying ["] I am glad you like that picture so much for it is mine" – which you know it is, for you gave it me . . .'[10] Here was a further Scottish link; Thomas Graham (1748–1843), later Baron Lynedoch, had come to Cadiz as lieutenant-governor to command the British defence troops.

Mrs Byron died in 1811, but when Newstead Abbey was sold abortively to Thomas Claughton in 1812–4 the Sanders painting was still there, since Byron's butler, 'old Joe' Murray, packed it up and kept it dry and 'in great perfection'.[11] In due course Hobhouse received the picture, of which a greatly inferior copy still hangs at Newstead. It was inherited in 1869 by his eldest daughter Charlotte, Lady Dorchester, who bequeathed it in 1914 to King George V. Part of the Royal

Collection since then, the Sanders portrait was lent by H.M. The Queen for the important Byron exhibition at the Victoria and Albert Museum in 1974.[12]

NOTES

1. See also David Piper, *The Image of the Poet: British Poets and their Portraits* (1982), 127.
2. Letter to John Murray, 23 October 1812. *Byron's Letters and Journals*, ed. Leslie A. Marchand (1973–82), vol. 2, 234.
3. Letter to Mrs Catherine Gordon Byron, 24 May 1810. *BLJ*, vol. 1, 243. The Sanders portrait is the colour frontispiece in this volume.
4. The Revd John Grant Michie, M.A., Minister of Dinnet, *The Records of Invercauld 1547–1827* (Aberdeen, 1901), 389.
5. Letter to Elizabeth Bridget Pigot, 11 August 1807. *BLJ*, vol. 1, 131–2.
6. Geoffrey Wills, 'A Forgotten Scottish Painter', *Country Life* (8 October 1953), 1120.
7. From Falmouth, 22 June 1809 (*BLJ*, vol. 1, 206); from Constantinople, 24 May and 28 June 1810 (ibid., 243, 251); from Athens, 20 July, and Patras, 2 October 1810 (*BLJ*, vol. 2, 4,18).
8. Doris Langley Moore, *Lord Byron: Accounts Rendered* (1974), 129.
9. *BLJ*, vol. 1, 272.
10. *Byron's Bulldog: The Letters of John Cam Hobhouse to Lord Byron*, ed. Peter W. Graham (1984), 52.
11. D. L. Moore, op. cit., 473.
12. Anthony Burton and John Murdoch, *Byron: An exhibition to commemorate the 150th anniversary of his death* (1974), 9. (Exhibit A41.)

On Singing 'Dark Lochnagar'

SHEENA BLACKHALL

I must be one of the few Scots who came to learn of the connection between Dark Lochnagar and George Gordon Byron by reading it on a tea-towel in a Ballater gift shop. Until then, I had assumed (by the number of Deesiders who sang it, who'd never heard of Byron) that it was a traditional North-East ballad, rising from the common ancestral pool, best sung at 5 a.m. on a Hogmanay morning, on looking proudly and blearily at the mountain from any one of a number of Deeside cottar windows, mid-down a bottle of the whisky of that name.

My father could never cross the Border without breaking into a rendition of Dark Lochnagar, with the verve of Bruce accosting de Bohun from the stirrups, axe raised, with a heavy emphasis on 'England, thy beauties are tame and domestic' – lest we were in danger of forgetting the fact. For of course, Lochnagar is our family property, grudgingly shared with others of Deeside extraction, Byron included (his maternal bloodline dips back to Aboyne), and nobody else's business.

Beside it, other mountains wither into inconsequence. I know this, having travelled – not extensively, but travelled none the less – in places as diverse as Europe, England and Fraserburgh.

I found the Eiger as cold and moth-eaten as Miss Havisham's inedible wedding-cake. I found it as populous as downtown Shanghai, its *pistes* traffic-jammed with tourists, its snow as shop-soiled as Woolworths' floor after a sale. I thought it crude Kentucky moonshine, chocolate-box saccharine, a phallic meringue of a mountain, which promised great things but whose ecstasies were small. On its lower slopes, the noblest beast it could boast was a curly-topped Swiss cow,

looking remarkably like a Heidi-Shirley-Temple, ringing its bell like a till opening and shutting incessantly.

In Ireland, try as I might, I couldn't find *any* mountains, merely a quilt of soggy green mammaries. 'The Irish', Father remarked as the guide extolled in song the umpteenth supposed beauty spot, 'will sing about *onything!*' Whereas we, of course, were particular. We sang about Lochnagar. Norman MacCaig once said that God was Mozart when he wrote Cul Mor. I have news for him. Lochnagar *is* Mozart.

But fancy George Gordon writing the song, Mistress Gordon of Gight's 'Crooked Deevil'! For years Crutchie Geordie, or Lord Byron, to give him his English title, was the moody, frozen man on the plinth outside my brother's school, whom my mother nodded to, never omitting to mutter 'Byron' conversationally as we trotted past. Not once did he have the good manners to reply. As a very small child, I remember thinking that extremely rude, and wondering why he let pigeons sit on his head, and did he ever get down to stretch his legs and walk about a bit?

Later, I discovered that Byron climbed my Lochnagar in the summer of 1803, from Invercauld, going by the Garrawalt Glen, by the crags of Loch an Uan, resting frequently but, typically, rejecting help. It wasn't his first sight of the mountain, he was familiar with it from the age of eight, from holidays at Ballaterach.

From the age of nothing onwards, I too spent summers on Deeside. By the time I was eight, I had officially adopted Lochnagar as my big brother cum playmate. I was always pleased to see him, and I was sure the joy was reciprocal; we spent hours of fun together in all weathers. Every night, before dropping off to sleep, I would say my round of 'goodnights' as only children and 'The Waltons' do, and always ended with 'Goodnight Lochnagar, goodnight Dee', because you should never omit your friends.

I grew very possessive of my mountain, and was rude and discourteous to the assortment of Japanese, American and people from Middlesbrough who asked directions to him. I would feign complete ignorance of English, and give incom-

prehensible replies in deepest Doric, throwing in a string of Gaelic oaths I had learned from an Inverness cousin of poor moral fibre. Later still, when behind the wheel of a car, I would pretend to be ignorant of passing places, and block the winding, tendril-thin road to give a quick paean of song to my friend, a chorus or two of Dark Lochnagar in his honour, thus obliging the Dutch, Hindi, Zulu, English or whatever to veer off the road into a bog, or reverse eight miles and go home, wherever home was.

For of course he is no mere mountain, he's also a panacea, the family medicine-chest. When any of us fell ill, my father would listen indulgently to the doctor's diagnosis and as soon as he left would say, 'Whit they *really* need's a wee whiff o' Lochnagar!' And a whiff of Lochnagar we would get. And what's more, it worked!

I like a piece of my mountain by me. It's nice to have friends near at hand. I squeeze my little Lochnagar stone when I'm anxious, and I squeeze it for comfort when I'm sad. It's the most valuable thing I own.

So, when I sing Dark Lochnagar, I don't sing for an audience. I don't sing for Byron, and I certainly don't sing for myself. When I sing Dark Lochnagar, you see, I'm singing for a friend.

Afterword

J. DRUMMOND BONE

Perhaps there is room for an afterword of gentle Byronic scepticism. The papers of this volume have as their reason for being a project to characterise Byron's Scottishness. I rather wonder if in the event they even try to do that in any very serious way. What I really wonder is whether this project could ever be anything more than a 'design-governing-posture', a useful fiction with which to compose one's thoughts. If there is a connecting tissue in what has gone before, and I do sense one, it seems to me rather to be the issue of Byron's political seriousness. National identity is a difficult and slippery concept, and there are dangers for both the object of thought and the thinker in its invocation. Byron's idea of nationality had not had to be tested by the consequences of the organic mysticism which was to invoke his name only a few years after his death. Perhaps too these consequences are never tested by the nationalism of a relatively small country such as Scotland, which can therefore afford to indulge itself – it is nationalism by opposition rather than by assumption. But all this is truism. The point which this collection begins to make almost against its own will is that Byron's characteristic way of being Byron is never serious enough to be dangerous by way of abstract identities, whether these are national or political. It *is* unserious enough to be wildly dangerous to those who think in terms of discrete identities, essences, purities, and polarities. Its unseriousness consists, for a start, in its inability not to be serious. Byron and his poetry refuse the dialectic of serious and unserious – and this is deeply worrying for those who believe they have a privileged access either to the serious or to the cynical.

It is entirely characteristic that Byron's so-called 'Luddite Song' has not yet been published *as it was written* outside of

157

editions of Byron's letters. For *pace* David Craig it was not suppressed until six years after his death – for fear of its Luddite subversiveness – but has effectively been suppressed until this day, and regrettably one has to include McGann's edition in this suppression, because it subverts a deeper decorum of seriousness and cynicism, of the politically committed and the personally scurrilous. The three stanzas of the Luddite song are bracketed by two other stanzas in the letter in which they occur, but in order to leave this poem as distinct and unambiguous they are conveniently displaced in all 'editions'. These are the stanzas beginning: 'What are you doing now,/Oh Thomas Moore . . . Sighing or suing now,/ Rhyming or wooing now . . . /Which Thomas Moore?' It is really 'not on' to appropriate Byron to a political identity, and then to blame him for not living up to it. But this stubborn refusal to be appropriated is uncomfortable – highly so.

But of course there is a different kind of unseriousness too, the opportunistic unseriousness of which Andrew Noble writes – the Byron cynically exploiting the market Scott had created for the sensationalist tale. That there is an element of escapist mannerism in these is difficult to deny. There have been recent attempts, notably by Daniel Watkins in *Byron's Eastern Tales* and by Marilyn Butler and Jerome McGann to read these as political allegories, or near allegories. Even if one is convinced by these arguments it would be impossible to say that the poems were redeemed by this however, since it has gone almost totally unrecognized from their publication until now. But are we not being misled by another impulse to impose unity on the man who 'must contradict himself on every page'? If these contradictions were easy, if they were always in the control of the writer, they would not really *be* contradictions. The 'mobility' of which Byron writes in *Don Juan* XVI is a 'most painful and unhappy attribute' – though witnesses to it can be misled by its seeming 'facility'. The honest writer, Byron seems to mean, does not even write his own contradictoriness, rather he is written by it. And that 'honest' can be misleading too, for the honest writer will also be written by his own dishonesties. To be committedly in bad

faith is part of the cant of consistency. This is not to defend the Turkish Tales from a charge of literary expediency. On the contrary, I feel that to attempt the defence is to interpret in bad faith if the defence is the discovery of a unified allegorical politics, while on the other hand, if the attack on these poems is seen as a description of some 'core' failing in Byron or his poetry, then this too is to avoid the difficulty of living inconsistently. It is difficult again to accept what Byron enacts – that consistency and total moral seriousness is *easier* and more cynical than inconsistency and moral dubiety.

Am I attributing a Romantic view of unity to my colleagues in this book when in fact they would disown it, and claim that particular politics or particular nationality have no essential quality in their usage? Perhaps. It would indeed be a Romantic view of language if we said that words themselves were reductionist. But it is also a characteristic of Romanticism to will the infinite into the particular, and thereby to envalue the word (while arguably silencing it). It might be inaccurate therefore to say simply that a national identity reduces the person – this would only be true if we believed the person to be a being greater than his or her historical self, and if we believed *that* we are also likely to believe that national identity is greater than the historical nation. The failure of words depends on the expectations laid on them. On the other hand not to realise the reductionist possibility is not to have any expectations of insight. But then once again is it perhaps wrong to feel that the project of identity is surreptitiously always transcendent? But if it is not, why is it so important to those who follow it? *Is* there a way out of all this? Perhaps the whole game of paths through identities to higher levels of being, or of the reduction of these higher levels through the discriminations of words, is not one which Byronists should really play, either for or against. If we love him, we should let him be – certainly not ourselves in nation or politics, not accountable to our understanding of the serious or the cynical – 'only' someone whose voice we fondly hear.

Index

Aberdeen, xvii, 7
Aberdeen, Lord, 94
Aberdeenshire, xi–xii
Albania, Byron and, 36–7, 38

Baillie, Joanna, 55
ballads, xii, xvii
Barbour, John, 81
Beattie, James, 82
Beatty, Bernard, 114
Beyle, Henri (Stendhal), 57
Blackstone, Bernard, 135
Blake, William, 18, 39, 44
Blessington, Lady, 67–8, 69, 84
Bligh, William, 136–7, 144
Boswell, James, 38, 40
Brougham, Henry, 108–9, 111–12, 117
Bounty mutiny, 135, 136–9, 143–4
Burns, Robert, 19–20, 38, 39–40, 83, 115–16, 134
Butler, Marilyn, 34–5, 158
Byron, Allegra, 47
Byron, Catherine Gordon, xv, 97, 152
Byron, Lady (Annabella Milbanke), 53, 55
Byron, Lord (George Gordon)
  and morality, 4, 39, 49, 95, 135, 136
  and politics, xiv–xviii, 2–3, 24, 33–34, 36–8, 40
  and religion, xiii–xiv, 3–4, 39, 40, 44–9, 59, 79, 134, 144
  and Scotland, xi, xvii, 2, 37–9, 58–62, 71, 79–84, 132–5, 139, 144–8, 151–2, 154, 155, 157
  assessments of, 16–18, 26–30, 35, 41, 57, 75, 112–18, 134, 158–9
  relations with Elgin, 86, 93–4, 96–107
  relations with Galt, 65–8, 74, 84
  relations with Scott, 51–62, 81, 133, 134

reviews of, 56, 65, 108–12
sources and influences, xii, 2, 30–1, 34, 41, 69, 80, 81–2, 95, 126, 134, 142 .
works: 'Address Intended to be Recited at the Caledonian Meeting', 38; *The Age of Bronze*, 133; *Beppo*, 82, 83–4, 111; *Cain*, xiv; *Childe Harold*, xii–xiii, 7, 30–1, 33–4, 52–3, 56, 72–3, 74, 76, 81, 95, 96–100, 107, 113, 123, 125–6, 142, 151; *The Curse of Minerva*, 2, 95, 101–5; 'Dark Lochnagar'; xii, 7, 154, 155; *Don Juan*, xiii, xvi–xvii, 8–9, 11–13, 24–6, 40–1, 49, 69–70, 72, 75–8, 82, 112, 113, 114, 117, 127–9, 133, 135–6, 145, 148–149; *English Bards and Scotch Reviewers*, 2, 52, 53, 65, 69, 80–1, 93–4, 109, 110; 'Harmodia', 122–123, 127; *Hebrew Melodies*, 123–5, 126; *Hours of Idleness*, 65, 108–9, 111–12, 113; *The Island*, 132–3, 135–49; *Manfred*, xiv; *Marino Faliero*, xiv–xv, 37; *Mazeppa*, 116; *The Siege of Corinth*, 74–5; 'So we'll go no more a-roving', xii; 'Song for the Luddites', 9, 15, 21, 157–8; *The Vision of Judgement*, 9–11, 21, 72, 78, 79, 82

Calder, Angus, 45
Calton Hill, 101
Calvinism, 45–7, 48, 59, 79, 134
  see also religion
Campbell, Thomas, 116
Castlereagh, Viscount, 8, 106
Catholicism, 46, 47–9
  see also religion
Cato Street conspiracy, 13, 36
censorship, xviii, 9, 11, 18, 21, 157–8

161